Photonic Sensing

WILEY SERIES IN MICROWAVE AND OPTICAL ENGINEERING

KAI CHANG, Editor
Texas A&M University

A complete list of the titles in this series appears at the end of this volume.

Photonic Sensing

Principles and Applications for Safety and Security Monitoring

Edited by

GAOZHI XIAO

WOJTEK J. BOCK

A JOHN WILEY & SONS, INC., PUBLICATION

Cover Photographs: courtesy of Wojtek Bock

Published by John Wiley & Sons, Inc., Hoboken, New Jersey.
Published simultaneously in Canada.

Library of Congress Cataloging-in-Publication Data:

Xiao, Gaozhi.
Photonic sensing : principles and applications for safety and security monitoring / Gaozhi Xiao, Wojtek J. Bock.
p. cm.
Includes bibliographical references and index.
ISBN 978-0-470-62695-5 (hardback)
1. Optical fiber detectors. 2. Optical fiber detectors–Safety measures. 3. Optical fiber detectors–Security measures. I. Bock, Wojtek J. II. Title.
TA1815.X53 2012
681′.25–dc23

2011044215

Printed in the United States of America.

10 9 8 7 6 5 4 3 2 1

Contents

Preface

Twenty-first century society places high priority on health, environment, and security; the threats of terrorism, climate change, and pollution are of increasing concern for both governments and citizens. While many of these threats can never be completely eliminated, their impact can be significantly mitigated by the presence of effective early-detection and early-warning systems. To provide such protection requires flexible, cost-effective sensing, and monitoring systems in areas such as structural integrity, environmental health, human security and health, and industrial process control. Photonics (the use of light) has the potential to provide highly effective solutions tailored to meet a broad range of specific sensing requirements, particularly, as it can leverage many of the technology platforms that were successfully developed for the communications industry. However, optimal sensing solutions require the development of specialized materials, novel optical devices, and new networking algorithms and platforms. As a sensor technology, photonics offers low power requirements, high sensitivity and selectivity, and immunity from electromagnetic interference. A single optical fiber can be used both to detect disturbances at multiple locations and to transit the information to a central point for data processing. Over the last two decades, we have seen the rapid development of photonic sensing technologies and their application in fields including food bacteria detection, oil/gas pipe structure health monitoring, bio chips, explosives detection, defense platform health monitoring, etc.

This book comprises a series of chapters contributed by leading experts in the field of photonic sensing, with target applications to safety and security. The objective is to provide a most comprehensive, though by no means complete, review of this exciting field. This book aims for multidisciplinary readership. The editors intend that the book serve as an invaluable reference that aids research and development of those areas that concern safety and security. Another aim of the book is to stimulate the interest of researchers from physics, chemistry, biology, medicine, mechanics, electronics, defense and others, and foster collaboration through multidisciplinary programs.

Each chapter of the book deals with a specific area of safety and security monitoring using the photonic technique. It provides discussions on background, operation principles, and applications. Chapter 1 is on surface plasmons and their applications to biodetection, in particular to food pathogen detection. Chapter 2 is on microchip-based flow cytometry and its application in bacteria detection and analysis. Chapter 3 is on optofluidic techniques and their application in bioanalysis. Chapter 4 is on optical fiber sensors and their application in explosives detection. Chapter 5 is on photonic liquid crystal fiber sensors and their application in safety and security monitoring. Chapter 6 is on optic fiber sensor systems targeting air vehicle structural health monitoring applications. Chapter 7 is on optical coherence tomography and its application in document security and biometrics. Chapter 8 is on photonics-assisted instantaneous frequency measurement and its potential for electronic warfare applications.

We are grateful to all the authors for their contributions. Without their assistance and cooperation, this book would not have been possible. We also like to thank Ms. Aleksandra Czapla for the cover design, Ms. Kari Capone and Ms. Lucy Hitz of Wiley for their continual support during the course of this book.

Gaozhi Xiao and Wojtek Bock

Contributors

Pavel Adam, Institute of Photonics and Electronics, Academy of Sciences of the Czech Republic, Prague, Czech Republic

Wojtek J. Bock, Centre de recherche en photonique, Département d'informatique et d'ingénierie, Université du Québec en Outaouais, Gatineau, Québec, Canada

Shoude Chang, Institute for Microstructural Sciences, National Research Council Canada, Ottawa, ON, Canada

Costel Flueraru, Institute for Microstructural Sciences, National Research Council Canada, Ottawa, ON, Canada

Honglei Guo, Microwave Photonics Research Laboratory, School of Information Technology and Engineering, University of Ottawa, Ottawa, ON, Canada

Jiří Homola, Institute of Photonics and Electronics, Academy of Sciences of the Czech Republic, Prague, Czech Republic

Jianjun Ma, Centre de recherche en photonique, Département d'informatique et d'ingénierie, Université du Québec en Outaouais, Gatineau, Québec, Canada

Youxin Mao, Institute for Microstructural Sciences, National Research Council Canada, Ottawa, ON, Canada

Nezih Mrad, Air Vehicles Research Section, Defense R&D Canada, Department of National Defense, National Defense Headquarters, Ottawa, ON, Canada

Shilong Pan, Microwave Photonics Research Laboratory, School of Information Technology and Engineering, University of Ottawa, Ottawa, ON, Canada

Marek Piliarik, Institute of Photonics and Electronics, Academy of Sciences of the Czech Republic, Prague, Czech Republic

Hana Šípová, Institute of Photonics and Electronics, Academy of Sciences of the Czech Republic, Prague, Czech Republic

Tomáš Špringer, Institute of Photonics and Electronics, Academy of Sciences of the Czech Republic, Prague, Czech Republic

Milan Vala, Institute of Photonics and Electronics, Academy of Sciences of the Czech Republic, Prague, Czech Republic

Benjamin R. Watts, Department of Engineering Physics, McMaster University, Hamilton, ON, Canada

Tomasz Wolinski, Faculty of Physics, Warsaw University of Technology, Warszawa, Poland

Gaozhi Xiao, Institute for Microstructural Sciences, National Research Council Canada, Ottawa, ON, Canada

Chang-Qing Xu, Department of Engineering Physics, McMaster University, Hamilton, ON, Canada

Jianping Yao, Microwave Photonics Research Laboratory, School of Information Technology and Engineering, University of Ottawa, Ottawa, ON, Canada

Zhiyi Zhang, Institute for Microstructural Sciences, National Research Council of Canada, Ottawa, ON, Canada

CHAPTER ONE

Surface Plasmons for Biodetection

PAVEL ADAM, MAREK PILIARIK, HANA ŠÍPOVÁ, TOMÁŠ ŠPRINGER, MILAN VALA, and JIŘÍ HOMOLA
Institute of Photonics and Electronics, Academy of Sciences of the Czech Republic, Prague, Czech Republic

1.1 INTRODUCTION

The diffusion of inorganic and biological worlds represents an important paradigm of modern science and technology (1). Biophotonics stands as an emerging field of research at the crossroads of physical, chemical, and life sciences. The integration of photonics, biology, and nanotechnology is leading to a new generation of devices that makes it possible to characterize chemical and other molecular properties and to discover novel phenomena and biological processes occurring at the molecular level. Biophotonics is widely regarded as the key science on which the next generation of clinical tools and biomedical research instruments will be based.

The last two decades have witnessed an increasing effort devoted to the research and development of optical biosensors and biochips worldwide. Recent scientific and technological advances have demonstrated that such devices hold tremendous potential for applications in areas such as genomics, proteomics, medical diagnostics, environmental monitoring, food analysis, agriculture, and security (2–4). Label-free optical biosensors present a unique technology that enables the direct observation of molecular interaction in real-time and allows for the study of molecular systems, which cannot be labeled and studied by fluorescence spectroscopy (2).

Photonic Sensing: Principles and Applications for Safety and Security Monitoring, First Edition.
Edited by Gaozhi Xiao and Wojtek J. Bock.

Optical label-free biosensors measure binding-induced refractive index changes and are typically based on interferometric transducers, such as the integrated optical Mach–Zehnder interferometer (5), the integrated Young interferometer (6), and the white light interferometer (7), and transducers based on spectroscopy of guided modes of dielectric waveguides, such as the resonant mirror sensor (8) and the grating coupler sensor (9), or metal-dielectric waveguides, such as the surface plasmon resonance (SPR) sensor.

Since the first demonstration of the SPR method for the study of processes at the surfaces of metals (10) and sensing (11) in the early 1980s, SPR sensors have received a great deal of attention and allowed for great advances both in terms of technology and applications (12). Thousands of research papers on SPR biosensors have been published and SPR biosensors have been extensively featured in books (1, 2, 4, 13) and reviews (3, 12, 14–18). SPR biosensors have become a crucial tool for characterizing and quantifying biomolecular interactions. SPR biosensors have also been increasingly developed for the detection of chemical and biological species and numerous SPR biosensors for the detection of analytes related to medical diagnostics, environmental monitoring, food safety, and security have been reported as well.

This chapter describes the principles of SPR biosensors and discusses the advances that SPR biosensors have made both in terms of technology and applications over the last decade. The first part (Section 1.2) describes the fundamentals of SPR biosensors. Sections 1.3 and 1.4 are concerned with the optical configurations and immobilization methods used in current SPR sensors. The last part (Section 1.5) presents examples of applications of SPR biosensors for the detection of chemical and biological species with an emphasis on food safety and security applications.

1.2 PRINCIPLES OF SPR BIOSENSORS

1.2.1 Surface Plasmons

Surface plasmons (SPs) are electromagnetic modes guided by metallic waveguides. The simplest geometry supporting SPs comprises a planar boundary between a semi-infinite metal and a semi-infinite dielectric. The optical properties of the metal are characterized by a complex permittivity ε_m ($\varepsilon_m = \varepsilon'_m + i\varepsilon''_m$, where ε'_m and ε''_m are the real and imaginary parts of ε_m) and the dielectric is characterized by the refractive index n_d. Analysis of Maxwell's equations with appropriate boundary conditions suggests that this structure can only support a single guided mode of electromagnetic field—an SP (19). The vector of intensity of the magnetic field of SP lies in the plane of the metal–dielectric interface and is perpendicular to the direction of propagation. Such a mode of the electromagnetic field is referred to as the transversally magnetic (TM) mode. A typical profile of the magnetic field of an SP is shown in Figure 1.1(a). The intensity of the magnetic field reaches its maximum at the metal–dielectric interface and decays into both the metal and the dielectric. The field decay in the direction perpendicular to the metal–dielectric

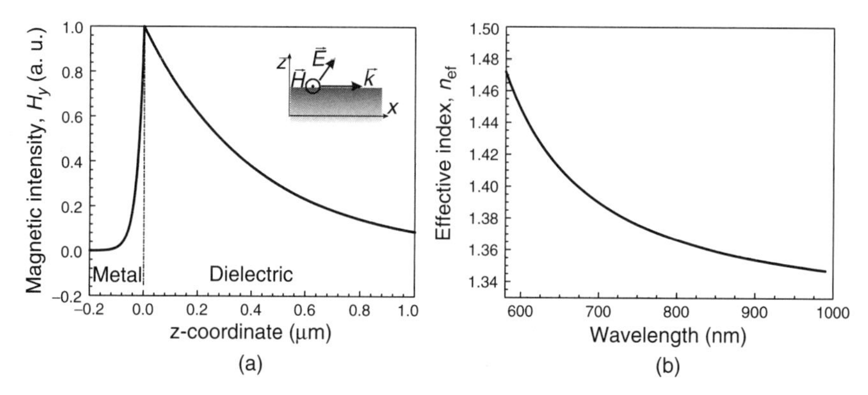

FIGURE 1.1 (a) Spatial distribution of the magnetic intensity for SP at the interface between gold and a dielectric ($n_d = 1.32$) in the direction perpendicular to the interface, $\lambda = 850$ nm. (b) Effective index of SP propagating along the interface between a dielectric (refractive index—1.32) and metal (gold) as a function of the wavelength.

interface is characterized by the penetration depth. The penetration depth depends on the wavelength and permittivities of the materials involved. The penetration depth into the dielectric for an SP propagating along the interface of gold and a dielectric with $n_d = 1.32$ increases with a wavelength and ranges from 100 to 600 nm in the wavelength region 600–1000 nm (19).

Propagation constant of SP β_{SP} at the metal–dielectric interface can be expressed as

$$\beta_{SP} = \frac{\omega}{c}\sqrt{\frac{n_d^2 \varepsilon_m}{n_d^2 + \varepsilon_m}}, \qquad n_{ef} = \frac{c}{\omega}\mathrm{Re}\,\{\beta_{SP}\}, \tag{1.1}$$

where c is the speed of light in a vacuum, ω is the angular frequency, and n_{ef} is the effective index of the SP (20, 21). If the structure is lossless ($\varepsilon_m'' = 0$), Equation 1.1 represents a guided mode only if the metal permittivity ε_m' is negative and $\varepsilon_m' < -n_d^2$. Metals such as gold, silver, and aluminum exhibit a negative real part of permittivity in the visible and near-infrared region of the spectrum. Figure 1.1b depicts the wavelength dependence of the effective index of SP n_{ef} for the gold waveguide. The imaginary part of the propagation constant is associated with the imaginary part of the metal permittivity ε_m'' and determines attenuation of the SP in the direction of propagation (20).

A special example of the metallic waveguide is a symmetric dielectric–metal–dielectric planar structure. When the metal film thickness is much larger than the SP penetration depth into the metal, an independent SP may propagate at each metal–dielectric boundary. If the thickness of the metal film is decreased, coupling between the SPs at opposite sides of the metal film can occur, giving rise to mixed modes of electromagnetic field—symmetric and antisymmetric SPs (22, 23). The profiles of magnetic intensity of symmetric and antisymmetric SPs are symmetric or antisymmetric with respect to the plane of symmetry of the structure. The field of the symmetric SP penetrates much deeper into the dielectric medium than the field of the antisymmetric SP or the field of a conventional SP at a single metal–dielectric interface. Moreover, the symmetric SP exhibits a lower attenuation than its antisymmetric counterpart and therefore it is referred to as a long-range surface plasmon (LRSP) while the antisymmetric mode is referred to as a *short-range surface plasmon* (22).

1.2.2 Excitation of Surface Plasmons

1.2.2.1 Prism Coupling The most common approach to the excitation of SPs is by means of a prism coupler and the attenuated total reflection method (ATR). In the Kretschmann geometry of the ATR method (24), a high refractive index prism with refractive index n_p is interfaced with a metal–dielectric waveguide consisting of a metal film with permittivity ε_m and a semi-infinite dielectric with a refractive index n_d ($n_d < n_p$), Figure 1.2.

When a light wave propagating in the prism totally reflects on the prism base, an evanescent electromagnetic wave decays exponentially in the direction perpendicular to the prism–metal interface (25). If the metal film is sufficiently thin (less than 100 nm for light in the visible and near-infrared part of spectrum), the evanescent wave penetrates through the metal film and couples with an SP at the outer boundary of the metal film. In terms of the effective index, this coupling condition can be written as follows:

$$n_p \sin\theta = n_{ef} = \mathrm{Re}\left\{\sqrt{\frac{n_d^2 \varepsilon_m}{n_d^2 + \varepsilon_m}}\right\} + n_{ef}^{(1)}, \tag{1.2}$$

where n_{ef} is the effective index of the SP, and the perturbation in effective index $n_{ef}^{(1)}$, and the respective propagation constant of SP $n_{ef}^{(1)}$ describe the effect of the presence of the prism.

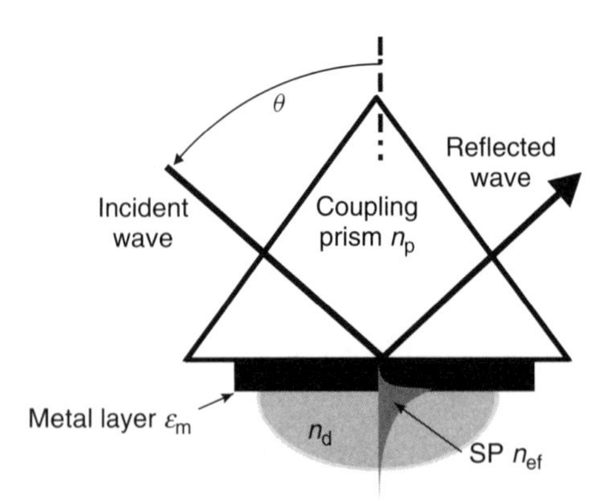

FIGURE 1.2 Excitation of surface plasmons in the Kretschmann geometry of the attenuated total reflection (ATR) method.

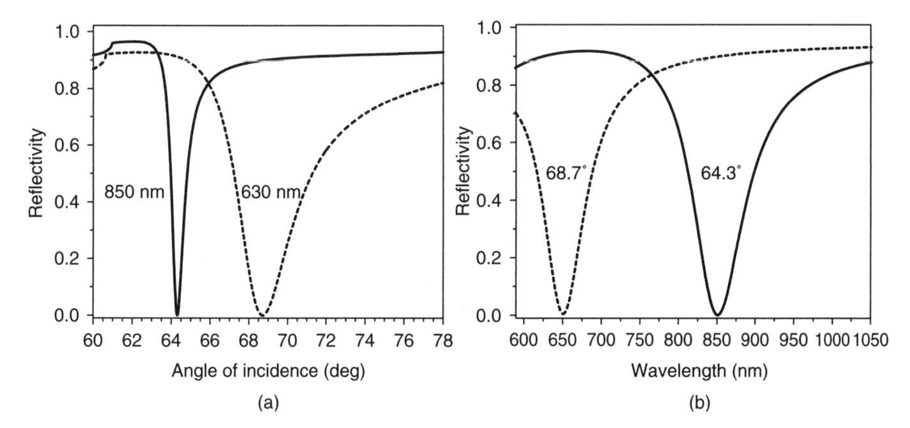

FIGURE 1.3 TM reflectivity as a function of the angle of incidence (a) and the wavelength (b) calculated for two different angles of incidence using the rigorous Fresnel reflection theory. Configuration: BK7 glass, gold film (thickness—48 nm for the wavelength of 650 nm and 50 nm for the wavelength of 850 nm), water.

Figure 1.3 shows the angular and wavelength spectra calculated using a rigorous Fresnel model of light reflection on a multilayer structure calculated at two different wavelengths and for two angles of incidence, respectively. The reflectivity spectra exhibit distinct dips in TM polarization, which are associated with the transfer of energy from the incident light wave into an SP and its subsequent dissipation in the metal film.

The reflectivity spectra can be rigorously calculated using Maxwell equations and the boundary condition of the planar multilayer structure. Assuming that the permittivity of metal ε_{m} obeys $|\varepsilon'_{\mathrm{m}}| \gg n_{\mathrm{d}}^2$ and $|\varepsilon'_{\mathrm{m}}| \gg \varepsilon''_{\mathrm{m}}$, a Lorentzian (with respect to n_{ef}) approximation of the reflectivity can be used as follows (20):

$$R(\theta, \lambda) \approx 1 - \frac{4\gamma_{\mathrm{i}}\gamma_{\mathrm{rad}}}{(n_{\mathrm{p}} \sin\theta - n_{\mathrm{ef}})^2 + (\gamma_{\mathrm{i}} + \gamma_{\mathrm{rad}})^2}, \tag{1.3}$$

where $\gamma_{\mathrm{i}} = \mathrm{Im}\{\beta_{\mathrm{SP}}\}\lambda/2\pi$ and $\gamma_{\mathrm{rad}} = \mathrm{Im}\{\beta^{(1)}\}\lambda/2\pi$ denote the attenuation coefficients of SPs owing to absorption and radiation, respectively. As follows from Equation 1.3, the minimum of the dip in the reflectivity spectrum occurs when the coupling condition (Eq. 1.2) is matched and the shape of the reflectivity dip depends strongly on the strength of the coupling between the excitation wave and SP represented by γ_{rad}. This approximation has been shown to provide a good estimate of the position of the reflectivity dip and to predict the shape of the reflectivity curve in the neighborhood of the minimum (19). In addition, the Lorentzian curve exhibits the same width as the dips calculated using the rigorous approach (26).

1.2.2.2 Grating Coupling Another approach to optical excitation of SPs is based on the diffraction of light on a diffraction grating. In this method, a light wave

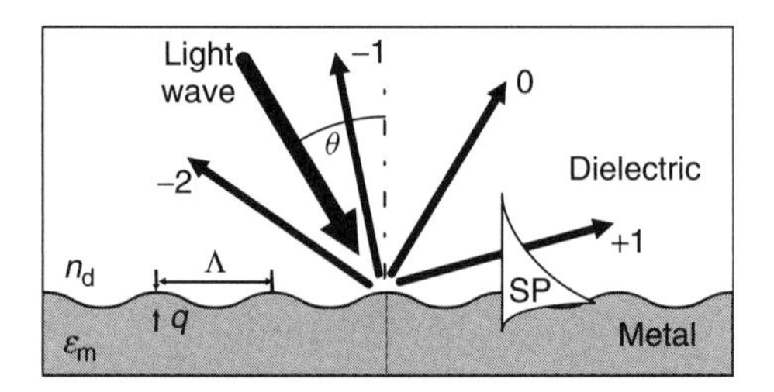

FIGURE 1.4 Excitation of surface plasmons by the diffraction of light on a diffraction grating.

is incident at an angle of incidence θ from a dielectric medium with the refractive index n_d on a metal grating with the dielectric constant ε_m, the grating period Λ, and the grating depth q (Fig. 1.4). The diffracted wave in m—the diffraction order can couple with SPs when their propagation constants are closely phase-matched. In terms of the effective index, the coupling condition can be written as

$$n_\mathrm{d}\sin\theta + m\frac{\lambda}{\Lambda} = \pm\left(\mathrm{Re}\left\{\sqrt{\frac{n_\mathrm{d}^2\varepsilon_\mathrm{m}}{n_\mathrm{d}^2+\varepsilon_\mathrm{m}}}\right\} + n_\mathrm{ef}^{(1)}\right), \tag{1.4}$$

where $n_\mathrm{ef}^{(1)}$ accounts for the presence of the grating.

The grating-moderated interaction between a light wave and an SP can be modeled by solving Maxwell's equations in differential form with a grating profile approximated by a stack of layers (27, 28), or in an integral form by solving the Helmholtz–Kirchhoff integral (29).

Figure 1.5 shows the reflectivity spectra (angular and wavelength) for light incident from water onto a gold grating (calculated using the integral method). The

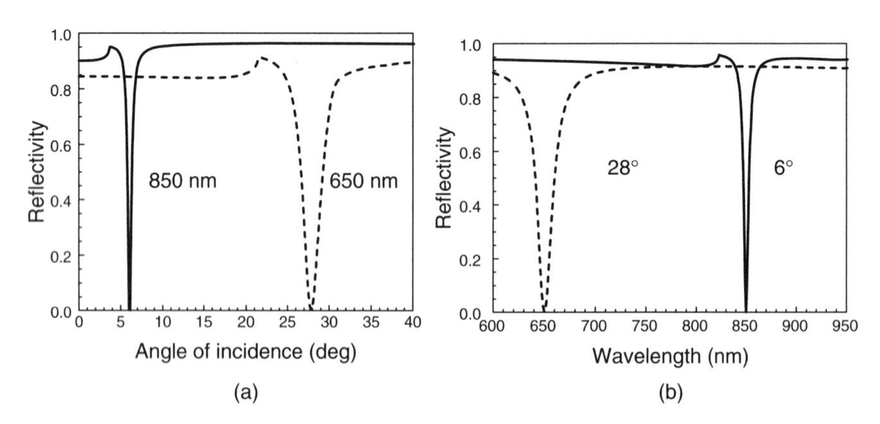

FIGURE 1.5 Reflectivity as a function of the angle of incidence (a) and the wavelength (b) calculated for two wavelengths and angles of incidence, respectively (denoted in the label). Configuration: gold–water interface; grating period 672 nm, grating depth 30 nm, angle of incidence taken in the air.

angular as well as wavelength reflectivity spectra exhibit a characteristic dip caused by the transfer of energy of thc incidcnt light into an SP. Figure 1.5 shows resonant spectra narrower by a factor of 10 compared to resonant dips obtained with a prism coupler in Figure 1.3. On shallow diffraction gratings, SPs are excited at the angles of incidence close to the coupling angles predicted from the matching condition, neglecting the effect of the grating. The depth of the reflectivity dips depends on the depth of the grating. For the structure considered in Figure 1.5 (gold–water interface; grating period 672 nm), the strongest excitation of an SP ($R = 0$) occurs with a grating depth of about 30 nm.

1.2.2.3 Waveguide Coupling SP can also be excited by modes of dielectric waveguides, such as planar or channel-integrated optical waveguides and optical fibers. Typically, coupling between a dielectric waveguide mode and an SP propagating along a metal layer in the proximity of the dielectric waveguide is achieved by coupling the evanescent tails of the two waves. The evanescent wave of the dielectric waveguide mode can couple with SP when the propagation constant of the mode β_{M} is equal to the real part of the propagation constant of the SP β_{SP}:

$$\beta_{\mathrm{M}} = \mathrm{Re}\{\beta_{\mathrm{SP}}\}. \tag{1.5}$$

As the coupling condition is fulfilled for only a narrow range of wavelengths, the excitation of SPs can be observed as a narrow dip in the spectrum of light transmitted through the waveguide structure.

1.2.3 Sensors Based on Surface Plasmons

A change in the refractive index of the dielectric medium produces a change in the propagation constant of SP at the interface of the metal and the dielectric. Thischange in propagation constant alters the coupling condition between the light wave and the SP, which can be observed as a change in the characteristics of the optical wave interacting with SP. The change in SP propagation is associated with the change in resonant coupling conditions, for example the resonant wavelength, the angle of incidence, or the strength of the SP coupling. Figure 1.6 illustrates the shift in the angular of wavelength SPR spectra because of the change in the refractive index of the dielectric medium n_d. Depending on which characteristics of the reflected light wave are measured in the SPR sensor, SPR sensors are classified as (i) SPR sensors with wavelength modulation (the angle of incidence is fixed and the coupling wavelength serves as a sensor output); (ii) SPR sensors with angular modulation (the coupling wavelength is fixed and the coupling angle of incidence serves as a sensor output); (iii) SPR sensors with intensity modulation (both the angle of incidence and the wavelength of incident light are fixed at nearly resonant values and the light intensity serves as a sensor output); and (iv) SPR sensors with phase modulation (both the angle of incidence and the wavelength of incident light are fixed at nearly resonant values and the phase of the reflected light serves as a sensor output).

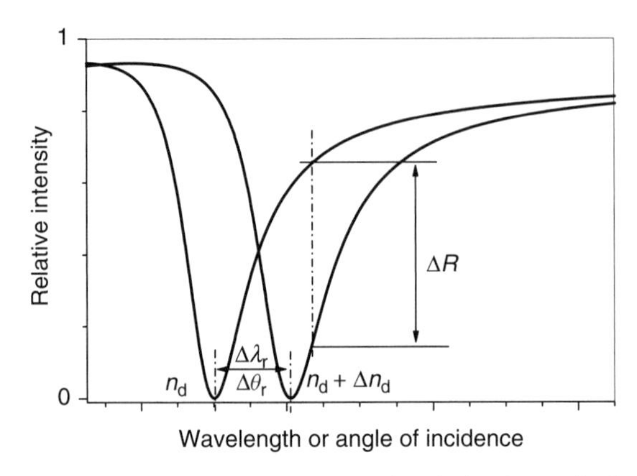

FIGURE 1.6 Change in the SPR spectra associated with the change in the refractive index of the dielectric media. Parameters $\Delta\lambda_r$, $\Delta\theta_r$, and ΔR denote the shift in the resonant wavelength, the resonant angle of incidence, and the change in light intensity.

A change in the refractive index Δn_d of the dielectric film with thickness h, produces a change in the effective index of the SP Δn_{ef}. If the thickness of the layer is much higher that the penetration depth L_{pd} of the SP field, the change in the effective refractive index of the SP can be calculated by differentiating the dispersion relation (1.1), which for the conventional SP yields (19):

$$\Delta n_{ef} = \frac{n_{ef}^3}{n_d^3} (\Delta n_d)_{h \gg L_{pd}}. \tag{1.6}$$

A change in the effective index of an SP owing to refractive index changes within a thin layer $h \ll L_{pd}$ can be estimated using the perturbation theory (19) as

$$\Delta n_{ef} = \frac{2h}{L_{pd}} \frac{n_{ef}^3}{n_d^3} (\Delta n_d)_{h \ll L_{pd}}. \tag{1.7}$$

1.2.4 SPR Affinity Biosensors

SPR biosensors employ special biomolecules—referred to as *biorecognition elements*—that can recognize and capture target analytes. Such biorecognition elements are immobilized in the form of a sensitive layer on the surface of the SPR metallic waveguide. When a solution containing analyte molecules is brought into contact with the SPR sensor, analyte molecules in solution bind to the molecular recognition elements, producing an increase in the refractive index of the sensitive layer (Fig. 1.7).

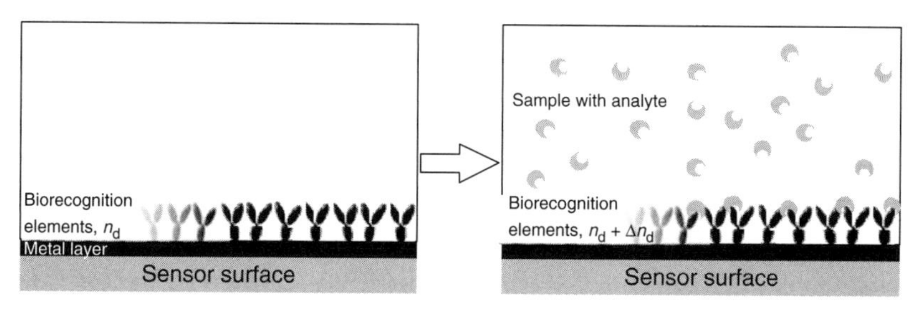

FIGURE 1.7 Principle of SPR affinity biosensing.

The change in refractive index Δn_d occurring within a layer of thickness h can be expressed as

$$\Delta n_d = \left(\frac{dn}{dc}\right)_{vol} \Delta c_b = \left(\frac{dn}{dc}\right)_{vol} \frac{\Delta\Gamma}{h}, \tag{1.8}$$

where $(dn/dc)_{vol}$ is the refractive index increment, Δc_b is the wt/vol concentration of bound molecules within the sensitive layer with the thickness h, and $\Delta\Gamma$ is the corresponding surface concentration (mass per surface area). The linear relation between the refractive index change and the surface concentration of bound molecular mass in Equation 1.4 is often referred to as the *de Feijter formula* (30). The refractive index increment $(dn/dc)_{vol}$ is a well-characterized property for most biochemical species and ranges typically from 0.1 to 0.3 cm^3/g (31). Proteins and nucleic acids exhibit quite a consistent refractive index increment value, which falls within 8% from the value of $(dn/dc)_{vol} = 0.18$ cm^3/g (31). As follows from Equations 1.6–1.8, a change in the effective index of the SP owing to the capture of analyte can be expressed as

$$\Delta n_{ef} = K \Delta\Gamma, \tag{1.9}$$

where K is a constant.

1.2.5 Performance Characteristics of SPR Biosensors

The performance of SPR biosensors is usually characterized in terms of the sensitivity, resolution, limit of detection (LOD), linearity, accuracy, reproducibility, and dynamic range (19, 26).

The *sensitivity* of an SPR sensor is the ratio of the change in sensor output to the change in the quantity to be measured (*e.g.*, the refractive index n_d). The sensitivity of an SPR sensor to a refractive index S can be written as

$$S = \frac{\Delta Y}{\Delta n_d} = \frac{\Delta Y}{\Delta n_{ef}} \frac{\Delta n_{ef}}{\Delta n_d}, \tag{1.10}$$

where Y denotes the sensor output and depending on the modulation approach usually represents the resonant angle of incidence, or the resonant wavelength, or the reflectivity. Equation 1.10 describes the decomposition of the sensor sensitivity in two parts: (i) the sensitivity of the sensor output to the change in the effective index of an SP and (ii) the sensitivity of the effective index of an SP to the change in the refractive index (19, 26). Therefore, the second term (ii) is independent of the modulation method and the method of excitation, and the first term (i) represents the instrumental factor which is independent of the measurand.

Analogously, the sensitivity of an SPR biosensor to a concentration of analyte c derives from the change in the refractive index of the sensitive layer n_{d} caused by the analyte binding:

$$S = \frac{\Delta Y}{\Delta c} = \frac{\Delta Y}{\Delta n_{\mathrm{ef}}} \frac{\Delta n_{\mathrm{ef}}}{\Delta n_{\mathrm{d}}} \frac{\Delta n_{\mathrm{d}}}{\Delta c}. \tag{1.11}$$

As with Equation 1.10, the first term is the instrumental factor and the second term characterizes the properties of SP mode. The third term in Equation 1.11 is derived from the relationship between the analyte concentration and the refractive index change described in Equation 1.8, the binding capabilities of biorecognition elements, and analyte transport to the sensor surface.

The instrumental factor $\Delta Y / \Delta n_{\mathrm{ef}}$ has been analyzed in detail for different combinations of the SPR couplers and modulation approaches in recent publications (26). Clearly, the instrumental factor depends on which method of excitation of the SPs and modulation approach are used. For the most common sensors based on SPR spectroscopy with a prism coupler, the instrumental factor can be calculated by differentiating coupling conditions (1.2) as follows:

$$\frac{\mathrm{d}\theta_{\mathrm{r}}}{\mathrm{d}n_{\mathrm{ef}}} = (n_{\mathrm{p}}^2 - n_{\mathrm{ef}}^2)^{-\frac{1}{2}}, \tag{1.12}$$

$$\frac{\mathrm{d}\lambda_{\mathrm{r}}}{\mathrm{d}n_{\mathrm{ef}}} = \left(\frac{\mathrm{d}n_{\mathrm{p}}}{\mathrm{d}\lambda} \frac{n_{\mathrm{ef}}}{n_{\mathrm{p}}} - \frac{\mathrm{d}n_{\mathrm{ef}}}{\mathrm{d}\lambda} \right)^{-1}, \tag{1.13}$$

where λ_{r} and θ_{r} denote the resonant wavelength and angle of incidence respectively. Instrumental factors are thus determined by the geometry and material constants of the SPR coupler and dispersion of the SP mode. In similar manner, the instrumental factor $\Delta R / \Delta n_{\mathrm{ef}}$ can be calculated for SPR sensors based on intensity modulation using the Lorentzian approximation Equation 1.3 as

$$\frac{\mathrm{d}R}{\mathrm{d}n_{\mathrm{ef}}} = \pm \frac{3\sqrt{3}}{2} \frac{\gamma_{\mathrm{i}} \gamma_{\mathrm{rad}}}{(\gamma_{\mathrm{i}} + \gamma_{\mathrm{rad}})^3}. \tag{1.14}$$

Unlike instrumental factors calculated for SPR spectroscopy (Eqs. 1.12 and 1.13), Equation 1.14 suggests a significant dependence of the sensitivity on the strength of SPR coupling represented by the radiation coefficient γ_{rad}. This behavior is associated with the effect of γ_{rad} term on the shape of the SPR dip as discussed in Section 1.2.2.

The *resolution* of an SPR sensor is defined as the smallest change in the refractive index that produces a detectable change in the sensor output. The magnitude of the change in sensor output that can be detected depends on the level of uncertainty of the sensor output—the output noise. Resolution of an SPR sensor σ_{n}, is typically expressed in terms of the standard deviation of noise of the sensor output σ_{Y} translated to the refractive index of the bulk medium:

$$\sigma_{\mathrm{n}} = \frac{\sigma_{\mathrm{Y}}}{S}. \quad (1.15)$$

The noise in the sensor output originates from the noise of individual light intensities involved in the calculation of the sensor output. The propagation of noise to the sensor output was investigated by Piliarik and Homola (26). Their study revealed that, independent of the SPR coupling principle, the refractive index resolution of an SPR sensor can be expressed as follows:

$$\sigma_{\mathrm{n}} = \frac{rK}{\sqrt{N}} \frac{(\gamma_{\mathrm{i}} + \gamma_{\mathrm{rad}})^3}{\gamma_{\mathrm{i}}\gamma_{\mathrm{rad}}} \left(\frac{\partial n_{\mathrm{ef}}}{\partial n}\right)^{-1} \frac{\sigma_{I\,(\max)}}{I_0} \quad (1.16)$$

where K is the noise distribution factor, r is the noise correlation factor, N is the number of intensities involved in the measurement, I_0 is the intensity of the incident light which corresponds to one detector (e.g., spectrometer pixel), and $\sigma_{I\,(\max)}$ is the standard deviation of intensity noise (for amplitude sensors this corresponds to the level of measured intensity and for spectroscopic sensors this corresponds to the level at the threshold of the SPR dip). The noise distribution factor is between $K = 0.38$ for amplitude SPR sensors and $K = 0.50$ for SPR spectroscopy with noise homogeneously distributed across the SPR dip ($K = 0.43$ for shot noise limited SPR spectroscopy being the most common case). The noise correlation factor $r = 1$ for uncorrelated intensities. The factor r is derived from the Pearson correlation coefficient ρ of individual pixels of the spectrum as $r = \sqrt{1 - \rho}$ for spectroscopic SPR sensors. For SPR sensors with intensity modulation, $r = \sqrt{1 + (N - 1)\rho}$, where N is the number of detector pixels.

As follows from the analysis presented above, the resolution of the SPR sensor depends on the noise of the used optoelectronic components, the strength of the coupling between the light wave and the SP, and material parameters of the metallic waveguide (26). In contrast, although the sensitivity of SPR sensors depends strongly on the coupling principle and modulation, a comparable resolution can be achieved regardless of the SPR coupling principle and modulation.

The LOD of an SPR biosensor represents the ability of a biosensor to detect an analyte. LOD is defined as the concentration of analyte derived from the smallest measure of the sensor output Y, which can be distinguished from the sensor output corresponding to a blank sample. The value of the sensor output corresponding to the LOD, Y_{LOD}, can be expressed as

$$Y_{\mathrm{LOD}} = Y_{\mathrm{blank}} + m\sigma_{\mathrm{blank}}, \quad (1.17)$$

where Y_{blank} is the mean of the blank measures, σ_{blank} is the standard deviation of the blank measures, and m is a numerical factor chosen for the required confidence level (typically $m = 3$).

1.3 OPTICAL PLATFORMS FOR SPR SENSORS

Since the first demonstration of SPR sensors in the early 1990s, optical platforms of SPR sensors have made substantial advances in terms of both performance (32–45) and new capabilities. SPR instruments have become compact enough to be used for routine bioanalytical tasks in the field (46–52) and have expanded to enable parallelized detection of hundreds of different analytes at a time (35, 53–56).

In this section, we present a brief overview of the advances in SPR optical platforms, in particular within the last decade. The section is organized into four subsections, based on the method of coupling of light to SPs (prism, grating, and waveguide). The last subsection presents examples of the available commercial SPR systems.

1.3.1 Prism-Based SPR Sensors

The use of prism couplers to couple light to SPs is straightforward, versatile, and does not require complex optical instrumentation. Therefore, SPR sensors based on the attenuated total reflection method and prism couplers have been the most widely used. All major types of modulation have been implemented in prism-based SPR sensors. In SPR sensors based on spectroscopy of SPs, the angular or wavelength spectrum of the optical wave coupled with the SP is measured. Alternatively, changes in the intensity or phase of the reflected wave can be measured at a fixed wavelength and an angle of incidence. While the spectroscopic SPR sensors usually offer higher resolution, they provide a rather limited number of sensing channels. By contrast, intensity or phase modulations can be adopted by SPR imaging configurations, where independent measurements are performed in as many as hundreds of sensing channels simultaneously.

1.3.1.1 Spectroscopic Prism-Based SPR Sensors The use of angular modulation in SPR sensors has a long history. In 1991, Sjolander et al. reported an angular modulation-based SPR sensor utilizing a light-emitting diode (LED), a glass prism with a gold layer, and a detector array with imaging optics to detect the angular spectrum of light reflected from the gold layer (Fig. 1.8) (32). A divergent beam emitted by the LED was collimated in the plane parallel to the gold layer and focused by a cylindrical lens in the perpendicular plane to produce a wedge-shaped beam illuminating a thin gold film. The imaging optics shaped the reflected beam in such a way that the angular spectrum of each sensor channel was projected on a separate row (or rows) of the array detector. This design has been further advanced by Biacore AB (Sweden) and has resulted in a variety of commercial SPR sensors (subsection 1.3.4).

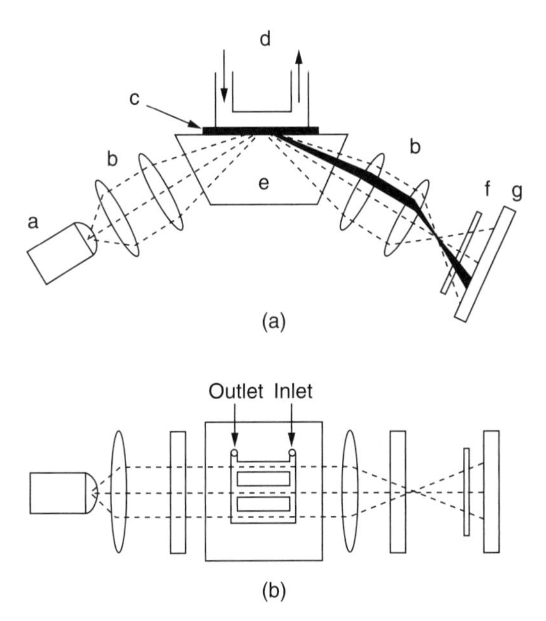

FIGURE 1.8 (a) Side view and (b) top view of a prism-based SPR sensor in angular configuration with three parallel channels. a—light-emitting diode, b—lenses, c—sensor chip, d—microfluidic cartridge, e—coupling prism, f—polarizer, g—photodiode array detector. *Source:* Reprinted, by permission, from Reference 32.

An interesting approach in the miniaturization of SPR sensors with angular modulation was reported by Thirstrup et al. (49). They developed a polymeric exchangeable sensor chip containing two diffractive optical coupling elements (DOCEs). The DOCEs (chirped relief gratings) were used to focus an incident parallel beam to a sensing spot and project the reflected beam onto a detector array (Fig. 1.9). The sensor provided an angular sensitivity of 140°/RIU and a resolution of 5×10^{-7} RIU. The SPR chips with DOCEs were produced in plastic by injection molding, which provides a low-cost method for the mass production of SPR chips. However, the limited quality of the diffraction coupling elements caused background illumination, which had a negative influence on the performance of the sensor.

Another compact SPR sensor based on angular modulation was proposed by the researchers at Texas Instruments and the University of Washington in the 1990s. In recent years, this design (referred to as *Spreeta*), has been further refined. Spreeta SPR sensor had specific dimensions of only $3 \times 1.5 \times 4$ cm^3 and consisted of a plastic prism molded to a printed circuit board (PCB). The PCB contained a microelectronic circuit containing an LED and a linear diode array detector. The divergent light beam emitted by the LED passed through a polarizer and hit the sensor surface at a range of angles above the critical angle. Therefore, different areas on the sensor chip were illuminated under different angles (Fig. 1.10). The light reflected from the mirror placed on the top of the sensor was captured by the detector. The sensor was showed to have a resolution of 1.8×10^{-7} RIU within a time interval

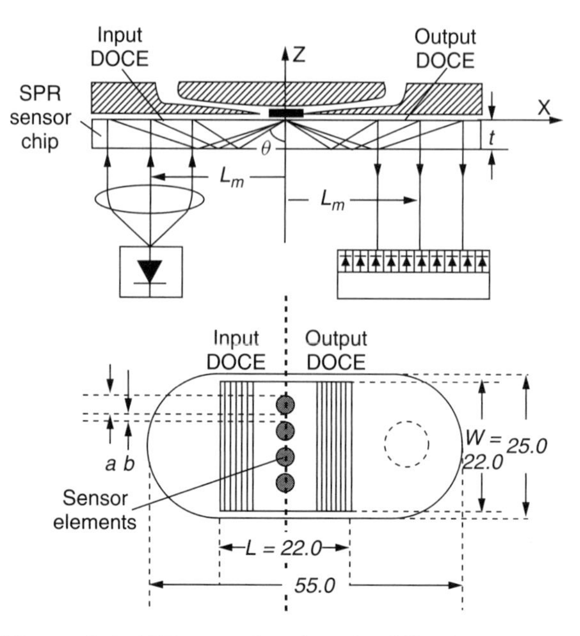

FIGURE 1.9 Scheme of the SPR sensor based on the diffractive optical coupling element (DOCE). *Source:* Reprinted, by permission, from Reference 49.

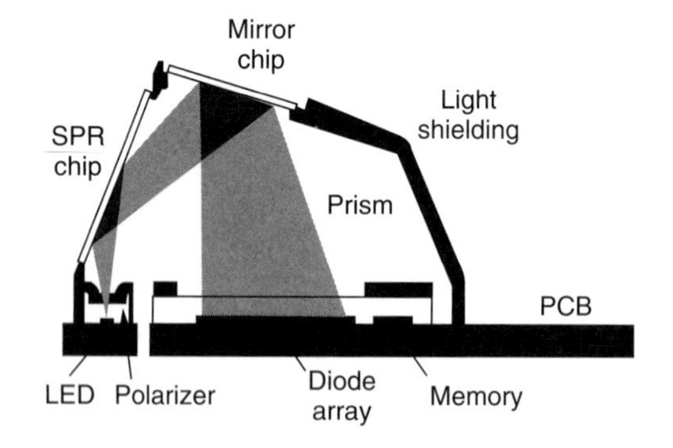

FIGURE 1.10 Scheme of the Spreeta sensor from Texas Instruments. *Source:* Reprinted, by permission, from Reference 47.

of 0.8 s. However, the sensor exhibited deviations from a smooth response of 0.2% during the measurement of a change in the refractive index of 0.04 RIU, which corresponds to a resolution of 8×10^{-5} RIU (47). Subsequently, Chinowsky et al. demonstrated a portable 24-channel SPR sensor integrating eight 3-channel Spreeta sensors into a single suitcase (48). The sensor contained supporting electronics and a microfluidic system with a temperature-controlled flow cell and was demonstrated to provide a resolution of $1\text{–}3 \times 10^{-6}$ RIU (48). Currently, the Spreeta sensor is also commercially available from Sensata Technologies (Section 1.3.4).

Kim et al. reported a portable SPR platform based on angular modulation of SPR in eight parallel sensing channels (50). A special gold-coated replaceable prism served as a coupling prism and a sensing chip at the same time. This made it possible to avoid the use of refractive index matching oil. The system was reported to exhibit a baseline noise of 0.7 mdeg (50), which for a typical angular sensitivity of about 95°/RIU (19) corresponds to a resolution of 7×10^{-6} RIU.

In 2008, Feltis et al. reported a fully integrated handheld SPR device with a size of only 15×8 cm^2 and a weight of 600 g (51) (Fig. 1.11). The sensor was based on angular modulation of SPR using a laser diode and a photodiode array as a light source and detector, respectively. The sensor only measured in a single channel and was demonstrated to achieve a resolution of about 3×10^{-6} RIU (51).

In 2002, Song et al. reported an SPR sensor based on angular interrogation of SPR, which utilized a bi-cell detector instead of a photodiode array (57). The intensity of the reflected light captured by the detector's two cells, A and B, was analyzed and the position of the SPR minimum was determined by dividing the differential and sum signals $(A - B)/(A + B)$. The presented baseline noise of about $5 \times 10^{-4\circ}$ corresponds to a resolution of 4×10^{-6} RIU (assuming an operating wavelength of 635 nm and a sensitivity of about 140°/RIU (19)). A similar approach was reported by Zhang et al. (36), who used a quadrant detector and thus were able to measure in two channels simultaneously. This sensor was reported to provide an angular resolution of 10^{-5}, which corresponded to a refractive index resolution of about 10^{-7} RIU (assuming a sensitivity of 130°/RIU (19)).

In 2007, researchers from Agilent Technologies (USA) reported an SPR system, which offers both angular modulation and SPR imaging (35). The sensor

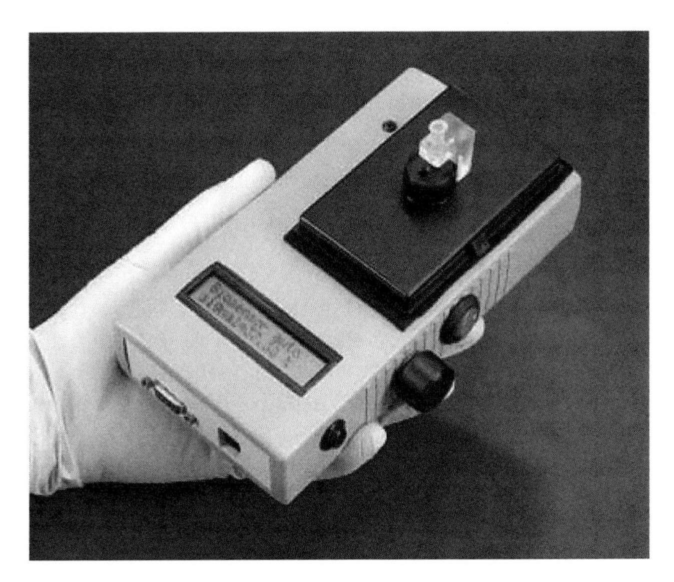

FIGURE 1.11 Photograph of a handheld SPR device based on a prism coupler and angular interrogation. *Source:* Reprinted, by permission, from Reference 51.

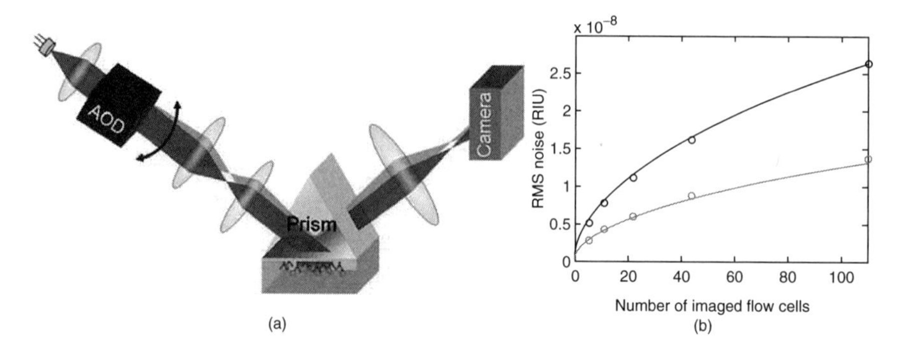

FIGURE 1.12 (a) Configuration of the high-resolution and high-throughput SPR imaging sensor based on fast angular scanning. (b) System performance as a function of the number of cells and dynamic range. Black data (experimental datapoints and theoretical extrapolations) corresponds to a dynamic range of 0.06 RIU, while the gray data were obtained for a 0.015 RIU dynamic range. *Source:* Reprinted, by permission, from Reference 35.

configuration with angular is depicted in Figure 1.12a. The light from a laser diode was collimated and passed through an acousto-optic deflector (AOD) where the direction of the beam was modulated within a defined angular span. The deflected light was then imaged by a two-lens telescope onto a sensing chip so that the illuminated region remained at the same place during the angular scanning. The surface of the sensing chip was then imaged onto a high-speed CMOS-based camera. In the performed experiment, the camera frame-rate was 1.1 kHz and scanning through the full angular range was done every 100 ms, which allowed for the acquisition of more than 100 spectra each second. The combination of a powerful light source and a high-speed camera allowed for massive averaging of angular spectra and therefore offered spectra with extremely low noise. The resulting performance of the sensor as a function of the number of imaged spots is shown in Figure 1.12b for two different values of dynamic range. For example, the sensor imaging 30 sensing spots and offering a dynamic range of 0.015 RIU provided a resolution of about 7×10^{-9} RIU (35).

Wavelength modulation has also found its way to spectroscopic SPR sensors. For instance, a high-resolution prism-based SPR sensor with wavelength modulation was reported by Homola et al. (Fig. 1.13) (33, 58). Light from a halogen lamp was transmitted through a multimode optical fiber and after collimation and polarization it was made incident on an SPR chip with a gold film interfaced with a coupling prism. The reflected polychromatic beam was then coupled to output optical fibers and transmitted to a multichannel spectrometer. With the use of an advanced data processing algorithm (34), a baseline noise as low as 1.5×10^{-3} nm was established. This low noise level makes it possible to achieve a refractive index resolution of 2×10^{-7} RIU (assuming an operating wavelength of 750 nm and sensor sensitivity of 7500 nm/RIU (19)).

In order to increase the amount of information in SPR sensors with wavelength modulation, wavelength division multiplexing (WDM) was introduced to

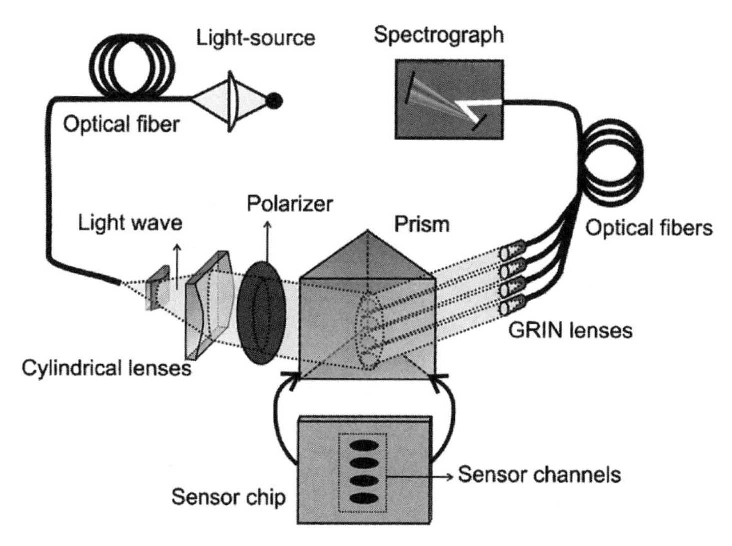

FIGURE 1.13 Four-channel sensor based on a wavelength interrogation of SPR in a Kretschmann geometry. *Source:* Reprinted, by permission, from Reference 58.

SPR sensors with wavelength modulation. Two different approaches to WDM were proposed by Homola's group (39, 40). In the first approach, a sensing surface consisting of a thin gold film partly coated with a thin dielectric film was used. As the presence of the thin dielectric film red-shifted the resonant wavelength (compared to the bare gold), the reflected light exhibited two dips associated with the excitation of SPs in the area with and without the dielectric overlayer (39). In the second configuration, a special skewed prism was used in which polychromatic light was sequentially made incident on two sensing channels under different angles (40). This WDM approach was combined with the previously described 4-channel SPR sensor with four parallel beams yielding an 8-channel SPR sensor. The resolution was 1×10^{-6} RIU and 7×10^{-7} RIU for the short-wavelength (640 nm) channel and the long-wavelength (790 nm) channel, respectively (40).

Bardin et al. (41) reported an SPR sensor combining high resolution of spectroscopy of SPs with a high number of sensing channels. In this spectro-imaging SPR sensor, polychromatic light reflected from a linear array of sensing spots was dispersed by a diffraction grating and projected onto a 2D detector. The spectra of light reflected from individual channels appeared as narrow bands on the detector. The sensor was demonstrated to attain a resolution of 3.5×10^{-7} RIU in up to 20 channels (41).

More complex plasmonic modes, such as LRSPs, have also been exploited in spectroscopic SPR sensors. In 2001, Nenninger et al. (37) reported an SPR sensor utilizing LRSPs. LRSP is a guided mode of an optical structure consisting of a very thin metal film sandwiched between two dielectrics with similar refractive indices (Section 1.2.1). Nenninger et al. investigated LRSP sensing structures employing two dielectrics as a buffer layer—one with refractive indices slightly below (Teflon

AF) and above (MgF_2) that of water. In the wavelength-modulation-based SPR sensor, the MgF_2 chips were found to perform better and to achieve a resolution of 2×10^{-7} RIU (37). In 2007, Slavík and Homola improved the design and implementation of the LRSP-based sensor (with Teflon AF as a buffer layer) and achieved a resolution of 2.5×10^{-8} RIU (38).

1.3.1.2 Prism-Based Sensors with Intensity or Phase Modulation Intensity modulation has been increasingly used in SPR sensors to enable spatially resolved measurements (59, 60). In typical SPR imaging sensors, a parallel TM-polarized beam of monochromatic light is launched into a prism coupler and made incident on a thin metal film at an angle of incidence close to the coupling angle for the excitation of SPs. The intensity of reflected light depends on the strength of the coupling between the incident light and the SP and therefore can be correlated with the distribution of the refractive index along the metal film surface (61–63). In order to increase the sensor stability and optimize the contrast of SPR images, Fu et al. proposed an SPR imaging employing a white light source and a bandpass interference filter (64). This SPR sensor was demonstrated to provide a refractive index resolution of 3×10^{-5} RIU (61).

In 2007, Piliarik et al. (53) reported an SPR imaging sensor based on polarization contrast and an array of patterned multilayers. In this configuration, each sensing channel consisted of a pair of sensing spots (type I and II; Fig. 1.14) with different multilayers. The multilayers were designed in such a way that with an increasing refractive index at the sensor surface, the intensity of the reflected light increased for spot type I and decreased for spot type II. When the sensor response is defined as a ratio of the intensities of light reflected from the spots I and II, the sensitivity is

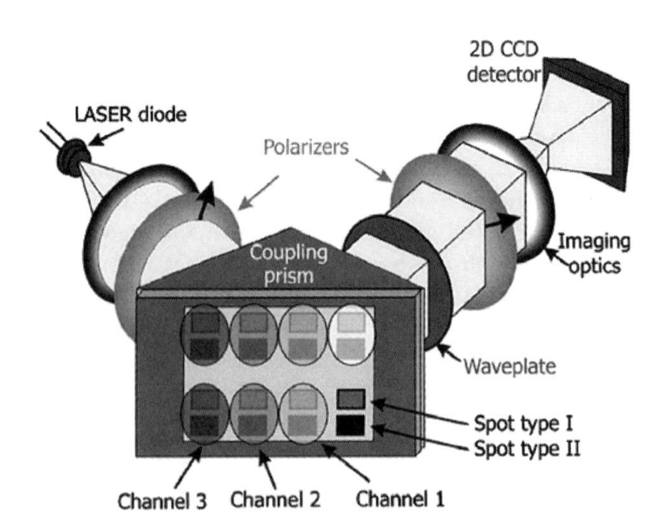

FIGURE 1.14 Scheme of an SPR imaging device based on polarization contrast and internal referencing using two types of multilayer structures on each sensing channel. *Source:* Reprinted, by permission, from Reference 53.

higher than the sensitivity for each individual spot. In addition, the defined sensor output is therefore insensitive to light level fluctuations, which are correlated across the sensing area. The sensor was demonstrated to provide an extended dynamic range of 0.012 RIU and a resolution of 2×10^{-6} RIU. Later, the same group reported an SPR imaging sensor combining the polarization contrast with a different concept of internal referencing. This concept is based on two mirrors enabling real-time compensation of fluctuations of the dark current and intensity of incident light (54). Resolution as low as 2×10^{-7} RIU in 120 sensing spots was achieved with this type of sensor (54).

In 2007, Law et al. reported an SPR sensor with phase modulation and an extended dynamic range (65). The sensor was based on temporal modulation of the phase and detection of the second and third modulation harmonics. While the detection of the third modulation harmonics resulted in a good resolution (3×10^{-7} RIU) and a limited operating range (0.004 RIU), detection of the second harmonic modulation offered a worse resolution (10^{-5} RIU) and a larger operating range (0.012 RIU) (65).

A high-resolution SPR sensor was demonstrated by Wu and Ho (42). They demonstrated a single-beam self-referenced phase-sensitive SPR sensor with an ultrahigh resolution of 5×10^{-9} RIU. The sensor was based on a differential phase-detection scheme, in which the phase of the incident beam was periodically modulated using a liquid crystal modulator. The modulated beam was split into a TM-polarized probe beam passing through the sensor head and a TE-polarized reference beam. Signals from both arms were then compared using a digital oscilloscope. However, the sensor response was strongly nonlinear and this level of resolution was only available within a refractive index range of 6×10^{-5} RIU. In the linear portion of the operating range, the resolution of the sensor was much poorer (10^{-6} RIU) (42).

Another approach in the development of SPR sensors with phase modulation was proposed by Nikitin's group. In 2000, Nikitin et al. reported two novel approaches to SPR imaging based on interferometry (43). In the first configuration, the Mach–Zehnder interferometer with TM-polarized beams in both the signal and reference arms was used. The second approach was based on the interference of the TM-polarized signal beam with the TE-polarized reference beam. This configuration was observed to be less sensitive to vibrations as both beams passed through the same optical elements. The second method was demonstrated in a "phase contrast mode" and in a "fringe mode" in which a small angle was introduced between the combined interfering beams resulting in a pattern of interference fringes superimposed over the image of the surface. The sensor operating in "fringe mode" was able to measure variations in the refractive index as small as 10^{-7} RIU. Expandability of this approach to high-throughput measurements was demonstrated by imaging multiple spots (diameter—50 μm) coated with a molecular monolayer (43).

An SPR sensor based on an interferometric approach allowing for multiple passes of the signal beam through the sensor was reported by Ho et al. (44). In this sensor, a coupling prism with an SPR chip was placed inside one arm of the interferometer and phase variations induced by SPR were measured through the interference of the beams on a detector. The sensor was tested in three configurations based on Mach–Zehnder, Michelson, and Fabry–Perot interferometers, which resulted in single-pass, double-pass, and multiple-pass of the optical beam through the SPR sensor head, respectively. The resolution of the tested configurations were 1.5×10^{-6} RIU and about 8×10^{-7} RIU for single-pass and double-pass, respectively (44).

A high-throughput SPR sensor based on differential phase SPR imaging using the Mach–Zehnder interferometer was reported by Wong et al. (45). In the reference arm of the interferometer, the optical path was periodically changed using a movable mirror mounted on a piezoelectric transducer (PZT). With the PZT continuously shifting the reference optical phase, intensity variations were captured by a detector pixel after recombining the beams from both arms of the interferometer, which followed a truncated sine function. The fast Fourier transform (FFT) was combined with appropriate signal processing, which then converted the raw data into a two-dimensional SPR phase map that has a direct correlation with the refractive index changes on the sensor surface. The sensor was demonstrated to achieve a resolution of 9×10^{-7} RIU in a 5×5 array of sensing spots (45).

1.3.2 SPR Sensors Based on Grating Couplers

Although grating couplers were initially used in SPR sensor much less than prism couplers, in the last decade, the popularity of grating couplers has grown substantially owing to several unique features and benefits. Replacing bulky and costly prism couplers with grating couplers formed on planar substrates allows for a reduction of the size and costs of SPR instruments. In addition, the use of grating couplers eliminates the need for refractive index matching fluids, which are needed in order to make an optical contact between the prism coupler and the SPR chip. Grating-based SPR sensors based on all the main modulation approaches have been developed. Nevertheless, the majority of grating-based SPR sensors are based on angular or wavelength spectroscopy of SPs.

1.3.2.1 Spectroscopic Grating-Coupled SPR Sensors Various SPR sensors based on the classical approach to spectroscopy of SPs have been developed. In this approach, light is made incident on several areas of a coupler and angular or wavelength spectrum of light coupled to SPs in these distinct areas is measured in parallel. For instance, recently, Vala et al. reported a mobile SPR sensor based on angular spectroscopy of SPs in 10 independent sensing channels (52). In this sensor, a light from a narrow band laser diode is shaped by special lens optics to a wedge beam and made incident on a disposable SPR cartridge incorporating a grating coupler. SPs were excited through the -1 diffraction order of the grating and the reflected light was projected on a CCD detector. A resolution as low as 6×10^{-7} RIU was observed.

A high-throughput SPR sensor based on angular spectroscopy of SPs was reported by Dostálek et al. (55). In this sensor, a TM-polarized monochromatic light was focused on a sensor chip cartridge (Fig. 1.15) and the back-reflected

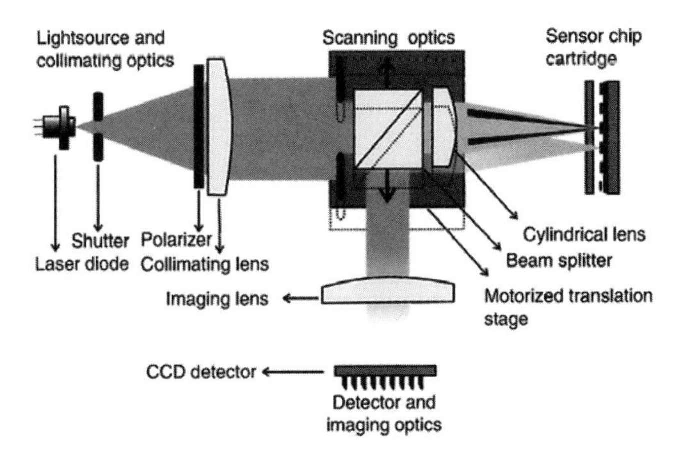

FIGURE 1.15 High-throughput grating-based SPR sensor with angular modulation. *Source:* Reprinted, by permission, from Reference 55.

light was imaged on a CCD detector. The cartridge consisted of a two-dimensional array of SPR diffraction gratings and a microfluidic system with six independent flow-chambers. The imaging optics was motorized, which made scanning of the grating array possible. The design of the sensor was further optimized with respect to the operating wavelength, the parameters of the diffraction grating coupler, and the supporting optical system, which altogether yielded a resolution of 5×10^{-7} RIU in more than a hundred sensing channels (66). However, the motorization and large size of some of the components limited the measurement speed and potential for miniaturization of the sensor.

Another interesting approach to high-throughput SPR sensing was reported by Kastl et al. (56). Their sensor was based on scanning multiple variably shaped (bar-coded) microgratings. These grating particles were settled at the bottom of a polydimethylsiloxane (PDMS) flow cell and their type and orientation were identified by optical imaging. Simultaneously, SPs were excited on the gratings and the angular distribution of monochromatic light coupled to the SPs was measured. The readout time was 6 s per grating and the resolution was 3×10^{-5} RIU. The main advantage of this approach is that it allows for the convenient introduction of different functionalizations and biorecognition elements into the measurement area.

Recently, Piliarik et al. (46) demonstrated a compact high-resolution SPR sensor based on a new approach to spectroscopy of SPs on a special diffraction grating referred to as a surface plasmon resonance coupler and disperser (SPRCD). In this sensor, polychromatic light was made incident on the SPRCD element and while one of the diffraction orders of the grating was used to excite SPs, the light diffracted away from the grating is dispersed across a CMOS detector (Fig. 1.16). The SPRCD element was integrated into a miniature cartridge with six independent microfluidic channels. The size of the sensor unit was 15×15 cm^2. A resolution as low as 3×10^{-7} RIU was observed with this type of sensor (46).

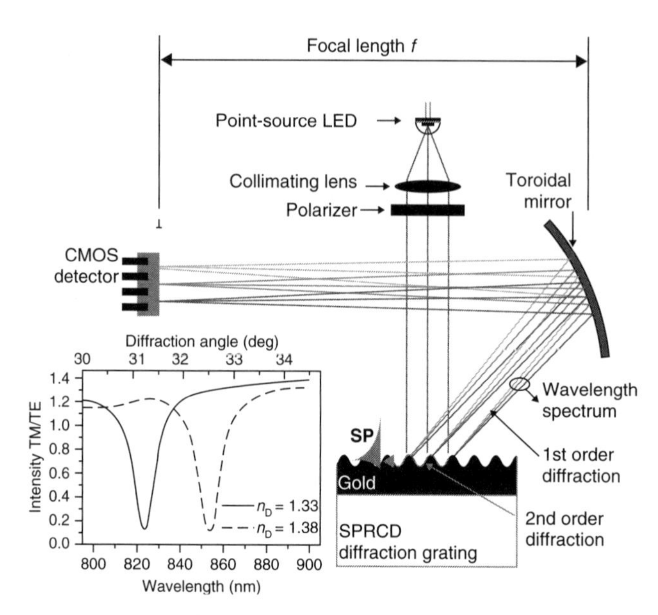

FIGURE 1.16 Principle of operation of the surface plasmon resonance coupler and disperser. *Source:* Reprinted, by permission, from Reference 46.

Advanced diffractive structures allow for the simultaneous excitation of multiple SPs by different spectral components of a polychromatic light, making multiple SP spectroscopy possible. These structures include, for instance, multidiffractive gratings on which SPs can be excited through diffraction on different harmonics of the multidiffractive grating (67), a diffraction grating with extremely thin metal film supporting short-range and long-range SPs (68), and bi-diffractive structures supporting Bragg-scattered SPs (69). As SPs excited at different wavelengths exhibit different field profiles, probing the same binding event with multiple plasmons can reveal more information about the system under study, for instance, in discriminating surface refractive index changes due to the binding from background refractive index variations. However, this process requires accurate calibration and sophisticated data processing (70).

1.3.2.2 Grating-Based Sensors with Intensity Modulation Brockman and Fernandez (71) demonstrated a grating-based SPR imaging device. In this sensor, a collimated beam of monochromatic light was made incident onto a plastic chip with a gold-coated grating (Fig. 1.17). The reflected light with spatially distributed SPR information was projected onto a CCD detector. The system was demonstrated to perform simultaneous measurements in 400 sensing spots with a resolution in the order of 10^{-6} RIU. This design was commercialized by HTS Biosystems (Germany) as a FLEX-chip system and later acquired by Biacore (72).

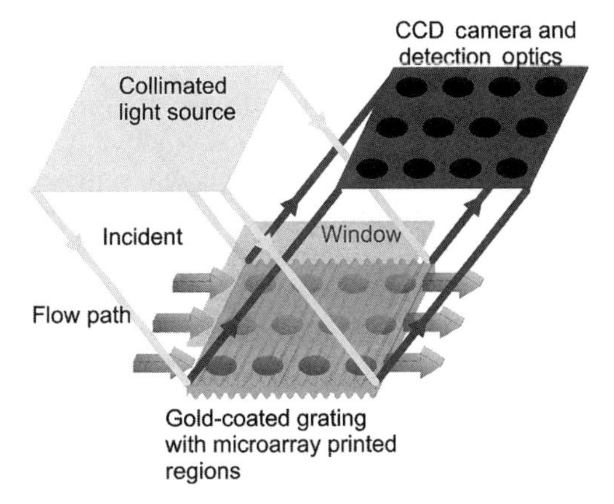

FIGURE 1.17 Concept of an SPR sensor based on a grating coupling and intensity modulation.

1.3.3 SPR Sensors Based on Optical Waveguides

Optical waveguides have been exploited in SPR sensors to miniaturize the sensing element and to create a connection between the SPR instrument and the sensing element. The waveguide-based SPR sensors can be divided into two groups on the basis of whether they are based on optical fibers or integrated optical waveguides.

1.3.3.1 Fiber Optic SPR Sensors There are several configurations of fiber optic SPR sensors. A fiber-based SPR sensor can be, for example, fabricated by locally removing the cladding (e.g., by side-polishing) and coating the exposed area of the fiber with a thin metal film. Recently, an SPR spectroscopic sensor based on a side-polished multimode optical fiber was reported with a resolution in the order of 10^{-6} RIU (73). Pollet et al. (74) presented a compact SPR sensor based on SPR spectroscopy employing a reusable SPR multimode fiber sensing element. In this sensor design, the sensing area was produced at the tip of an optical fiber and light reflected by the metal-coated end of the fiber was interrogated by a spectrometer. However, the performance of SPR sensors based on multimode fibers is limited by modal and polarization noise owing to perturbations of the fiber (deformations, thermal effect, etc.). To overcome this limitation, fibers guiding only a few modes (75) or a single mode (73, 76) were demonstrated with resolutions as low as 5×10^{-7} RIU (76). However, the performance of even single-mode optical-fiber-based SPR sensors could be negatively affected by deformations of the fiber. To suppress sensitivity of the SPR fiber optic sensor, Piliarik et al. developed a wavelength modulation-based fiber optic SPR sensor using a polarization-maintaining fiber (77). The sensor was demonstrated to achieve a resolution better than 4×10^{-6} RIU under moderate deformations of the fiber.

Another approach to the development of SPR sensors is based on tapered fibers. Chang et al. developed an SPR sensor based on intensity modulation and a single-mode optical fiber, the tip of which was tapered and coated with a metal film around the tip of the fiber (78). Another SPR sensor was reported by Monzon-Hernandez et al. who developed a sensing element based on a uniform-tapered asymmetrically metal-coated fiber supporting hybrid SP modes. A light transmitted through the sensing element exhibited multiple resonance peaks observed in the spectrum of transmitted light. The resolution of the sensor was estimated to be about 7×10^{-7} RIU (79). A similar uniform-tapered sensing probe was described by Diaz-Herrera et al. (80). The sensing element was illuminated by a polychromatic light. Two different fiber Bragg gratings (FBGs) located behind the sensing area reflected light of two selected wavelengths and these were detected by two photodiodes. These two wavelengths were tuned to correspond with the slopes of the SPR dip. From the intensity changes of the light at these two wavelengths, the SPR position changes were deduced. The observed resolution was 2×10^{-5} RIU.

Recently, Bragg gratings inscribed in the fiber have been used to excite SPs coupled to cladding modes of the fiber. Špacková et al. proposed a fiber optic sensor with two different FBGs inscribed into two different regions of the sensing area. In this sensor, polychromatic light propagated in the fiber and through diffraction on the grating, the light of distinct spectral components excited different cladding modes including cladding modes coupled to SPs on the outer boundary of metal film deposited around the optical fiber. The coupling gave rise to narrow dips in the spectrum of transmitted light. As the response from two different gratings is encoded in different parts of the spectrum, the two gratings form two independent sensing channels. The resolution of the sensor was estimated to be around 2×10^{-6} RIU (81). Shao et al. proposed an SPR sensor using an optical fiber with a tilted FBG in the core of the fiber (82). In contrast to the cladding mode resonances, which are sensitive to fluctuations in the temperature outside the fiber structure, the core mode back reflection resonance is only sensitive to the temperature of the fiber. Therefore, such a structure can be used to compensate for fluctuations in the temperature of the sample.

1.3.3.2 Integrated Optical SPR Sensors The first SPR sensors based on integrated optical waveguides were developed in the 1990s. The sensors typically consisted of a waveguide formed in a glass substrate and a thin metal layer deposited on the waveguide. The combination of materials and operating wavelengths in the visible and near infrared spectra determined the resonant refractive index of the medium under study above 1.4 RIU, which is much larger than the values of the refractive index of samples typically encountered in SPR biosensing. Several approaches have been proposed to shift the operating range of the sensor to an aqueous environment. Dostálek et al. used a high refractive index dielectric overlayer deposited on top of the metal film and demonstrated an integrated optical SPR sensor for aqueous environment with a resolution better than 2×10^{-6} RIU (83). Other integrated optics SPR sensors were aimed at compact sensors as presented by Suzuki et al. (84). In this sensor, monochromatic light emitted from two different

LEDs was coupled to a waveguide with a sensing area. The intensity of output light modulated by SPR was measured by photodiodes. The reported sensor system was light and compact and powered by a 9 V battery. Another compact and integrated sensor was presented by Johnston et al. (85) who exploited a photodetection system that integrates optical computing with each pixel of the detector. An electro-optical SPR sensor was proposed by Wang et al. (86). In this sensor, the SPR characteristics can be modulated by the applied voltage. Therefore, the SPR position can be changed. Such a feature can be used to cover a wide refractive index range, even when using a high-resolution spectrograph with a rather limited operating range. Jette-Charbonneau presented an integrated optical sensor utilizing LRSPs (87). The sensor consists of a metal slab of a constant thickness with a periodic change of the width of the slab. This design allows the excitation of LRSPs with an extended field, which makes the sensor suitable for the detection of large analytes.

1.3.4 Commercial SPR Sensors

In the last two decades, rapid advances in SPR technology and a growing number of possible applications have led to the development of a number of commercial solutions to SPR biosensing. Nowadays, tens of companies produce SPR sensors, which range from portable SPR sensing devices to large laboratory SPR instruments offering high-resolution and/or high-throughput. This section presents examples of commercial SPR sensors.

The first commercial SPR biosensor was launched in 1990 by Pharmacia Biosensor AB (since 1996 Biacore, now a part of GE Healthcare). This sensor was based on the attenuated total reflection method, with a prism coupler and angular modulation (Section 1.3.1.1, Fig. 1.8). Biacore have further improved this configuration and in the following years launched a wide range of laboratory SPR instruments offering different combinations of features and performance (72). Biacore currently offers eight different SPR instruments (72). Among those, Biacore 4000 offers 5 sensing spots in each of its 4 microfluidic channels for increased throughput measurements and Biacore T200 provides a resolution as low as 3×10^{-8} RIU. More than 80% of SPR studies have been performed using Biacore systems (88, 89), making Biacore instruments the most widespread SPR biosensor technology.

An SPR instrument for protein interaction analysis is offered by Bio-Rad (USA). Their SPR instrument is based on a prism coupler and angular modulation. The sensor chip is illuminated using a diode array under 15 distinct angles. The criss-crossed microfluidic channel array allows for simultaneous measurement in 36 spots (6 analytes with 6 ligands). The resolution of the instrument is 1×10^{-6} RIU (90).

IBIS Technologies (The Netherlands) has developed and markets an SPR sensor combining the robustness of angular spectroscopy of SPs with the high-throughput of SPR imaging. This feature was achieved by using a movable mirror that deflects the light beam impinging on the sensor chip in such a way that angular spectra are measured from each sensing spot (91). The sensor is designed to enable measurements in more than 500 spots with a baseline noise of 0.3 mdeg (92). This corresponds to a resolution of 3×10^{-6} RIU.

High-throughput SPR imaging systems based on intensity modulation are available from Genoptics (part of the Horiba Jobin Yvon group, France). Their sensors are able to monitor interactions in up to 400 spots on the sensor surface (93). The minimum detectable surface coverage specified for the SPRi-PlexII system is 5 pg/mm^2 (93). This corresponds to a refractive index resolution of about 5×10^{-6} RIU.

Compact SPR sensor Spreeta is offered by Sensata Technologies (USA). The Spreeta sensor is based on an integrated prism—light source—detector module and angular modulation (Fig. 1.10). The sensor allows for measurement in three independent sensing channels. The specified sensor resolution determined by the baseline noise is 3×10^{-7} RIU, the drift of the baseline is about 1×10^{-6} RIU per minute. The dimensions of the Spreeta module are $3 \times 1.5 \times 4$ cm^3 (94).

1.4 FUNCTIONALIZATION METHODS FOR SPR BIOSENSORS

The development of SPR sensor instrumentation and applications of SPR biosensors proceeds hand in hand with the need for advanced biointerfaces. The immobilization of biorecognition elements presents one of the key steps in the development of SPR biosensors with direct implications for sensor characteristics such as sensitivity, specificity, and LOD. The direct and label-free nature of SPR biosensors poses unique challenges in the development of procedures for the immobilization of biorecognition elements. For instance, the biorecognition elements are required to be oriented on the surface, which makes them accessible for the analyte and to be immobilized in such a manner that preserves their active structure. Moreover, the growing number of analytical applications in complex media creates an increasing demand for the development of advanced coatings, which can resist adsorption of nontarget molecules from a complex sample to the SPR sensor surface.

Functionalization strategies used in SPR biosensors are mainly concerned with the functionalization of gold, which is the predominant material used as a support for SPs in SPR sensors. It is worth noting that in addition to the traditional methods for functionalization of gold surfaces, the deposition of nanometer-thick layers on top of gold and the subsequent functionalization of these layers is emerging as a promising strategy. These layers can be made of various materials (such as silicates (95)) and allow the established protocols optimized for other analytical techniques to be applied in the SPR biosensors (96–98). This approach may significantly extend the range of surface functionalization methods available in SPR sensors in the future.

The gold surface of SPR chips offers a convenient method for the immobilization of molecules via thiol chemisorption. Indeed, the simplest method for the immobilization of biorecognition elements is their adsorption on the gold surface via thiol groups available on the receptor (Fig. 1.18a). However, this approach often results in a loss of receptor activity. Moreover, the proteins and other molecules tend to bind nonspecifically to the reactive metal surface, which may create large background sensor response and thus limit performance of the SPR sensor. Therefore,

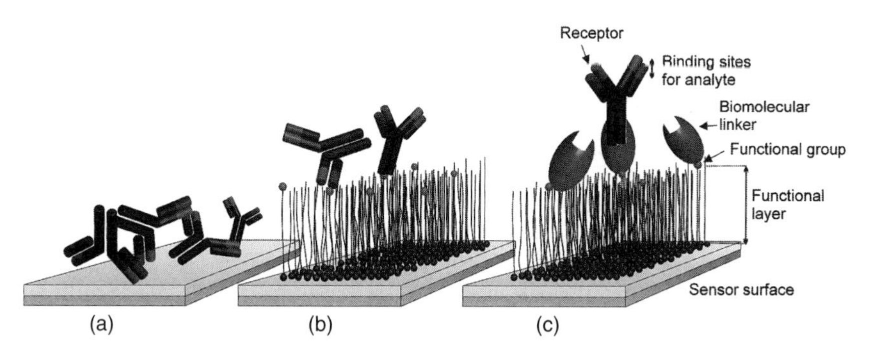

FIGURE 1.18 Scheme of the most common immobilization strategies: (a) adsorption of the receptor directly on gold, (b) coupling to the functional groups of a functional layer, and (c) immobilization via biomolecular linkers.

the gold surface is commonly covered with a functional layer, which contains functional groups for receptor immobilization and ensures low-fouling properties of the sensor surface (Fig. 1.18b). In addition, biomolecular linkers with specific recognition sites for the receptor may be attached to the functional layer (Fig. 1.18c), resulting in surfaces with better ordered receptors. This section describes traditional methods for functionalization of sensing surfaces in SPR sensors as well as advances in surface chemistries in the last decade. The methods suitable for immobilization of particular types of receptors are summarized at the end of this section.

1.4.1 Functional Layers

Substantial effort has been invested in the development of sophisticated functional layers that bear functional groups for the defined attachment of biorecognition elements and can resist nonspecific adsorption from complex samples. The most popular approaches include the well-established technology of self-assembled mono-molecular film. When a high surface capacity for receptors is desired, hydrogels, such as modified dextrans, are often employed. In addition, various polymer coatings aimed at providing functional surfaces with a high resistance to nonspecific adsorption have emerged in the last decade.

1.4.1.1 Self-Assembled Monolayers Gold surfaces can be readily functionalized with self-assembled monolayers (SAMs) of ω-functionalized thiols, sulfides, or disulfides. The SAMs are traditionally formed by alkanethiols, but may also be composed from other molecules such as aminoacids (99) and peptides (100). Monolayer formation is driven by the strong coordination of sulfur with metal atoms accompanied by van der Waals interactions between the alkyl chains (Fig. 1.19). With a sufficient chain length, the resulting monolayer forms a densely packed structure oriented more or less perpendicularly to the metal surface. The terminal groups of alkanethiols are important for the attachment of the receptors as well as for surface resistance toward nonspecific adsorption (101). A variety of the alkanethiols are

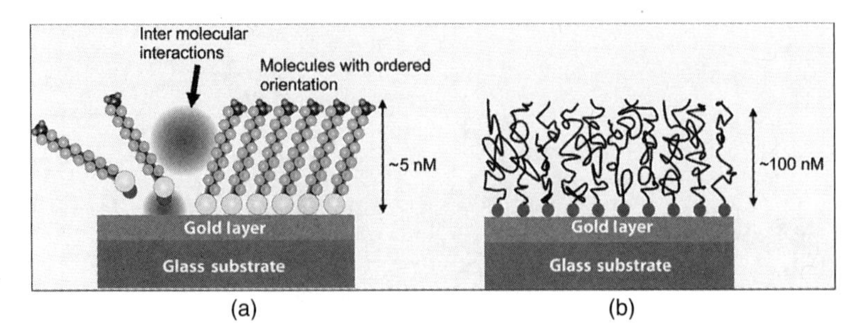

FIGURE 1.19 Formation of a self-assembled monolayer of alkanethiols (a) and carboxymethylated dextran matrix (b).

therefore commercially available or have been synthesized (102). The most widely used are alkanethiols with hydroxyl, carboxyl, and amine terminal groups; however, other types of alkanethiols that include 2-(biotinamido)ethanethiol terminal group for the immobilization of streptavidin (103) have also been reported. Various modifications to the alkenthiol chains have been introduced in order to minimize nonspecific adsorption. For example, the incorporation of oligo(ethylene glycol) (OEG) has been found to decrease adsorption of the plasma proteins. Recently, fluorinated SAMs with improved fouling properties have also been reported (104).

1.4.1.2 Polymers The electromagnetic field of an SP extends over several hundred nanometers. Therefore, the SPR method can detect binding events, which occur much further from the metal surface, than the thickness of a single molecular monolayer. In order to take advantage of the full field of the SP, polymer structures with thicknesses of hundreds of nanometers have been developed, which provide enhanced capacity for the immobilization of biorecognition elements compared to a single functional layer. The polymers are often grafted using initiator groups conjugated with SAM, which enable radical, ionic, or atom-transfer growth mechanisms. Such polymer-grafting reactions offer numerous methods for producing polymer layers with different properties and functions. One of the most widely used polymers is dextran, which is composed of linear chains of glucose units with a size ranging from 10 kDa to over a million daltons. The dextran hydrogel forms a highly hydrophilic environment with a 3D structure, which provides a large number of sites for immobilization of biorecognition elements. However, their accessibility may be limited owing to a slower diffusion of target molecules in the matrix. The detection conditions and dextran properties must therefore be tailored to each application (105). The most widely used derivative is carboxymethylated dextran employed in the Biacore systems. The general disadvantage of dextran layers is the overall negative charge of the surface, which makes the dextran polymer prone to nonspecific electrostatic binding from complex solutions. Other hydrophilic polymers have been developed as alternatives to dextran. For example, poly(vinyl alcohol) and poly(acryl acid), and their graft combination have been demonstrated for use in SPR applications. The methacrylate allows for the formation of UV-initiated gradient matrices (106), or *in situ* polymerization induced by the evanescent field of the SPR sensor (107). Poly (L-lysine) has become popular for DNA arrays owing to its positive charge.

Considerable effort has been invested in the development of surfaces with low-fouling properties. Although the use of SAMs and dextran-coated surfaces significantly reduces nonspecific binding compared to the bare gold surface (108), the nonspecific adsorption of proteins in complex media still remains a tough challenge. In recent years, improvements in the performance of these approaches have been demonstrated by incorporating various functional terminal groups (101, 109), by optimizing detection conditions (110) or by employing blocking agents, such as bovine serum albumin (BSA), casein and milk (111). A reduced nonspecific adsorption of plasma proteins was also achieved by grafting polyethylene glycol (PEG) to the SAM, poly (L-lysine) surfaces (112) or directly to the gold (113, 114). PEG is a water soluble, electrically neutral polyether that has been used as a material with significant inertness to cell and protein adhesion (112, 115). However, the main limitation of PEG surfaces lies in their susceptibility to oxidation. In recent years, zwitterionic layers grafted on SAMs have attracted a great deal of attention as an alternative to conventional PEG surfaces. Zwitterions are molecules with a neutral total charge, which results in a reduction of nonspecific electrostatic interactions between the proteins and the sensor surface. Zwitterionic groups used with SPR sensors include phosphorylcholine (PC) (109), sulfobetaine (SB), and carboxybetaine (CB) (112). The CB allows for the incorporation of carboxylic groups and thus the employment of amine coupling chemistry (116), Fig. 1.20. Surfaces coated with poly(carboxybetaine acrylamide) were demonstrated to exhibit very low fouling when exposed to undiluted plasma (97, 117).

1.4.2 Attachment of Receptors to Functional Surfaces

The immobilization of biorecognition elements to functional surfaces usually involves a more or less specific interaction between the biorecognition element

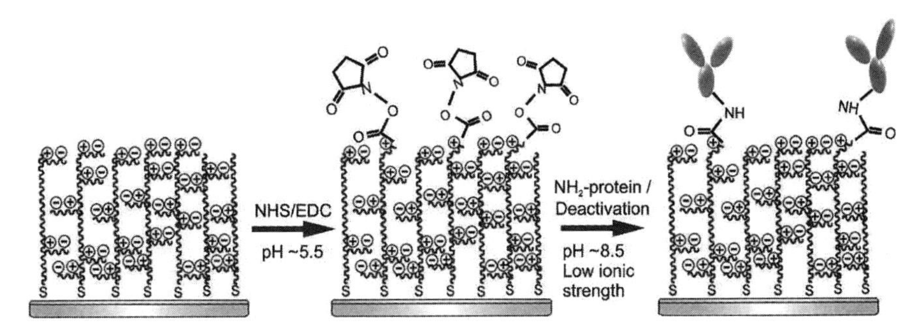

FIGURE 1.20 Scheme of the surface activation, protein immobilization, and surface deactivation of a poly(carboxybetaine acrylamide)-coated surface. *Source:* Reprinted, by permission, from Reference 116.

and a functional group located on the sensor surface. The main interactions employed in the immobilization of biorecognition elements include physical adsorption, the formation of a covalent bond, and specific interactions involving functional molecules.

1.4.2.1 Physical Adsorption Physical adsorption is suitable for biosensors in which no directional orientation of biomolecules is necessary. The immobilization is performed via electrostatic or hydrophobic interactions. Hydrophobic interaction is an alternative method for the immobilization of biorecognition elements with lipophilic properties, such as membrane proteins or liposomes (118). The sensor chip is modified with hydrophobic residues that are inserted in the membrane and create a stable lipid surface. Immobilization via electrostatic interaction takes advantage of the electrostatic attraction between the charged molecules and the oppositely charged surface. Surfaces containing positively charged amines or negatively charged carboxylic groups are most suitable for this approach. However, there is a drawback in this approach, that the resulting surfaces are prone to nonspecific adsorption of charged molecules and are therefore not well suited for the analysis of complex samples.

1.4.2.2 Covalent Immobilization Covalent immobilization presents a more defined alternative to physical adsorption. In this approach, the functional group on the biorecognition molecule is coupled to the functional group on the surface. This usually requires transformation of one of the functional groups into a more reactive form.

Amine Coupling The most versatile and widely used approach involves carboxylic moieties and reactive nucleophile functional groups. For the formation of a covalent bond, the activation of carboxylic groups with carbodiimide is often performed (Fig. 1.21a). In aqueous solution, 1-ethyl-3-(3-dimethylopropyl)3-(3-Dimethylopropyl) (EDC) is used to transform the carboxylic group into reactive O-acyl isourea intermediate, which can then readily react with a nucleophile. To prevent the reactive groups from transforming back to carboxylic groups, a reactive hydroxyl compound is usually added forming an active ester that is stable for several minutes to hours. N-hydroxy succinimide (NHS) serves this purpose well and is therefore often added to EDC solution.

The common scenario for protein immobilization is that carboxylic groups are introduced to the functionalized surface and lysine residues on the protein provide the nucleophile groups. The immobilization is often performed below the pK of the protein taking advantage of the electrostatic attraction between the positively charged protein and the negatively charged surface containing unreacted carboxylic groups. In such cases, a much lower concentration of the protein and shorter incubation times can be used. In some instances, the reverse order is preferred in which the protein carries the activated carboxylic groups and the surface contains functional amine groups. This situation arises when the biorecognition element lacks reactive amines, or they are suspected to be situated too close to its binding site.

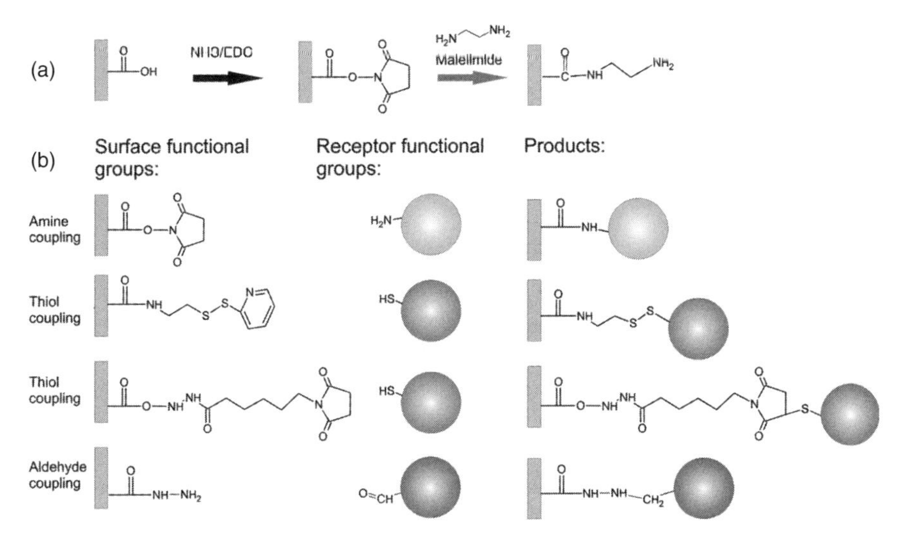

FIGURE 1.21 Covalent immobilization of receptors. (a) Activation of the carboxylic groups with NHS/EDC mixture resulting in reactive succinimide esters. (b) Scheme of reactions giving rise to a covalent bond between the surface and the receptors.

Thiol Group Coupling Thiol coupling is one of the most commonly used immobilization methods. It is based on the reaction of thiol groups and thiol-reactive functional groups, such as pyridyl disulfides and their derivatives, maleiimides, or acyl halide derivatives (119) (Fig. 1.21b). The thiol group may be located on the receptor and the thiol-reactive group on the surface or vice versa. The multiple commercially available reagents for receptor modifications with thiol-reactive groups are typically directed toward amine groups.

The disulfide bonds can be selectively coupled under very mild conditions with little or no interference with other nucleophiles. In addition, they may be easily disrupted in solutions containing thiols (e.g., mercaptoethanol), which can be exploited for regeneration of the sensor surface and its repeated use. The interaction of thiols and maleimides or acryl halide produces a thioether linkage, which is more stable than a disulfide bond. However, it is also less specific and subject to interference from other nucleophilic groups.

Aldehyde Coupling The formation of a covalent bond can also be achieved with Schiff-base condensation of aldehyde groups to amines and hydrazines (Fig. 1.21b). This method is particularly useful for the immobilization of glycoproteins, where an aldehyde functional group is formed on carbohydrate residues under mild oxidative conditions. The carbohydrates are not usually situated near the active site of the protein and therefore this approach allows for the immobilization of proteins without compromising their activity. Although the Schiff-base formation can be performed with amine groups, hydrazine is a better option as it is more stable in aqueous environments. It can be introduced to the surface by coupling hydrazine or carbohydrazine to activated carboxylic groups after activation with NHS/EDC. The Schiff-base reaction presents an alternative to the amine coupling and is well suited for the immobilization of small proteins, the binding sites of which may deteriorate during random coupling.

1.4.2.3 Coupling Via Functional Molecules Although immobilization of biorecognition elements via nonspecific interactions or covalent bonds is rapid, simple, and cheap, such an approach often leads to a random orientation of the immobilized molecules, which may result in a partial loss of the receptor activity. Immobilization via functional molecules and linkers based on noncovalent interactions allows for the creation of well-defined and oriented biomolecular assemblies.

Avidin–Biotin Interaction A popular strategy for receptor immobilization inspired by affinity chromatography is the specific binding of biotin to the tetramer avidin (or streptavidin) that is anchored to the sensor surface. Streptavidin is usually preferred over glycosylated avidin, which may increase the undesired nonspecific adsorption. The streptavidin–biotin interaction is considered one of the strongest and almost irreversible bonds with an affinity constant $K_D \sim 10^{-15}$ M. Streptavidin monolayers are commonly formed on the sensor surface by amine coupling or by the use of a biotinylated functional surface. The latter allows for the creation of a multilayer structure: the biotin–streptavidin–biotinylated biorecognition element, which yields a high degree of organization of the immobilized biorecognition elements (120). In some instances, lower affinity ligands toward streptavidin may be preferred. Biorecognition elements conjugated with such ligands may be immobilized in an oriented manner and their bonds to the streptavidin can be disrupted with a regeneration solution, allowing for repeated use of the sensor (121, 122).

Immobilization Via a Complementary Nucleic-Acid Strand Site-directed immobilization of biorecognition elements via sequence-specific hybridization takes advantage of well-established methods for the immobilization of nucleic acids and DNA chip technology (123). In this approach, one DNA strand is incorporated in the biorecognition element (typically protein). Taking advantage of the specificity of a Watson-Crick base pairing of two complementary single-stranded nucleic acids, the protein is site-directly immobilized to the complementary DNA strand on the sensor surface (124, 125). This method produces a well-oriented and stable functional surface, which can be easily regenerated by solvents of low or high pH. The main challenge for this method consists in the incorporation of the DNA strand to the complex protein molecule as discussed by C. Niemeyer (126). The immobilization of proteins in the form of DNA conjugates has become quite popular in high-throughput SPR sensors, where it can be exploited to create an array of sensing areas, each containing different biorecognition elements (127, 128).

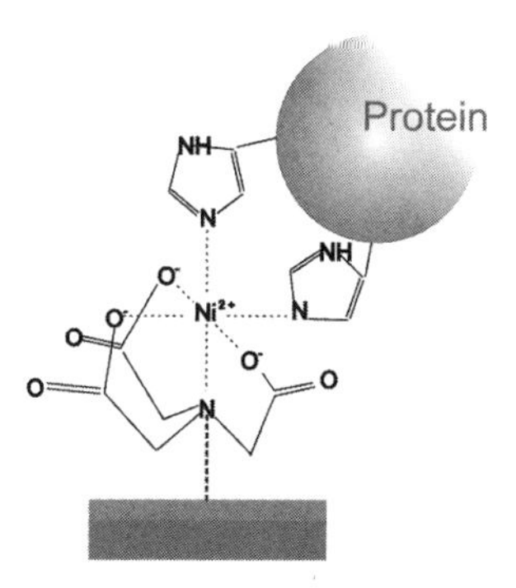

FIGURE 1.22 Interaction between His-tagged protein and NTA–Ni functionalized gold surface.

Nickel Nitrilotriacetic Acid Nickel nitriloacetate (Ni–NTA) chemistry can be used for the immobilization of His-tagged biorecognition elements (129). This approach is particularly useful for the immobilization of recombinant proteins bearing His tags, which are produced with genetic engineering. The NTA surfaces may be produced with a covalent attachment of NTA derivatives to dextran or SAM surfaces. The NTA forms a hexagonal complex with divalent metal ions, especially Ni^{2+}, with two binding positions available for binding to the His tag (Fig. 1.22). The His-tagged receptor can be easily removed with ethylenediaminetetraacetate (EDTA) or imidazole. The main drawback of the Ni–NTA immobilization method is the metal-dependent nonspecific adsorption to the surface and the low affinity of His tag to the Ni–NTA complex (K_D is in the order of μM) potentially leading to dissociation of the complex (130). The affinity may be significantly enhanced by the incorporation of multiple His tags into the biorecognition element and increasing the surface density of NTA groups, while preserving the reversibility of the complex formation (131).

Protein A/G Specific antibody-binding proteins such as protein A and protein G can be used for oriented immobilization of antibodies (132, 133). Antibodies consist of one Fc fragment and two Fab subunits, which contain antigen-binding regions. Protein A or protein G specifically binds the Fc region of certain kinds of antibodies. Therefore, the immobilization of antibodies to proteins A or G immobilized on the sensor surface (e.g., via covalent immobilization or thiol chemisorption to gold) can lead to the favorable orientation of antibodies and a high binding capacity of the sensing surface (134). To form a covalent bond between the G protein and the antibody, the binding site of the protein G may be mutated and modified with photo cross-linker benzophenone (135). On selective binding of the antibody, UV light is used to create the covalent bond between the Fc fragment and the protein G.

1.4.3 Molecular Recognition Elements

Various biorecognition elements have been employed in SPR biosensors. These include proteins, peptides, nucleic acids, carbohydrate structures and small organic or inorganic molecules. Typically, biorecognition elements are attached to the functional groups available on functional layers via interactions described in Section 1.4.2. Owing to the diverse chemical structures of the receptors, no single strategy can be used as a universal approach. Sometimes, even chemical modifications are necessary to introduce functional groups. This section summarizes the most common immobilization strategies used for the main types of biorecognition clements.

1.4.3.1 Proteins and Peptides In SPR sensors for bioanalytical applications, proteins are the most widely used biorecognition elements. They include antibodies, small peptides consisting of a few amino acids, as well as large proteins. As proteins are heterogeneous in terms of amino acid composition and a secondary structure, no universal immobilization method appropriate for all the proteins is available. Most frequently, the immobilization is performed via amino acids that contain a wide range of functional groups, such as $-NH_2$ (lysine), –SH (cystein), or –COOH (asparagine and glutamine).

The covalent immobilization is most often realized via amine coupling to activated carboxylic groups on the surface. This approach may be impaired by an abundance of lysines (over 10%) in the proteins, which often leads to heterogeneity and restricted protein flexibility owing to the multipoint attachment of the protein. Protein-containing cystein residues can be immobilized directly on the gold surface. A passivation step is then often necessary in order to block the remaining areas of the surface. Cysteins may also be used for the thiol coupling procedure as described above. As their availability is rather low (<1%), the cysteins may be introduced to the protein through site-specific mutation or the attachment of oligo-Cys tag (132). When covalent immobilization is not an option, alternative immobilization strategies include the introduction of recombinant tags (oligohistidines, biotin, etc.) or the attachment of the protein directly to the surface via conjugated linkers (136).

1.4.3.2 Nucleic Acids DNA probes (ssDNA, single-stranded DNA) are traditionally used for the detection of complementary RNA and DNA oligonucleotides. Alternatives include nucleic-acid derivatives, such as peptide nucleic acids (PNAs) or locked nucleic acids (LNAs), which have a higher affinity toward the complementary DNA/RNA strands than unmodified ssDNA. Short DNA or RNA molecules of various secondary structures (aptamers) are used for the detection of other molecules, such as proteins, whole cells, or small organic compounds (137, 138). Aptamers can be tailored to self-assemble to lattice structures on surface of a sensor (139) and can be used in microarrays as an alternative to antibodies (140).

Nucleic acids are composed of nitrogenous bases and negative phosphate groups, which are relatively resistant toward covalent coupling under mild aqueous conditions. Similar rules apply to DNA derivatives, such as peptide and locked nucleic acids. The immobilization of ssDNA oligonucleotide to a gold surface is therefore combined with chemical modification of one of the strand's ends. The common approach to the immobilization of nucleic acids is direct assembly of thiol-terminated ssDNA on gold (119), usually in combination with SAM (141, 142). This approach is also compatible with SPR sensor arrays and high-throughput applications (143). Another quite common strategy is the use of biotinylated derivatives, which can be obtained in the 3′ and 5′-end position (120). Recently, the coupling of N-terminated ssDNA strands to a monolayer of ply-L glutamic acid has been reported (144). Rewritable DNA arrays may be obtained by using plasma deposited poly(allylmercaptan) and the immobilization of DNA strands via disulfide chemistry (145).

1.4.3.3 Other Biorecognition Elements Other molecules often immobilized in SPR biosensors include carbohydrates and small organic molecules. The carbohydrates are present in glycoproteins and can be used for their immobilization. The primary coupling sites for sugars are hydroxyl groups, although they may play a crucial role in the binding to the analyte. Alternative strategies include the use of anomeric aldehyde groups for direct attachment to the surface, or for modification with linker molecules (thiol or biotin).

The use of small organic molecules, such as toxins, hormones, vitamins, and drug candidates with a low molecular weight (lower than 700 Da) in SPR biosensors has been growing. However, immobilization of these molecules is still rather challenging, as these molecules are often insoluble in aqueous media and often require the addition of organic solvents (e.g., DMSO or DMF). The number and position of functional groups is also limited. The molecules with aliphatic amines, thiols, and carboxylic groups can be covalently linked to the surface, as described previously. However, the use of those groups for immobilization may significantly alter the binding activity of the immobilized molecule. Synthesis of derivatives, conjugation with linkers (146), or larger molecules (typically proteins) (147) is therefore often necessary.

1.5 APPLICATIONS OF SPR BIOSENSORS

SPR biosensors have been applied to the detection of various analytes from small organic molecules, proteins, hormones, cancer markers to even large bacteria (18). With the advances in surface chemistries, SPR biosensors are used to tackle bioanalytical tasks in increasingly complex sample matrices. This section discusses applications of SPR biosensors for the detection of chemical and biological species with an emphasis on applications in food quality and safety.

1.5.1 Detection Formats

In SPR sensors, the detection of target molecules is performed in several formats. The choice of format depends on the details of the application, in particular on the size of the target molecule. Figure 1.23 shows four of the most commonly used detection formats: the direct detection format, the sandwich detection format, the competition detection format, and the inhibition detection format.

In the direct detection of the analyte (Fig. 1.23a), the analyte molecule is bound to the biorecognition element (e.g., antibody) immobilized on the surface of an SPR sensor, which gives rise to the SPR sensor response. To enhance the sensor response to the bound analyte, the sandwich detection format is often used (Fig. 1.23b). In this format, the secondary antibody is bound to the analyte which was previously captured on the sensor. To further enhance the amplification effect, the second antibody can be labeled with a mass tag (e.g., gold nanoparticles). The competitive detection format (Fig. 1.23c) is based on two analytes competing for the binding site on the chip surface. One of the analytes is free and the other is typically conjugated with a large molecule. The SPR sensor response is irreversibly proportional to the concentration of free analyte in the solution. In the inhibition detection format (Fig. 1.23d), the analyte is typically immobilized on the sensor surface. Subsequently, the mixture of an antibody and a free analyte in the solution is brought in contact with the sensor surface. The unreacted antibody is bound to the analyte immobilized on the sensor surface. The sensor response is then irreversibly proportional to the analyte concentration in the solution.

1.5.2 Medical Diagnostics

Modern health care demands the rapid, sensitive, and specific detection of various markers of health or diseases (e.g., cancer markers, cardiac markers) in bodily fluids (e.g., human plasma or serum, cerebrospinal fluid) (148). The detection of most of the biomarkers is demanding as the biomarkers are present at very low concentrations. Moreover, the biological samples are rather complex and interfere with SPR measurements. In order to address these challenges, low-fouling coatings and amplification approaches are used.

One of the most important areas in medical diagnostics is the early detection of cancer (148). In the last decade, various cancer markers have been detected by SPR biosensors such as prostate-specific antigen (prostate cancer) (149, 150), activated leukocyte cell adhesion molecule (pancreatic carcinoma) (97, 116, 151, 152), carcinoembryonic antigen (lung and breast cancer, epithelial ovarian tumors) (153), and cancer antigen 125 (ovarian cancer) (154). SPR biosensors for the detection of heart infarction have also been developed on the basis of the detection of cardiac markers such as myoglobin (155, 156), cardiac troponin (155, 157, 158), and brain natriuretic peptide (159). Other important biomarkers targeted by SPR biosensor technology include cholesterol (160–162), insulin autoimmune antibodies, and insulin (163).

1.5.3 Environmental Monitoring

Various chemical compounds (herbicides, pesticides, endocrine-disrupting chemicals, polychlorinated biphenyls) produced in industry and agriculture that are

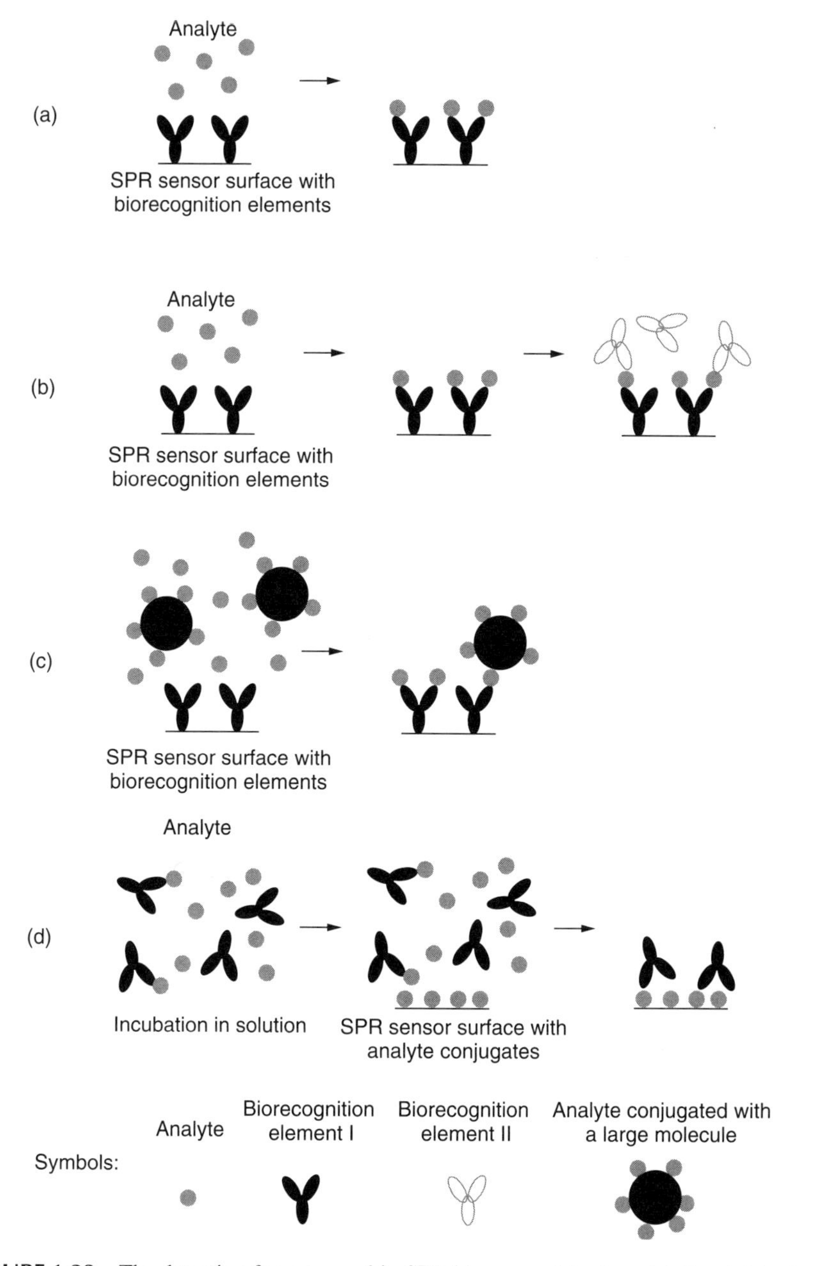

FIGURE 1.23 The detection formats used in SPR biosensor detection: (a) direct detection, (b) sandwich detection format, (c) competition detection format, and (d) inhibition detection format. *Source:* Reprinted, by permission, from Reference 18.

released into the environment are harmful for humans and animals. As such, they have been targeted by modern bioanalytical technologies, including SPR biosensors. Many of these chemicals have a rather low molecular weight and therefore SPR biosensors for the detection of these analytes usually use the inhibition and competition detection formats.

SPR biosensors have been widely applied for the detection of herbicides and pesticides such as 2,4-dichlorophenoxyacetic acid, isoproturon, DDT, chlorpyrifos, and carbaryl (164–169). Other environmental analytes targeted by the SPR biosensors include bisphenol A (170–173) and polychlorinated biphenyls (PCBs) (169, 174). SPR biosensors have also been used for the detection of explosives such as 2,4,6-trinitrotoluene (TNT) and 2,4-dinitrotoluene (DNT) (175–179).

1.5.4 Food Quality and Safety

Food contaminants may pose a considerable risk to human health (180, 181). For instance, the number of cases of gastroenteritis associated with food is estimated to be between 68 and 275 millions per year. There are many analytes that belong to this class, including low-molecular-weight toxins, toxic peptides, or pathogenic bacteria. Direct, sandwich, competitive, and inhibition detection formats have been used in the detection of these contaminants. As will be demonstrated, SPR biosensors offer a successful tool that enables fast and sensitive detection of these molecules. Moreover, the recent progress in surface chemistries of SPR biosensors has extended the efficient detection of these contaminants in real samples.

1.5.4.1 Low-Molecular-Weight Toxins This section reviews recent applications of SPR biosensors for the detection of low molecular weight (typically below 1 kDa). As the size of these compounds is too small to produce a strong sensor response in direct detection format, competitive or inhibition detection formats are typically used.

Mycotoxins Mycotoxins are fungal metabolites (e.g., aflatoxin B_1 (AFB_1), deoxynivalenol, zearalenone, ochratoxin A, and fumonisin B_1), which are often found in foods and have harmful effects on human health.

AFB_1 is a mycotoxin that is produced mainly by *Aspergillus flavus* and *Aspergillus parasiticus*. AFB_1 was detected by Dunne et al. using a Biacore SPR system and an inhibition assay involving single-chain antibody fragments (scFv) (182). AFB_1 derivate was immobilized on a CM5 chip surface and the mixture of aflatoxin and monomeric or dimeric scFvs was injected into the sensor flow cell. AFB_1 was detected in the concentration range from 375 pg/ml to 12 ng/ml and 190 pg/ml to 24 ng/ml for monomeric and dimeric scFvs, respectively. The sensing surface was regenerated and used in more than 70 detection cycles. In 2008, Cuccioloni et al. reported a sensitive approach, which was based on the interaction of neutrophil porcine elastase and AFB_1 to detect AFB_1 in maize extract (183). Elastase was immobilized on the sensor surface and AFB_1 was directly detected. The limit of quantification for this assay was determined to be 3.1 ppb (=ng/g) in maize extract, which is slightly above the EU limit for aflatoxin in food (<2 ppb).

Deoxynivalenol is a mycotoxin that is predominantly found in grains such as wheat, barley, oats, rye, and maize. Tudos et al. used a Biacore system and inhibition assay to detect deoxynivalenol in wheat samples (184). Deoxynivalenol was conjugated to casein and covalently immobilized on the surface of a CM5 chip. Three different monoclonal antibodies were evaluated in this work. Deoxynivalenol in buffer was detected in concentrations from 1 to 100 ng/ml. Ten contaminated wheat samples with concentrations of deoxynivalenol from 170 to 1400 ppb were analyzed. The results obtained using an SPR biosensor were shown to be consistent with those obtained via liquid chromatography–mass spectrometry.

Bram van der Gaag et al. demonstrated the simultaneous detection of multiple mycotoxins (aflatoxin B_1, zearalenone, ochratoxin A, deoxynivalenol, and fumonisin B_1) using a Biacore 2000 system and an inhibition assay (Fig. 1.24) (185). The toxins were separately immobilized of the Biacore 2000 apparatus. The solutions of toxins and antibodies were then flowed through the serially connected channels, which enabled the simultaneous detection of all the toxins. The LODs were determined as 0.2, 0.01, 0.1, 50, and 0.5 ppb for aflatoxin B_1, zearalenone, ochratoxin A, deoxynivalenol, and fumonisin B_1, respectively.

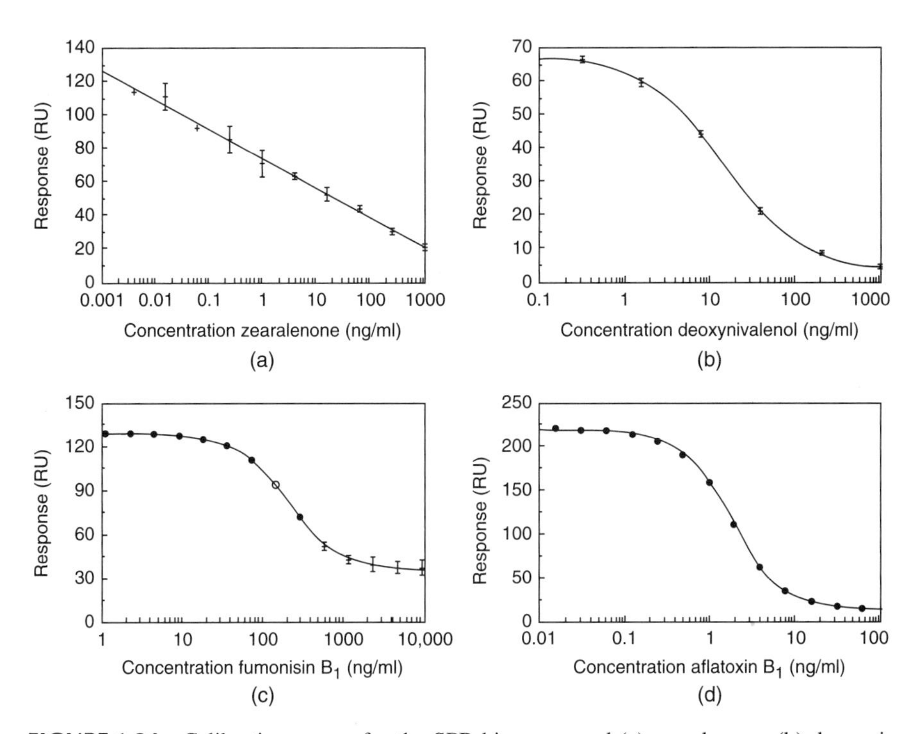

FIGURE 1.24 Calibration curves for the SPR biosensor and (a) zearalenone, (b) deoxynivalenol, (c) fumonisin B_1, and (d) aflatoxin B_1 measured with an inhibition assay. *Source:* Reprinted, by permission, from Reference 185.

Paralytic Shellfish Poisoning Toxin Paralytic shellfish poisoning (PSA) toxins are a group of more than 20 different neurotoxins found in both freshwater and marine environments. Several SPR biosensors for the detection of PSA have been reported in the last decade (186–188). One of the most dangerous toxins, saxitoxin, was targeted by Campbell et al. who used Biacore Q and an inhibition assay to detect PSA in shellfish samples (187). In their study, saxitoxin was immobilized on a dextran chip surface and the mixture of PSA toxin and antibody was injected to the surface of an SPR sensor. A total of 52 real shellfish samples were investigated and the LOD was determined to be 370 or 700 ppb depending on the extraction methods used. The cross-sensitivity of the developed SPR biosensor to nine other toxins was also studied and was found to be rather low for most of the toxins. A similar method based on the inhibition format was used by Fonfria et al. (186) who detected saxitoxin in mussels, cockles, clams, scallops, and oysters. Both these studies demonstrated that SPR biosensors are capable of detecting saxitoxin at concentrations below the regulatory limit of EU (800 ppb) (186).

Domoic Acid Domoic acid (DA) is a neuroexcitatory toxin that is typically produced by planktonic algae. Several SPR biosensors for the detection of DA have been reported (189–192). Yu et al. detected DA in buffer using a dual-channel SPR sensor (developed at the Institute of Photonics and Electronics, Prague) and an inhibition assay (189). DA was immobilized on the surface of the SPR sensor and antibodies were flowed along the chip surface. The LOD was established at 0.1 ng/ml. This LOD was found to be better by a factor of 4 than that provided by the reference method—ELISA. In 2006, the detection of DA in extracts of shellfish species was reported by Traynor et al. (190). In this work, to reduce the nonspecific adsorption, the extracts were filtered and diluted. The LOD was in the parts per million (in the order of microgram per gram) range. The achieved limits of detection were somewhat higher than those previously achieved in buffer; however, it was still under the EU's official action limit for DA of 20 ppm. In 2007, Stevenson et al. used a portable SPR biosensor to detect DA in buffer and diluted clam extracts down to 3 ng/ml (192).

Okadaic Acid Okadaic acid (OA) is a toxin in bivalves, which causes diarrhetic shellfish poisoning. SPR biosensors have been used to detect OAs in several studies (193, 194). Prieto-Simons utilized a Biacore SPR system, an antibody against DA covalently immobilized on a surface of SPR sensor and two competitive assays (193). Free OA and OA-albumin conjugates or OA-avidin complex was mixed and injected to the chip surface. The detection limits of 0.1 and 1 ng/ml were achieved in buffer for the OA-albumin conjugate assay and the OA-avidin complex assay, respectively. The more sensitive assay (employing OA-albumin conjugates) was employed to detect OA in mussel extracts. To decrease the complex matrix effect, the extracts were filtered before the analysis. The LOD for the spiked samples was determined to be 0.24 ng/ml.

Tetrodotoxin Tetrodotoxin (TTX) is a highly dangerous neurotoxin found in many animal species, including pufferfish, newts, and toads. Taylor et al. used an inhibition assay to detect TTX (195). TTX was covalently bound to the chip surface and a mixture of antibodies and TTX was flowed along the chip surface. The concentration of anti-TTX on the chip surface was optimized and the LOD in buffer was demonstrated to be 0.3 ng/ml. Vaisocherová et al. demonstrated an SPR biosensor for the detection of TTX in pufferfish samples with the LOD of 1.5 ng/ml (196).

1.5.4.2 Toxic Peptides This section discusses the use of SPR biosensors for the detection of highly toxic peptides, including botulinum neurotoxins, staphylococcal enterotoxins, and ricin. The weight of these compounds ranges from 5 to 150 kDa and therefore they are typically detected using direct detection or the sandwich format.

Botulinum Neurotoxins Botulinum neurotoxin, which is produced by the bacterium *Clostridium botulinum*, is considered to be one of the most toxic biological substances. While botulinum neurotoxin presents an important concern for food safety, it is also a potential security threat. Botulinum neurotoxin has been investigated in several studies by SPR biosensors (197, 198). In the study of Ladd et al., the botulinum neurotoxin serotypes A, B, and F were detected using the respective antibodies and a sandwich assay (197). In buffer, concentrations as low as 0.5–1 ng/ml were detected. The detection of neurotoxins was also performed in 20% honey and the LOD of about 5 ng/ml was achieved.

Staphylococcal Enterotoxins Staphylococcal enterotoxin B (SEB) is a foodborne toxin that is produced by the bacterium *Staphylococcus aureus*. Several SPR biosensors for the detection of staphylococcal enterotoxins have been developed in the last decade (199–202). In 2005, Medina et al. detected SEB in milk using the inhibition detection format (199). SEB was immobilized on the SPR sensor chip surface. Then SEB and anti-SEB antibodies in milk samples were injected to the sensor flow cell. This approach allowed for the detection of SEB in milk at concentrations ranging from 0.3 to 25 ppb. A similar approach was used for the detection of staphylococcal enterotoxin A (SEA) in whole eggs (200). Eggs were homogenized and spiked with SEA. Then the anti-SEA antibody was added and the mixture was centrifuged. Subsequently, SEA was successfully detected at concentrations from 1 to 40 ng/ml using the inhibition detection format. SEB was also detected by Naimushin et al. who demonstrated a portable two-channel SPR biosensor (201). A two-step amplification format and long incubation times were used to detect low levels of SEB in buffer. The lowest concentration, which was successfully detected, was 2.9 pg/ml.

Ricin Ricin is a highly toxic protein that is extracted from castor beans (*Ricinus communis*). Ricin toxin has also been targeted with SPR biosensors (51, 203). Tran et al. evaluated 10 monoclonal antibodies against commercial ricin and horticultural ricin variants from 6 different cultivars of *Ricinus communis* (203). The most sensitive antibody was chosen to detect commercial ricin and the LOD of 0.5 ng/ml was achieved. Further amplification with the secondary antibody was demonstrated to improve the LOD to 0.1 ng/ml. Ricin detection was also performed in six spiked environmental samples and the results were consistent with those obtained in buffer.

1.5.4.3 Bacteria Bacteria present an important problem for food safety and security (180). In the last decade, numerous SPR biosensors for the detection of bacterial pathogens have been developed. The main bacteria targeted by SPR biosensors include *Escherichia coli, Staphylococcus aureus, Salmonella, Listeria monocytogenes, Vibrio cholerae*, and *Bacillus anthracis*. SPR detection of bacteria poses several unique challenges. The dimensions of bacteria (typically in the range of micrometers) exceed the penetration depth of the evanescent field of conventional SPs. In addition, the large size of bacteria leads to rather low diffusion coefficients, which limits the transport of bacteria to the sensing surface. Therefore, various amplification approaches are often employed to enhance the sensor response and thus the sensitivity of the method.

Escherichia coli *Escherichia coli* (*E. coli*) have been a frequent target of SPR biosensors. *E. coli* was detected by Su et al. who used protein A to immobilize antibodies against *E. coli* on the surface of an SPR sensor (204). The bacteria were detected directly in the concentration range from 10^6 to 10^8 cfu/ml. Taylor et al. used the sandwich detection format with primary and secondary antibodies to detect *E. coli* in orange juice (205). The calibration curve is depicted in Figure 1.25. This approach has demonstrated a significant improvement in the sensitivity of SPR-based detection of *E. coli* and has achieved an LOD as low as 10^4 cfu/ml. The sandwich detection format was also employed by Linman et al. who detected *E. coli* in fresh spinach (206). The primary antibody was immobilized on an SPR chip. Then *E. coli* in the spinach sample was captured and the secondary antibody modified with horse radish peroxidase (HRP) was bound to the captured bacteria. Subsequently, tetramethylbenzidine in solution was allowed to react with HRP to produce insoluble precipitate on the chip surface. The SPR sensor response to the formation of the precipitation was monitored. This method also enabled the detection of *E. coli* down to 10^4 cfu/ml. Another approach to *E. coli* detection was presented by Eum et al. (207) who used the sandwich detection format and a secondary antibody conjugated with gold nanorods. This approach offered a fourfold improvement in the sensitivity when compared to the assay using a nonlabeled secondary antibody. Recently, Vala et al. used LRSPs to detect *E. coli* (208). It was demonstrated that this approach provided up to a sixfold improvement in sensitivity compared to detection using conventional SPs.

Staphylococcus aureus The detection of *Staphylococcus aureus* was performed by Chen et al. who immobilized the human protein IgG, which is specific against protein A on the surface of *S. aureus*, to a dextran matrix on an SPR chip (209).

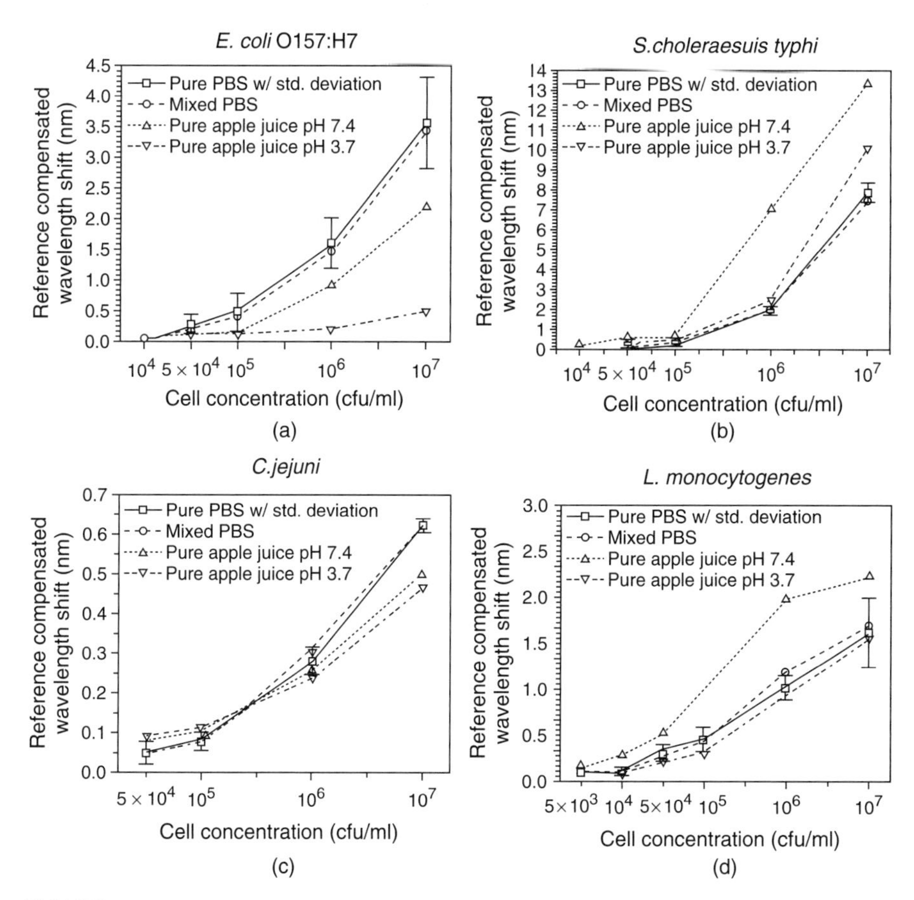

FIGURE 1.25 SPR detection of four different bacteria: (a) *Escherichia coli*, (b) *Salmonella choleraesuis* serotype typhimurium, (c) *Campylobacteur jejuni*, and (d) *Listeria monocytogene*. Experiments were performed in buffer and in apple juice samples. *Source:* Reprinted, by permission, from Reference 205.

Using this approach, *S. aureus* was detected directly at concentrations down to 10^6 cfu/ml. Moreover, the experiment with *E. coli, S. epidermidis*, and *B. subilis* showed that this method is specific only for *S. aureus*. The detection of *S. aureus* was also performed by Subramanian et al. (210) who used the sandwich detection format and achieved an LOD of 10^5 cfu/ml.

Salmonella Biosensors for the detection of *Salmonella* have also been reported. *Salmonella choleraesuis* serotype *typhimurium* in buffer and in apple juice was detected by Taylor et al. (205). They used the sandwich detection format and demonstrated an LOD of 10^5 cfu/ml, both in buffer and in apple juice. Baren et al. used the sandwich assay to sequentially detect two *Salmonella* serovars in milk using the same sensing channel (211). The limits of detection for *Salmonella typhimurium* and *Salmonella enteritidis* were demonstrated to be 2.5×10^5 cfu/ml and 2.5×10^8 cfu/ml, respectively.

Listeria Several studies have reported SPR biosensors for the detection of *Listeria monocytogenes* (205, 212, 213). Nanduri et al. physically adsorbed bacteriophages with single-chain antibodies against *L. monocytogenes* on the sensing surface, which resulted in the direct detection of *L. monocytogenes* down to 2×10^6 cfu/ml (212). An alternative approach, as shown in Figure 1.25, was described by Taylor et al. who used the sandwich detection format to detect *L. monocytogenes* in apple juice (205). This study reported the LOD as low as 5×10^4 cfu/ml.

Vibrio cholerae O1 In 2006, Joung et al. performed the direct detection of *Vibrio cholerae* O1 (214). They employed monoclonal antibodies against *V. cholerae*, which were immobilized on the sensor surface via a protein G layer. Detectable sensor response was observed for *V. cholerae* concentrations between 10^5 and 10^9 cfu/ml.

Bacillus anthracis *Bacillus anthracis*, an important biological warfare agent, has also been targeted by SPR biosensors. A mouse monoclonal antibody was pre-incubated with Bacillus spores. Then the free antibodies were separated by centrifugation and detected with an SPR sensor functionalized with anti-mouse antibody. By this method, a concentration of whole spores of *B. anthracis* as low as 10^4 cfu/ml was determined within 40 min (215). The low cross-reactivity of the method with other related Bacillus spores (at a concentration of 10^7 cfu/ml) was demonstrated (Fig. 1.26).

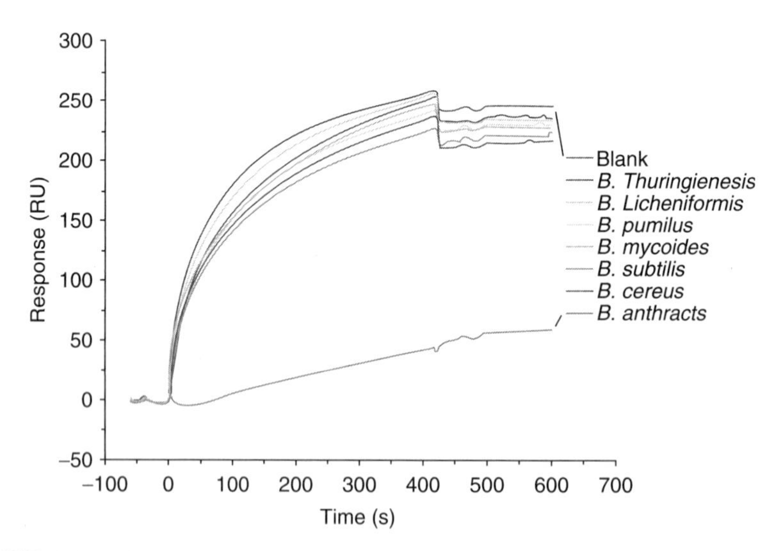

FIGURE 1.26 Cross-reactivity of the monoclonal antibody against *Bacillus anthracis*. *Source:* Reprinted, by permission, from Reference 215.

1.6 SUMMARY

SPR biosensors represent one of the most advanced label-free biophotonic sensing technologies. SPR biosensors have become a mainstay of both life science and pharmaceutical research and have enabled a huge number of studies investigating biomolecules and their interactions. In order to facilitate parallelized monitoring of a large number (>100) of molecular interactions, high-throughput SPR biosensors have emerged. These sensors are expected to expand the capabilities of conventional SPR biosensors and meet the growing need of pharmaceutical research. SPR biosensors are also increasingly applied in bioanalytics. These applications are supported by substantial recent advances in the development of portable SPR instruments for field use and the development of functionalization methods suppressing nonspecific adsorption from complex real-world samples. It is envisioned that these advances will facilitate the penetration of SPR biosensor technology into various areas including medical diagnostics, environmental monitoring, food safety, and security.

REFERENCES

1. Vo-Dinh T. *Nanotechnology in Biology and Medicine: Methods, Devices, and Applications*. Boca Raton (FL): CRC Press; 2007.
2. Ligler FS, Taitt CAR. *Optical Biosensors: Present and Future*. Amsterdam, Oxford: Elsevier; 2002.
3. Nakamura H, Karube I. Current research activity in biosensors. Anal Bioanal Chem 2003;377(3):446–468.
4. Gorton L. *Biosensors and Modern Biospecific Analytical Techniques*. 1st ed. Amsterdam, Boston (MA): Elsevier; 2005.
5. Heideman RG, Lambeck PV. Remote opto-chemical sensing with extreme sensitivity: design, fabrication and performance of a pigtailed integrated optical phase-modulated Mach-Zehnder interferometer system. Sens Actuator B Chem 1999;61(1–3):100–127.
6. Schmitt K, Schirmer B, Hoffmann C, Brandenburg A, Meyrueis P. Interferometric biosensor based on planar optical waveguide sensor chips for label-free detection of surface bound bioreactions. Biosens Bioelectron 2007;22(11):2591–2597.
7. Schmitt H-M, Brecht A, Piehler J, Gauglitz G. An integrated system for optical biomolecular interaction analysis. Biosens Bioelectron 1997;12(8):809–816.
8. Cush R, Cronin JM, Stewart WJ, Maule CH, Molloy J, Goddard NJ. The resonant mirror: a novel optical biosensor for direct sensing of biomolecular interactions Part I: principle of operation and associated instrumentation. Biosens Bioelectron 1993;8(7–8):347–354.
9. Clerc D, Lukosz W. Direct immunosensing with an integrated-optical output grating coupler. Sens Actuator B Chem 1997;40(1):53–58.
10. Gordon JG, Ernst S. Surface-plasmons as a probe of the electrochemical interface. Surf Sci 1980;101(1–3):499–506.
11. Nylander C, Liedberg B, Lind T. Gas-detection by means of surface-plasmon resonance. Sens Actuator 1982;3(1):79–88.

12. Rebecca L, Rich DGM. Survey of the year 2005 commercial optical biosensor literature. J Mol Recognit 2006;19(6):478–534.
13. Malhotra BD, Turner APF. *Perspectives in Biosensors*. Amsterdam: JAI Press; 2003.
14. Ince R, Narayanaswamy R. Analysis of the performance of interferometry, surface plasmon resonance and luminescence as biosensors and chemosensors. Anal Chim Acta 2006;569(1–2):1–20.
15. Bally M, Halter M, Voros J, Grandin HM. Optical microarray biosensing techniques. Surf Interface Anal 2006;38(11):1442–1458.
16. Boozer C, Kim G, Cong S, Guan H, Londergan T. Looking towards label-free biomolecular interaction analysis in a high-throughput format: a review of new surface plasmon resonance technologies. Curr Opin Biotechnol 2006;17(4):400–405.
17. Phillips KS, Cheng Q. Recent advances in surface plasmon resonance based techniques for bioanalysis. Anal Bioanal Chem 2007;387(5):1831–1840.
18. Homola J. Surface plasmon resonance sensors for detection of chemical and biological species. Chem Rev 2008;108(2):462–493.
19. Homola J. *Surface Plasmon Resonance Based Sensors*. Berlin: Springer; 2006.
20. Raether H. Volume 111, *Surface Plasmons on Smooth and Rough Surfaces and On Gratings*, *Springer Tracts in Modern Physics*. Berlin (NY): Springer-Verlag; 1988. p 1–133.
21. Boardman AD. *Electromagnetic Surface Modes*. Chichester, New York: John Wiley and Sons; 1982.
22. Burke JJ, Stegeman GI, Tamir T. Surface-polariton-like waves guided by thin, lossy metal-films. Phys Rev B 1986;33(8):5186–5201.
23. Sarid D. Long-range surface-plasma waves on very thin metal-films. Phys Rev Lett 1981;47(26):1927–1930.
24. Kretschmann E, Raether H. Radiative decay of non radiative surface plasmons excited by light. Z Naturforsch [A] 1968;23:2135–2136.
25. Born M, Wolf E. *Principles of Optics: Electromagnetic Theory of Propagation, Interference and Diffraction of Light*. 7th (expanded) ed. Cambridge: Cambridge University Press; 1999.
26. Piliarik M, Homola J. Surface plasmon resonance (SPR) sensors: approaching their limits? Opt Express 2009;17(19):16505–16517.
27. Pang HS, Lee TW, Moharam MG, Likamwa PL, Cho HJ. Integrated optical SPR sensor based on mode conversion efficiency. Electron Lett 2008;44(16):971–972.
28. Li LF. Multilayer modal method for diffraction gratings of arbitrary profile, depth, and permittivity. J Opt Soc Am A 1993;10(12):2581–2591.
29. Goray LI, Seely JF. Efficiencies of master, replica, and multilayer gratings for the soft-x-ray-extreme-ultraviolet range: modeling based on the modified integral method and comparisons with measurements. Appl Opt 2002;41(7):1434–1445.
30. de Feijter JA, Benjamins J, Veer FA. Ellipsometry as a tool to study the ad-sorption of synthetic and biopolymers at the air-water interface. Biopolymers 1978;17:1759–1772.
31. Tumolo T, Angnes L, Baptista MS. Determination of the refractive index increment (dn/dc) of molecule and macromolecule solutions by surface plasmon resonance. Anal Biochem 2004;333(2):273–279.

32. Sjolander S, Urbaniczky C. Integrated fluid handling-system for biomolecular interaction analysis. Anal Chem 1991;63(20):2338–2345.
33. Homola J, Dostálek J, Chen SF, Rasooly A, Jiang SY, Yee SS. Spectral surface plasmon resonance biosensor for detection of staphylococcal enterotoxin B in milk. Int J Food Microbiol 2002;75(1–2):61–69.
34. Nenninger GG, Piliarik M, Homola J. Data analysis for optical sensors based on spectroscopy of surface plasmons. Meas Sci Technol 2002;13(12):2038–2046.
35. VanWiggeren GD, Bynum MA, Ertel JP, et al. A novel optical method providing for high-sensitivity and high-throughput biomolecular interaction analysis. Sens Actuator B Chem 2007;127(2):341–349.
36. Zhang HQ, Boussaad S, Tao NJ. High-performance differential surface plasmon resonance sensor using quadrant cell photodetector. Rev Sci Instrum 2003;74(1):150–153.
37. Nenninger GG, Tobiš P, Homola J, Yee SS. Long-range surface plasmons for high-resolution surface plasmon resonance sensors. Sens Actuator B Chem 2001;74(1–3): 145–151.
38. Slavík R, Homola J. Ultrahigh resolution long range surface plasmon-based sensor. Sens Actuator B Chem 2007;123(1):10–12.
39. Homola J, Lu HBB, Nenninger GG, Dostálek J, Yee SS. A novel multichannel surface plasmon resonance biosensor. Sens Actuator B Chem 2001;76(1–3):403–410.
40. Dostálek J, Vaisocherová H, Homola J. Multichannel surface plasmon resonance biosensor with wavelength division multiplexing. Sens Actuator B Chem 2005; 108(1–2):758–764.
41. Bardin F, Bellemain A, Roger G, Canva M. Surface plasmon resonance spectro-imaging sensor for biomolecular surface interaction characterization. Biosens Bioelectron 2009;24(7):2100–2105.
42. Wu SY, Ho HP. Single-beam self-referenced phase-sensitive surface plasmon resonance sensor with high detection resolution. Chin Opt Lett 2008;6(3):176–178.
43. Nikitin PI, Grigorenko AN, Beloglazov AA, et al. Surface plasmon resonance interferometry for micro-array biosensing. Sens Actuator A Phys 2000;85(1–3): 189–193.
44. Ho HP, Yuan W, Wong CL, et al. Sensitivity enhancement based on application of multi-pass interferometry in phase-sensitive surface plasmon resonance biosensor. Opt Commun 2007;275(2):491–496.
45. Wong CL, Ho HP, Suen YK, et al. Real-time protein biosensor arrays based on surface plasmon resonance differential phase imaging. Biosens Bioelectron 2008; 24(4):606–612.
46. Piliarik M, Vala M, Tichý I, Homola J. Compact and low-cost biosensor based on novel approach to spectroscopy of surface plasmons. Biosens Bioelectron 2009; 24(12):3430–3435.
47. Chinowsky TM, Quinn JG, Bartholomew DU, Kaiser R, Elkind JL. Performance of the Spreeta 2000 integrated surface plasmon resonance affinity sensor. Sens Actuator B Chem 2003;91(1–3):266–274.
48. Chinowsky TM, Soelberg SD, Baker P, et al. Portable 24-analyte surface plasmon resonance instruments for rapid, versatile biodetection. Biosens Bioelectron 2007;22(9–10):2268–2275.

49. Thirstrup C, Zong W, Borre M, Neff H, Pedersen HC, Holzhueter G. Diffractive optical coupling element for surface plasmon resonance sensors. Sens Actuator B Chem 2004;100(3):298–308.
50. Kim SJ, Gobi KV, Iwasaka H, Tanaka H, Miura N. Novel miniature SPR immunosensor equipped with all-in-one multi-microchannel sensor chip for detecting low-molecular-weight analytes. Biosens Bioelectron 2007;23(5):701–707.
51. Feltis BN, Sexton BA, Glenn FL, Best MJ, Wilkins M, Davis TJ. A hand-held surface plasmon resonance biosensor for the detection of ricin and other biological agents. Biosens Bioelectron 2008;23(7):1131–1136.
52. Vala M, Chadt K, Piliarik M, Homola J. High-performance compact SPR sensor for multi-analyte sensing. Sens Actuator B Chem 2010;148(2):544–549.
53. Piliarik M, Vaisocherová H, Homola J. Towards parallelized surface plasmon resonance sensor platform for sensitive detection of oligonucleotides. Sens Actuator B Chem 2007;121(1):187–193.
54. Piliarik M, Parová L, Homola J. High-throughput SPR sensor for food safety. Biosens Bioelectron 2009;24(5):1399–1404.
55. Dostálek J, Homola J, Miler M. Rich information format surface plasmon resonance biosensor based on array of diffraction gratings. Sens Actuator B Chem 2005;107(1):154–161.
56. Kastl KF, Lowe CR, Norman CE. Encoded and multiplexed surface plasmon resonance sensor platform. Anal Chem 2008;80(20):7862–7869.
57. Song FY, Zhou FM, Wang J, et al. Detection of oligonucleotide hybridization at femtomolar level and sequence-specific gene analysis of the Arabidopsis thaliana leaf extract with an ultrasensitive surface plasmon resonance spectrometer. Nucleic Acids Res 2002;30(14):e72.
58. Hegnerová K, Bocková M, Vaisocherová H, et al. Surface plasmon resonance biosensors for detection of Alzheimer disease biomarker. Sens Actuator B Chem 2009; 139(1):69–73.
59. Rothenhausler B, Knoll W. Surface-plasmon microscopy. Nature 1988;332(6165): 615–617.
60. Hickel W, Kamp D, Knoll W. Surface-plasmon microscopy. Nature 1989;339(6221): 186–186.
61. Fu E, Chinowsky T, Foley J, Weinstein J, Yager P. Characterization of a wavelength-tunable surface plasmon resonance microscope. Rev Sci Instrum 2004; 75(7):2300–2304.
62. Piliarik M, Vaisocherová H, Homola J. A new surface plasmon resonance sensor for high-throughput screening applications. Biosens Bioelectron 2005;20(10):2104–2110.
63. Bassil N, Maillart E, Canva M, et al. One hundred spots parallel monitoring of DNA interactions by SPR imaging of polymer-functionalized surfaces applied to the detection of cystic fibrosis mutations. Sens Actuator B Chem 2003;94(3):313–323.
64. Fu E, Foley J, Yager P. Wavelength-tunable surface plasmon resonance microscope. Rev Sci Instrum 2003;74(6):3182–3184.
65. Law WC, Markowicz P, Yong KT, et al. Wide dynamic range phase-sensitive surface plasmon resonance biosensor based on measuring the modulation harmonics. Biosens Bioelectron 2007;23(5):627–632.

66. Dostálek J, Homola J. Surface plasmon resonance sensor based on an array of diffraction gratings for highly parallelized observation of biomolecular interactions. Sens Actuator B Chem 2008;129(1):303–310.

67. Adam P, Dostálek J, Homola J. Multiple surface plasmon spectroscopy for study of biomolecular systems. Sens Actuator B Chem 2006;113(2):774–781.

68. Vala M, Dostálek J, Homola J. Diffraction grating-coupled surface plasmon resonance sensor based on Spectroscopy of long-range and short-range surface plasmons. In: Baldini F, Homola J, Lieberman RA, Miler M, editors. Volume 6585, *Optical Sensing Technology and Applications*. Bellingham: Spie-Int Soc Optical Engineering. 2007. p U547–U555.

69. Dostálek J, Adam P, Kvasnicka P, Telezhnikova O, Homola J. Spectroscopy of Bragg-scattered surface plasmons for characterization of thin biomolecular films. Opt Lett 2007;32(20):2903–2905.

70. Adam P, Dostálek J, Telezhniková O, Homola J. SPR sensor based on a bi-diffractive grating. In: Baldini F, Homola J, Lieberman RA, Miler M, editors. Volume 6585, *Optical Sensing Technology and Applications*. Bellingham: Spie-Int Soc Optical Engineering. 2007. p U528–U536.

71. Brockman JM, Fernandez SM. Grating-coupled surface plasmon resonance for rapid, label-free, array-based sensing. Am Lab 2001;33(12):37–41.

72. Official website of the Biacore AB. 2010. Available at http://www.biacore.com/lifesciences/products/systems_overview/index.html. Last access year: 2012.

73. Lin HY, Tsai WH, Tsao YC, Sheu BC. Side-polished multimode fiber biosensor based on surface plasmon resonance with halogen light. Appl Opt 2007;46(5): 800–806.

74. Pollet J, Delport F, Janssen KPF, et al. Fiber optic SPR biosensing of DNA hybridization and DNA-protein interactions. Biosens Bioelectron 2009;25(4):864–869.

75. Jang HS, Park KN, Kang CD, Kim JP, Sim SJ, Lee KS. Optical fiber SPR biosensor with sandwich assay for the detection of prostate specific antigen. Opt Commun 2009;282(14):2827–2830.

76. Slavík R, Homola J, Čtyroký J, Brynda E. Novel spectral fiber optic sensor based on surface plasmon resonance. Sens Actuator B Chem 2001;74(1–3):106–111.

77. Piliarik M, Homola J, Maníková Z, Čtyroký J. Surface plasmon resonance sensor based on a single-mode polarization-maintaining optical fiber. Sens Actuator B Chem 2003;90(1–3):236–242.

78. Chang YJ, Chen YC, Kuo HL, Wei PK. Nanofiber optic sensor based on the excitation of surface plasmon wave near fiber tip. J Biomed Opt 2006;11(1):014032.

79. Monzon-Hernandez D, Villatoro J. High-resolution refractive index sensing by means of a multiple-peak surface plasmon resonance optical fiber sensor. Sens Actuator B Chem 2006;115(1):227–231.

80. Diaz-Herrera N, Viegas D, Jorge PAS, et al. Fibre-optic SPR sensor with a FBG interrogation scheme for readout enhancement. Sens Actuator B Chem 2010;144(1):226–231.

81. Špacková B, Homola J. Theoretical analysis of a fiber optic surface plasmon resonance sensor utilizing a Bragg grating. Opt Express 2009;17(25):23254–23264.

82. Shao LY, Shevchenko Y, Albert J. Intrinsic temperature sensitivity of tilted fiber Bragg grating based surface plasmon resonance sensors. Opt Express 2010; 18(11):11464–11471.

83. Dostálek J, Čtyroký J, Homola J, et al. Surface plasmon resonance biosensor based on integrated optical waveguide. Sens Actuator B Chem 2001;76(1–3):8–12.
84. Suzuki A, Kondoh J, Matsui Y, Shiokawa S, Suzuki K. Development of novel optical waveguide surface plasmon resonance (SPR) sensor with dual light emitting diodes. Sens Actuator B Chem 2005;106(1):383–387.
85. Johnston MMW, Wilson DM, Booksh KS, Cramer J. Integrated optical computing: system-on-chip for surface plasmon resonance imaging. 2005 IEEE International Symposium on Circuits and Systems. New York: IEEE; 2005. p 3483–3486.
86. Wang TJ, Lin WS, Liu FK. Integrated-optic biosensor by electro-optically modulated surface plasmon resonance. Biosens Bioelectron 2007;22(7):1441–1446.
87. Jette-Charbonneau S, Charbonneau R, Lahoud N, Mattiussi GA, Berini P. Bragg gratings based on long-range surface plasmon-polariton waveguides: comparison of theory and experiment. IEEE J Quantum Electron 2005;41(12):1480–1491.
88. Rich RL, Myszka DG. Survey of the year 2007 commercial optical biosensor literature. J Mol Recognit 2008;21(6):355–400.
89. Rich RL, Myszka DG. Grading the commercial optical biosensor literature-Class of 2008: 'The Mighty Binders'. J Mol Recognit 2009;23(1):1–64.
90. Official website of Bio-Rad. 2010. Available at http://www.bio-rad.com. last access year: 2012.
91. Lokate AMC, Beusink JB, Besselink GAJ, Pruijn GJM, Schasfoort RBM. Biomolecular interaction monitoring of autoantibodies by scanning surface plasmon resonance microarray imaging. J Am Chem Soc 2007;129(45):14013–14018.
92. Official website of IBIS Technologies. 2010. Available at http://www.ibis-spr.nl/p_1.php. Last access year: 2012.
93. Oficial website of the Genoptics. 2010. Available at http://www.genoptics-spr.com. Last access year: 2012.
94. Official website of Sensata Technologies. 2010. Available at http://www.spreeta.com/sensors/spreeta-analytical-sensor-highlights.htm. Last access year: 2012.
95. Phillips KS, Han JH, Martinez M, Wang ZZ, Carter D, Cheng Q. Nanoscale glassification of gold substrates for surface plasmon resonance analysis of protein toxins with supported lipid membranes. Anal Chem 2006;78(2):596–603.
96. Lockett MR, Smith LM. Fabrication and characterization of DNA arrays prepared on carbon-on-metal substrates. Anal Chem 2009;81(15):6429–6437.
97. Brault ND, Gao CL, Xue H, et al. Ultra-low fouling and functionalizable zwitterionic coatings grafted onto SiO2 via a biomimetic adhesive group for sensing and detection in complex media. Biosens Bioelectron 2010;25(10):2276–2282.
98. Boozer C, Yu QM, Chen SF, et al. Surface functionalization for self-referencing surface plasmon resonance (SPR) biosensors by multi-step self-assembly. Sens Actuator B Chem 2003;90(1–3):22–30.
99. Bolduc OR, Masson JF. Monolayers of 3-mercaptopropyl-amino acid to reduce the nonspecific adsorption of serum proteins on the surface of biosensors. Langmuir 2008;24(20):12085–12091.
100. Bolduc OR, Clouthier CM, Pelletier JN, Masson JF. Peptide self-assembled monolayers for label-free and unamplified surface plasmon resonance biosensing in crude cell lysate. Anal Chem 2009;81(16):6779–6788.

101. Holmlin RE, Chen X, Chapman RG, Takayama S, Whitesides GM. Zwitterionic SAMs that resist nonspecific adsorption of protein from aqueous buffer. Langmuir 2001;17(9):2841–2850.
102. Chen HX, Lee MS, Lee JB, et al. Building a novel vitronectin assay by immobilization of integrin on calixarene monolayer. Talanta 2008;75(1):99–103.
103. Jung SH, Son HY, Yuk JS, et al. Oriented immobilization of antibodies by a self-assembled monolayer of 2-(biotinamido)ethanethiol for immunoarray preparation. Colloids Surf B Biointerfaces 2006;47(1):107–111.
104. Klein E, Kerth P, Lebeau L. Enhanced selective immobilization of biomolecules onto solid supports coated with semifluorinated self-assembled monolayers. Biomaterials 2008;29(2):204–214.
105. Monchaux E, Vermette P. Development of dextran-derivative arrays to identify physicochemical properties involved in biofouling from serum. Langmuir 2007;23(6):3290–3297.
106. Andersson O, Larsson A, Ekblad T, Liedberg B. Gradient hydrogel matrix for microarray and biosensor applications: an imaging SPR study. Biomacromolecules 2009;10(1):142–148.
107. Chegel V, Whitcombe MJ, Turner NW, Piletsky SA. Deposition of functionalized polymer layers in surface plasmon resonance immunosensors by in-situ polymerization in the evanescent wave field. Biosens Bioelectron 2009;24(5):1270–1275.
108. Masson JF, Battaglia TM, Davidson MJ, et al. Biocompatible polymers for antibody support on gold surfaces. Talanta 2005;67(5):918–925.
109. Ostuni E, Chapman RG, Liang MN, et al. Self-assembled monolayers that resist the adsorption of proteins and the adhesion of bacterial and mammalian cells. Langmuir 2001;17(20):6336–6343.
110. Situ C, Wylie ARG, Douglas A, Elliott CT. Reduction of severe bovine serum associated matrix effects on carboxymethylated dextran coated biosensor surfaces. Talanta 2008;76(4):832–836.
111. Vikholm-Lundin I, Piskonen R. Binary monolayers of single-stranded oligonucleotides and blocking agent for hybridisation. Sens Actuator B Chem 2008;134(1): 189–192.
112. Emmenegger CR, Brynda E, Riedel T, Sedlakova Z, Houska M, Alles AB. Interaction of blood plasma with antifouling surfaces. Langmuir 2009;25(11):6328–6333.
113. Lu HB, Campbell CT, Castner DG. Attachment of functionalized poly(ethylene glycol) films to gold surfaces. Langmuir 2000;16(4):1711–1718.
114. Feller L, Bearinger JP, Wu L, Hubbell JA, Textor M, Tosatti S. Micropatterning of gold substrates based on poly(propylene sulfide-bl-ethylene glycol), (PPS-PEG) background passivation and the molecular-assembly patterning by lift-off (MAPL) technique. Surf Sci 2008;602(13):2305–2310.
115. Blattler TM, Pasche S, Textor M, Griesser HJ. High salt stability and protein resistance of poly(L-lysine)-g-poly(ethylene glycol) copolymers covalently immobilized via aldehyde plasma polymer interlayers on inorganic and polymeric substrates. Langmuir 2006;22(13):5760–5769.
116. Vaisocherová H, Yang W, Zhang Z, et al. Ultralow fouling and functionalizable surface chemistry based on a zwitterionic polymer enabling sensitive and specific protein detection in undiluted blood plasma. Anal Chem 2008;80(20):7894–7901.

117. Vaisocherová H, Zhang Z, Yang W, et al. Functionalizable surface platform with reduced nonspecific protein adsorption from full blood plasma-material selection and protein immobilization optimization. Biosens Bioelectron 2009;24(7):1924–1930.
118. Erb EM, Chen XY, Allen S, et al. Characterization of the surfaces generated by liposome binding to the modified dextran matrix of a surface plasmon resonance sensor chip. Anal Biochem 2000;280(1):29–35.
119. Smith EA, Wanat MJ, Cheng YF, Barreira SVP, Frutos AG, Corn RM. Formation, spectroscopic characterization, and application of sulfhydryl-terminated alkanethiol monolayers for the chemical attachment of DNA onto gold surfaces. Langmuir 2001;17(8):2502–2507.
120. Yang N, Su XD, Tjong V, Knoll W. Evaluation of two- and three-dimensional streptavidin binding platforms for surface plasmon resonance spectroscopy studies of DNA hybridization and protein-DNA binding. Biosens Bioelectron 2007;22(11):2700–2706.
121. Li YJ, Bi LJ, Zhang XE, et al. Reversible immobilization of proteins with streptavidin affinity tags on a surface plasmon resonance biosensor chip. Anal Bioanal Chem 2006;386(5):1321–1326.
122. Garcia-Aljaro C, Munoz FX, Baldrich E. Captavidin: a new regenerable biocomponent for biosensing? Analysis 2009;134(11):2338–2343.
123. Teles FRR, Fonseca LR. Trends in DNA biosensors. Talanta 2008;77(2):606–623.
124. Chung JW, Park JM, Bernhardt R, Pyun JC. Immunosensor with a controlled orientation of antibodies by using NeutrAvidin-protein A complex at immunoaffinity layer. J Biotechnol 2006;126(3):325–333.
125. Jung Y, Lee JM, Jung H, Chung BH. Self-directed and self-oriented immobilization of antibody by protein G-DNA conjugate. Anal Chem 2007;79(17): 6534–6541.
126. Niemeyer CM. Semisynthetic DNA-protein conjugates for biosensing and nanofabrication. Angew Chem Int Ed 2010;49(7):1200–1216.
127. Ladd J, Boozer C, Yu QM, Chen SF, Homola J, Jiang S. DNA-directed protein immobilization on mixed self-assembled monolayers via a streptavidin bridge. Langmuir 2004;20(19):8090–8095.
128. Piliarik M, Bocková M, Homola J. Surface plasmon resonance biosensor for parallelized detection of protein biomarkers in diluted blood plasma. Biosens Bioelectron 2010;26(4):1656–1661.
129. Wegner GJ, Lee NJ, Marriott G, Corn RM. Fabrication of histidine-tagged fusion protein arrays for surface plasmon resonance imaging studies of protein-protein and protein-DNA interactions. Anal Chem 2003;75(18):4740–4746.
130. Yoshitani N, Saito K, Saikawa W, Asanuma M, Yokoyama S, Hirota H. NTA-mediated protein capturing strategy in screening experiments for small organic molecules by surface plasmon resonance. Proteomics 2007;7(4):494–499.
131. Khan F, He MY, Taussig MJ. Double-hexahistidine tag with high-affinity binding for protein immobilization, purification, and detection on ni-nitrilotriacetic acid surfaces. Anal Chem 2006;78(9):3072–3079.
132. Lee JM, Park HK, Jung Y, Kim JK, Jung SO, Chung BH. Direct immobilization of protein G variants with various numbers of cysteine residues on a gold surface. Anal Chem 2007;79(7):2680–2687.

133. Stenlund P, Babcock GJ, Sodroski J, Myszka DG. Capture and reconstitution of G protein-coupled receptors on a biosensor surface. Anal Biochem 2003; 316(2):243–250.
134. Bae YM, Oh BK, Lee W, Lee WH, Choi JW. Study on orientation of immunoglobulin G on protein G layer. Biosens Bioelectron 2005;21(1):103–110.
135. Jung YW, Lee JM, Kim JW, Yoon JW, Cho HM, Chung BH. Photoactivable antibody binding protein: site-selective and covalent coupling of antibody. Anal Chem 2009;81(3):936–942.
136. Mansuy-Schlick V, Delage-Mourroux R, Jouvenot M, Boireau W. Strategy of macromolecular grafting onto a gold substrate dedicated to protein-protein interaction measurements. Biosens Bioelectron 2006;21(9):1830–1837.
137. Lautner G, Balogh Z, Bardoczy V, Meszaros T, Gyurcsanyi RE. Aptamer-based biochips for label-free detection of plant virus coat proteins by SPR imaging. Analysis 2010;135(5):918–926.
138. Polonschii C, David S, Tombelli S, Mascini M, Gheorghiu M. A novel low-cost and easy to develop functionalization platform. Case study: aptamer-based detection of thrombin by surface plasmon resonance. Talanta 2010;80(5):2157–2164.
139. Lin CX, Katilius E, Liu Y, Zhang JP, Yan H. Self-assembled signaling aptamer DNA arrays for protein detection. Angew Chem Int Ed 2006;45(32):5296–5301.
140. Li Y, Lee HJ, Corn RM. Detection of protein biomarkers using RNA aptamer microarrays and enzymatically amplified surface plasmon resonance imaging. Anal Chem 2007;79(3):1082–1088.
141. Boozer C, Chen SF, Jiang SY. Controlling DNA orientation on mixed ssDNA/OEG SAMs. Langmuir 2006;22(10):4694–4698.
142. Lee CY, Gong P, Harbers GM, Grainger DW, Castner DG, Gamble LJ. Surface coverage and structure of mixed DNA/alkylthiol monolayers on gold: characterization by XPS, NEXAFS, and fluorescence intensity measurements. Anal Chem 2006;78(10):3316–3325.
143. Manera MG, Spadavecchia J, Leone A, et al. Surface plasmon resonance imaging technique for nucleic acid detection. Sens Actuator B Chem 2008;130(1):82–87.
144. Chen YL, Nguyen A, Niu LF, Corn RM. Fabrication of DNA microarrays with poly(L-glutamic acid) monolayers on gold substrates for SPR imaging measurements. Langmuir 2009;25(9):5054–5060.
145. Schofield WCE, McGettrick J, Bradley TJ, Badyal JPS, Przyborski S. Rewritable DNA microarrays. J Am Chem Soc 2006;128(7):2280–2285.
146. Mitchell JS, Wu YQ, Cook CJ, Main L. Sensitivity enhancement of surface plasmon resonance biosensing of small molecules. Anal Biochem 2005;343(1):125–135.
147. Gobi KV, Matsumoto K, Toko K, Ikezaki H, Miura N. Enhanced sensitivity of self-assembled-monolayer-based SPR immunosensor for detection of benzaldehyde using a single-step multi-sandwich immunoassay. Anal Bioanal Chem 2007; 387(8):2727–2735.
148. Wu J, Fu ZF, Yan F, Ju HX. Biomedical and clinical applications of immunoassays and immunosensors for tumor markers. Trends Anal Chem 2007;26(7):679–688.
149. Huang L, Reekmans G, Saerens D, et al. Prostate-specific antigen immunosensing based on mixed self-assembled monolayers, camel antibodies and colloidal gold enhanced sandwich assays. Biosens Bioelectron 2005;21(3):483–490.

150. Cao C, Sim SJ. Double-enhancement strategy: a practical approach to a femto-molar level detection of prostate specific antigen-alpha(1)-antichymotrypsin (PSA/ACT complex) for SPR immunosensing. J Microbiol Biotechnol 2007;17(6): 1031–1035.
151. Ladd J, Taylor AD, Piliarik M, Homola J, Jiang SY. Label-free detection of cancer biomarker candidates using surface plasmon resonance imaging. Anal Bioanal Chem 2009;393(4):1157–1163.
152. Gao CL, Li GZ, Xue H, Yang W, Zhang FB, Jiang SY. Functionalizable and ultra-low fouling zwitterionic surfaces via adhesive mussel mimetic linkages. Biomaterials 2010;31(7):1486–1492.
153. Su F, Xu C, Taya M, Murayama K, Shinohara Y, Nishimura SI. Detection of carcinoembryonic antigens using a surface plasmon resonance biosensor. Sensors 2008;8(7):4282–4295.
154. Suwansa-ard S, Kanatharana P, Asawatreratanakul P, Wongkittisuksa B, Limsakul C, Thavarungkul P. Comparison of surface plasmon resonance and capacitive immunosensors for cancer antigen 125 detection in human serum samples. Biosens Bioelectron 2009;24(12):3436–3441.
155. Masson JF, Battaglia TM, Khairallah P, Beaudoin S, Booksh KS. Quantitative measurement of cardiac markers in undiluted serum. Anal Chem 2007;79(2):612–619.
156. Masson JF, Obando L, Beaudoin S, Booksh K. Sensitive and real-time fiber-optic-based surface plasmon resonance sensors for myoglobin and cardiac troponin I. Talanta 2004;62(5):865–870.
157. Wei JY, Mu Y, Song DQ, et al. A novel sandwich immunosensing method for measuring cardiac troponin I in sera. Anal Biochem 2003;321(2):209–216.
158. Dutra RF, Mendes RK, da Silva VL, Kubota LT. Surface plasmon resonance immunosensor for human cardiac troponin T based on self-assembled monolayer. J Pharm Biomed Anal 2007;43(5):1744–1750.
159. Teramura Y, Arima Y, Iwata H. Surface plasmon resonance-based highly sensitive immunosensing for brain natriuretic peptide using nanobeads for signal amplification. Anal Biochem 2006;357(2):208–215.
160. Arya SK, Solanki PR, Singh SP, et al. Poly-(3-hexylthiophene) self-assembled monolayer based cholesterol biosensor using surface plasmon resonance technique. Biosens Bioelectron 2007;22(11):2516–2524.
161. Solanki PR, Arya SK, Nishimura Y, Iwamoto M, Malhotra BD. Cholesterol biosensor based on amino-undecanethiol self-assembled monolayer using surface plasmon resonance technique. Langmuir 2007;23(13):7398–7403.
162. Gehlot R, Sharma K, Mathew M, Kumbhat S. Surface plasmon resonance based biosensor for label free detection of cholesterol. Indian J Chem 2008;47(12):1804–1808.
163. Gobi KV, Iwasaka H, Miura N. Self-assembled PEG monolayer based SPR immunosensor for label-free detection of insulin. Biosens Bioelectron 2007;22(7): 1382–1389.
164. Gobi KV, Kim SJ, Tanaka H, Shoyama Y, Miura N. Novel surface plasmon resonance (SPR) immunosensor based on monomolecular layer of physically-adsorbed ovalbumin conjugate for detection of 2,4-dichlorophenoxyacetic acid and atomic force microscopy study. Sens Actuator B Chem 2007;123(1):583–593.

165. Kim SJ, Gobi KV, Tanaka H, Shoyama Y, Miura N. A simple and versatile self-assembled monolayer based surface plasmon resonance immunosensor for highly sensitive detection of 2,4-D from natural water resources. Sens Actuator B Chem 2008;130(1):281–289.
166. Dostálek J, Pribyl J, Homola J, Skladál P. Multichannel SPR biosensor for detection of endocrine-disrupting compounds. Anal Bioanal Chem 2007;389(6):1841–1847.
167. Gouzy MF, Kess M, Kramer PM. A SPR-based immunosensor for the detection of isoproturon. Biosens Bioelectron 2009;24(6):1563–1568.
168. Mauriz E, Calle A, Manclus JJ, Montoya A, Lechuga LM. Multi-analyte SPR immunoassays for environmental biosensing of pesticides. Anal Bioanal Chem 2007;387(4):1449–1458.
169. Shimomura M, Nomura Y, Zhang W, et al. Simple and rapid detection method using surface plasmon resonance for dioxins, polychlorinated biphenylx and atrazine. Anal Chim Acta 2001;434(2):223–230.
170. Hegnerová K, Piliarik M, Šteinbachová M, Flegelová Z, Černohorská H, Homola J. Detection of bisphenol A using a novel surface plasmon resonance biosensor. Anal Bioanal Chem 2010;398(5):1963–1966.
171. Marchesini GR, Meulenberg E, Haasnoot W, Irth H. Biosensor immunoassays for the detection of bisphenol A. Anal Chim Acta 2005;528(1):37–45.
172. Marchesini GR, Koopal K, Meulenberg E, Haasnoot W, Irth H. Spreeta-based biosensor assays for endocrine disruptors. Biosens Bioelectron 2007;22(9–10):1908–1915.
173. Matsumoto K, Sakai T, Torimaru A, Ishitobi S. A surface plasmon resonance-based immunosensor for sensitive detection of bisphenol A. J Fac Agric Kyushu Univ 2005;50(2):625–634.
174. Hong S, Kang T, Oh S, et al. Label-free sensitive optical detection of polychlorinated biphenyl (PCB) in an aqueous solution based on surface plasmon resonance measurements. Sens Actuator B Chem 2008;134(1):300–306.
175. Shankaran DR, Kawaguchi T, Kim SJ, Matsumoto K, Toko K, Miura N. Evaluation of the molecular recognition of monoclonal and polyclonal antibodies for sensitive detection of 2,4,6-trinitrotoluene (TNT) by indirect competitive surface plasmon resonance immunoassay. Anal Bioanal Chem 2006;386(5):1313–1320.
176. Matsumoto K, Torimaru A, Ishitobi S, et al. Preparation and characterization of a polyclonal antibody from rabbit for detection of trinitrotoluene by a surface plasmon resonance biosensor. Talanta 2005;68(2):305–311.
177. Larsson A, Angbrant J, Ekeroth J, Mansson P, Liedberg B. A novel biochip technology for detection of explosives - TNT: synthesis, characterisation and application. Sens Actuator B Chem 2006;113(2):730–748.
178. Singh P, Onodera T, Mizuta Y, Matsumoto K, Miura N, Toko K. Dendrimer modified biochip for detection of 2,4,6 trinitrotoluene on SPR immunosensor: fabrication and advantages. Sens Actuator B Chem 2009;137(2):403–409.
179. Nagatomo K, Kawaguchi T, Miura N, Toko K, Matsumoto K. Development of a sensitive surface plasmon resonance immunosensor for detection of 2,4-dinitrotoluene with a novel oligo (ethylene glycol)-based sensor surface. Talanta 2009;79(4):1142–1148.
180. Glynn B, Lahiff S, Wernecke M, Barry T, Smith TJ, Maher M. Current and emerging molecular diagnostic technologies applicable to bacterial food safety. Int J Dairy Technol 2006;59(2):126–139.

181. Patel PD. Overview of affinity biosensors in food analysis. J AOAC Int 2006;89(3): 805–818.
182. Dunne L, Daly S, Baxter A, Haughey S, O'Kennedy R. Surface plasmon resonance-based inummoassay for the detection of affatoxin B-1 using single-chain antibody fragments. Spectrosc Lett 2005;38(3):229–245.
183. Cuccioloni M, Mozzicafreddo M, Barocci S, et al. Biosensor-based screening method for the detection of aflatoxins B-1-G(1). Anal Chem 2008;80(23):9250–9256.
184. Tudos AJ, Lucas-van den Bos ER, Stigter ECA. Rapid surface plasmon resonance-based inhibition assay of deoxynivalenol. J Agric Food Chem 2003;51(20):5843–5848.
185. van der Gaag B, Spath S, Dietrich H, et al. Biosensors and multiple mycotoxin analysis. Food Control 2003;14(4):251–254.
186. Fonfria ES, Vilarino N, Campbell K, et al. Paralytic shellfish poisoning detection by surface plasmon resonance-based Biosensors in shellfish matrixes. Anal Chem 2007;79(16):6303–6311.
187. Campbell K, Huet AC, Charlier C, Higgins C, Delahaut P, Elliott CT. Comparison of ELISA and SPR biosensor technology for the detection of paralytic shellfish poisoning toxins. J Chromatogr B Analyt Technol Biomed Life Sci 2009;877(32): 4079–4089.
188. Rawn DFK, Niedzwiadek B, Campbell K, Higgins HC, Elliott CT. Evaluation of surface plasmon resonance relative to high pressure liquid chromatography for the determination of paralytic shellfish toxins. J Agric Food Chem 2009;57(21):10022–10031.
189. Yu QM, Chen SF, Taylor AD, Homola J, Hock B, Jiang SY. Detection of low-molecular-weight domoic acid using surface plasmon resonance sensor. Sens Actuator B Chem 2005;107(1):193–201.
190. Traynor IM, Plumpton L, Fodey TL, Higgins C, Elliott CT. Immunobiosensor detection of domoic acid as a screening test in bivalve molluscs: comparison with liquid chromatography-based analysis. J AOAC Int 2006;89(3):868–872.
191. Lotierzo M, Henry OYF, Piletsky S, et al. Surface plasmon resonance sensor for domoic acid based on grafted imprinted polymer. Biosens Bioelectron 2004;20(2): 145–152.
192. Stevens RC, Soelberg SD, Eberhart BTL, et al. Detection of the toxin domoic acid from clam extracts using a portable surface plasmon resonance biosensor. Harmful Algae 2007;6(2):166–174.
193. Prieto-Simon B, Miyachi H, Karube I, Saiki H. High-sensitive flow-based kinetic exclusion assay for okadaic acid assessment in shellfish samples. Biosens Bioelectron 2010;25(6):1395–1401.
194. Llamas NM, Stewart L, Fodey T, et al. Development of a novel immunobiosensor method for the rapid detection of okadaic acid contamination in shellfish extracts. Anal Bioanal Chem 2007;389(2):581–587.
195. Taylor AD, Ladd J, Etheridge S, Deeds J, Hall S, Jiang SY. Quantitative detection of tetrodotoxin (TTX) by a surface plasmon resonance (SPR) sensor. Sens Actuator B Chem 2008;130(1):120–128.
196. Vaisocherová H, Taylor AD, Jiang S, et al. Surface plasmon resonance biosensor for determination of tetrodotoxin: prevalidation study. J AOAC Int. 2008;94(2): 596–604.

197. Ladd J, Taylor AD, Homola J, Jiang SY. Detection of botulinum neurotoxins in buffer and honey using a surface plasmon resonance (SPR) sensor. Sens Actuator B Chem 2008;130(1):129–134.
198. Ferracci G, Miquelis R, Kozaki S, Seagar M, Leveque C. Synaptic vesicle chips to assay botulinum neurotoxins. Biochem J 2005;391:659–666.
199. Medina MB. A biosensor method for a competitive immunoassay detection of staphylococcal enterotoxin B (SEB) in milk. J Rapid Methods Autom Microbiol 2005;13(1):37–55.
200. Medina MB. A biosensor method for detection of staphylococcal enterotoxin A in raw whole egg. J Rapid Methods Autom Microbiol 2006;14(2):119–132.
201. Naimushin AN, Soelberg SD, Nguyen DK, et al. Detection of Staphylococcus aureus enterotoxin B at femtomolar levels with a miniature integrated two-channel surface plasmon resonance (SPR) sensor. Biosens Bioelectron 2002;17(6–7):573–584.
202. Tsai WC, Pai PJR. Surface plasmon resonance-based immunosensor with oriented immobilized antibody fragments on a mixed self-assembled monolayer for the determination of staphylococcal enterotoxin B. Microchim Acta 2009;166(1–2):115–122.
203. Tran H, Leong C, Loke WK, Dogovski C, Liu CQ. Surface plasmon resonance detection of ricin and horticultural ricin variants in environmental samples. Toxicon 2008;52(4):582–588.
204. Su XL, Li Y. Surface plasmon resonance and quartz crystal microbalance immunosensors for detection of Escherichia coli O157: H7. Trans ASAE 2005;48(1):405–413.
205. Taylor AD, Ladd J, Yu QM, Chen SF, Homola J, Jiang SY. Quantitative and simultaneous detection of four foodborne bacterial pathogens with a multi-channel SPR sensor. Biosens Bioelectron 2006;22(5):752–758.
206. Linman MJ, Sugerman K, Cheng Q. Detection of low levels of Escherichia coli in fresh spinach by surface plasmon resonance spectroscopy with a TMB-based enzymatic signal enhancement method. Sens Actuator B Chem 2010;145(2):613–619.
207. Eum NS, Yeom SH, Kwon DH, Kim HR, Kang SW. Enhancement of sensitivity using gold nanorods-antibody conjugator for detection of E. coli O157:H7. Sens Actuator B Chem 2010;143(2):784–788.
208. Vala M, Etheridge S, Roach JA, Homola J. Long-range surface plasmons for sensitive detection of bacterial analytes. Sens Actuator B Chem 2009;139(1):59–63.
209. Chen LL, Deng L, Liu LL, Peng ZH. Immunomagnetic separation and MS/SPR end-detection combined procedure for rapid detection of Staphylococcus aureus and protein A. Biosens Bioelectron 2007;22(7):1487–1492.
210. Subramanian A, Irudayaraj J, Ryan T. Mono and dithiol surfaces on surface plasmon resonance biosensors for detection of Staphylococcus aureus. Sens Actuator B Chem 2006;114(1):192–198.
211. Barlen B, Mazumdar SD, Lezrich O, Kampfer P, Keusgen M. Detection of salmonella by surface plasmon resonance. Sensors 2007;7(8):1427–1446.
212. Nanduri V, Bhunia AK, Tu SI, Paoli GC, Brewster JD. SPR biosensor for the detection of L-monocytogenes using phage-displayed antibody. Biosens Bioelectron 2007;23(2):248–252.
213. Koubová V, Brynda E, Karasová L, et al. Detection of foodborne pathogens using surface plasmon resonance biosensors. Sens Actuator B Chem 2001;74(1–3):100–105.

214. Jyoung JY, Hong SH, Lee W, Choi JW. Immunosensor for the detection of Vibrio cholerae O1 using surface plasmon resonance. Biosens Bioelectron 2006;21(12): 2315–2319.

215. Wang DB, Bi LJ, Zhang ZP, et al. Label-free detection of B-anthracis spores using a surface plasmon resonance biosensor. Analysis 2009;134(4):738–742.

CHAPTER TWO

Microchip-Based Flow Cytometry in Photonic Sensing: Principles and Applications for Safety and Security Monitoring

BENJAMIN R. WATTS
Department of Engineering Physics, McMaster University, Hamilton, ON, Canada

ZHIYI ZHANG
Institute for Microstructural Sciences, National Research Council of Canada, Ottawa, ON, Canada

CHANG-QING XU
Department of Engineering Physics, McMaster University, Hamilton, ON, Canada

2.1 INTRODUCTION

Flow cytometry is a powerful analysis tool whose power is derived from the speed of analysis and in its multiparameter detection capabilities. The measured characteristics of the sample analyzed by a flow cytometer give a very accurate picture of the sample while simultaneously eliminating the painstaking labor dictated by manual counting (1). Flow cytometry is accomplished by flowing a population of diversely

Photonic Sensing: Principles and Applications for Safety and Security Monitoring, First Edition.
Edited by Gaozhi Xiao and Wojtek J. Bock.

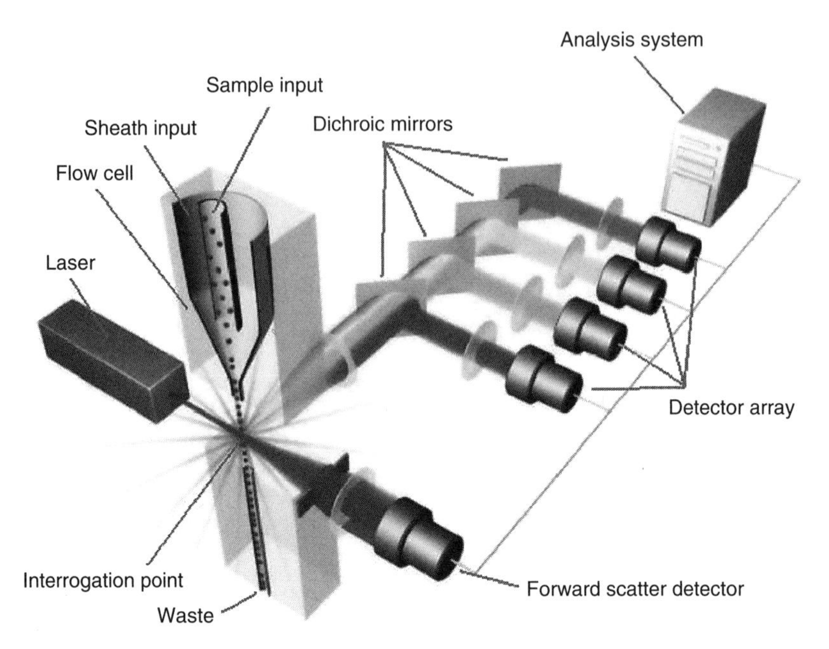

FIGURE 2.1 Diagram of a typical flow cytometer showing the flow cell, excitation optics, detection optics, and analysis system. *Source:* From Reference 2 Copyright © 2011 Life Technologies Corporation. Used under permission. www.lifetechnologies.com.

or similarly sized cells—in single file via hydrodynamic focusing—past a laser beam allowing individual interrogation of each cell in the population measuring properties such as size, shape, cell viability, cell DNA content, cell surface marker characteristics, or even cell life cycle distribution. A conventional cytometer consists of four basic systems integrated together, as shown in the schematic representation in Figure 2.1 (2). The four components are the flow cell for handling samples, the excitation system to provide a means for detection via scattering and fluorescence, the collection system to analyze the numerous parameters from labeled cells, and an analysis system that allows rapid storage and computation of raw data.

Conventional flow cytometers accomplish simultaneous detection of multiple parameters by tagging a different fluorescent dye to each individual feature of interest via very selective antigen–antibody chemical binding, making it possible to resolve up to 17 different features of a population of cells in a single-experimental run (3). By analyzing and comparing the light signature from each cell against the entire population, it is possible to see similarities and differences among cells via a multiparameter scatter plot (Fig. 2.2 (2)). This scatter plot is a two-parameter comparison of the cells' forward scattered light and side scattered light. It is easy to see the distinct cell types in this population based on grouping similar cell light signatures together.

Very accurate and specific characterization of the sample—because of specific and extensive labeling—allows flow cytometers to perceive very slight

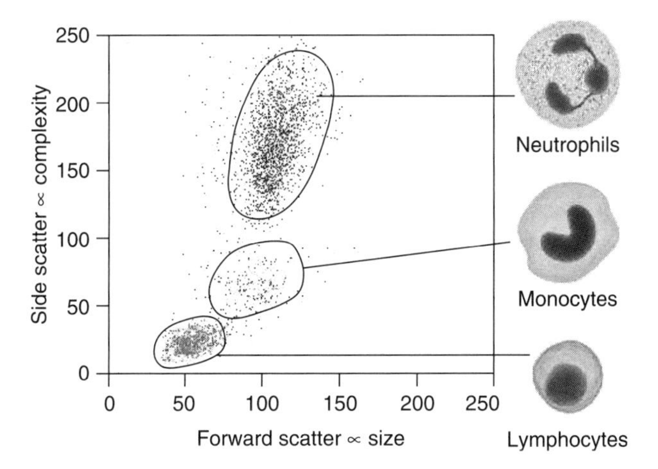

FIGURE 2.2 Plot of a typical data set obtained from a flow cytometric analysis showing the distinct populations of different cells based on the two-parameter (side scattered vs forward scattered light) collection. *Source:* From Reference 2, Copyright © 2011 Life Technologies Corporation. Used under permission. www.lifetechnologies.com.

heterogeneities within a perceived homogeneous sample (cells, bacteria, viruses, microbeads) or vice versa. Populations of cells in the sample can be subtyped and accurately classified in a single-experimental step, allowing a more accurate snap-shot of the current unaltered state of the biological sample in a very rapid time frame—an obvious advantage over other detection assay schemes reliant on amplification techniques that tend to average properties over the entire sample, hiding very important subtleties, and requiring multiple, expensive, time-consuming steps.

With the ability to resolve such subtleties within a cell population, flow cytometry has found widespread applications in both the clinic and within the laboratory, ranging from diagnostic applications in immunophenotyping and immunology, diagnosing, determining the viral loading; tracking the treatment of HIV infections, diagnosing and classifying lymphomas, analyzing cellular function, cycle; and apoptosis, diagnosing and monitoring residual disease in leukemia treatments, microbiology, drug discovery, classification and differentiation of T-cells, virology, sorting applications, genetics, environmental monitoring, and many others.

A microchip-based flow cytometer is a cytometer that has the flow cell miniaturized to form a low cost version of a conventional flow cytometer. It is suitable for handling small samples with low operation costs and has applications in important and novel areas that are unreachable by conventional flow cytometers.

2.2 MICROCHIP-BASED FLOW CYTOMETRY

Microchip-based flow cytometry aims to remove some of the restrictions that limit conventional flow cytometry by replacing the bulky flow cell with a microchannel

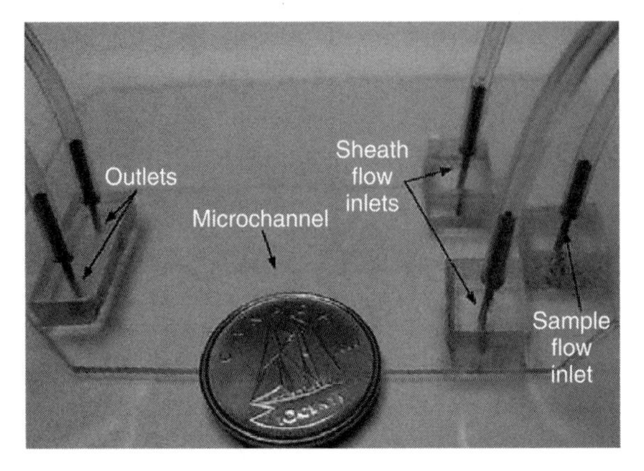

FIGURE 2.3 A packaged microchip-based flow cytometer device showing sample and sheath inlet ports for hydrodynamic focusing. *Source:* Reprinted from Reference 28, with permission from Elsevier.

that is fabricated monolithically on a substrate. This reduction in channel size means greater fluid-handling possibilities and increased efficiency through smaller analysis volumes and less dead zone volumes of liquid. Increased functionality is also possible through on-chip integration of conventional off-chip and free-space systems such as fluid-handling components and detection optics. These devices, such as the one shown in Figure 2.3 (28), create a smaller and more portable device capable of performing applications that are not suitable to conventional cytometry because of the bulky nature of conventional machines, large costs, and the large-scale facilities needed to operate them.

These miniaturized cytometers usually consist of a microfabricated flow cell coupled with a conventional free-space optical system. The microchip-based device can be fabricated with various materials through various processes, though all follow a fairly similar process. Careful selection of materials must be made to ensure that a couple of important considerations are met. First, the material must be transparent at optical frequencies to minimize losses to excitation and detection signals. Second, the material must be biocompatible, that is, it must be safe to use with live cells. Third, the material must be able to withstand large pressures that are generated by the small dimensions of the channels. To fabricate such a device, an open microfluidic channel is first created on a planar and optically transparent substrate, such as glass or polymer, through chemical etching (29–32), lithography-based micromolding processes (28, 33, 34), standard lithography processes, or laser ablation. The obtained open channels will be then sealed with a transparent lid and connected with external tubing (or a well in the cover material) to provide a fluidic interface to the external world. So far, device fabrication using polydimethylsiloxane (PDMS) through soft lithography (33) has been extensively used for research and development because of its low capital and materials cost, and easy and rapid fabrication (34). It uses a micromolding procedure to create a positive image from a negative mold with dimensional capabilities in the tens of microns. A micromolding

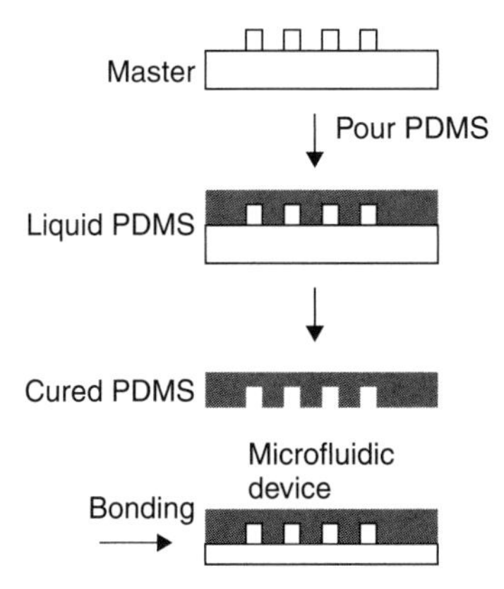

FIGURE 2.4 Diagram showing the fabrication process of a PDMS microfluidic device formed from a micromolding method, where a SU-8 master is used to form a pattern in PDMS as it is cured. *Source:* Reprinted from Reference 35, with permission from SPIE.

process (Fig. 2.4 (35)) requires that a master mold be made by applying a layer of photoresist of a specific thickness (which is the resulting channel depth) onto a wafer via a spin process. The photoresist is then patterned using a photomask that contains the desired designs creating a mold that is then used to cure a thick layer of PDMS, and removal of the cured PDMS results in a negative of the mold transferred to the elastomer. Bonding the patterned PDMS to a cover slip through various methods seals the channels and allows easy connection to external fluidic components.

Interfacing the fluidic platform with the external world is accomplished typically through an external pumping mechanism in the form of hydrodynamic flow (28, 35). Some systems utilize an integrated method by exploiting the phenomena of electroosmotic flow by electrokinetically forcing fluids down a microchannel via an applied electric field. With such small dimensions possible by microfabrication methods, a less applied voltage causes a large electric field that can move particles with great efficiency (36, 37) leading to easy integration on-chip with the flow cell.

To mimic the functionality of a conventional cytometer, these microchip-based devices must focus the introduced sample stream into a very thin stream in the detection zone during testing so that the cells will be detected one-by-one. As a planar device, hydrodynamic focusing in one dimension—a direction perpendicular to the flow of the liquids and the plane of the device—can be easily achieved in a microchip device, as seen in Figure 2.5 (28). The resulting fluid flow is commonly called *2D hydrodynamic focusing* despite the one dimension of control (38). Conventional cytometers achieve a full 3D hydrodynamic focusing by two concentric tubes to introduce the fluids to a single tube, the sheath forming an annulus shape around the central sample flow. With fabrication methods limiting devices to 2D design, different tricks have been implemented to achieve focusing in both the horizontal (in the plane of the chip, perpendicular to flow direction) and the vertical direction—called *3D hydrodynamical focusing* (39–43).

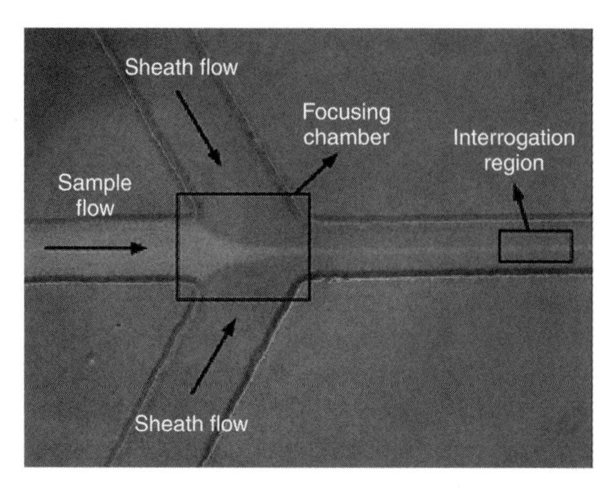

FIGURE 2.5 Device showing 2D hydrodynamic focusing of a sample liquid (center stream) between two buffer liquids in the plane of the device. *Source:* Reprinted from Reference 28, with permission from Elsevier.

Normally, microchip-based cytometers are coupled with a free-space optical detection scheme similar to that used in a conventional flow cytometer (Fig. 2.6 (29)). The device relies on beam shaping optics for excitation in order to form an ideal beam shape in the middle of the channel on the microfluidic device for targeted detection. On the collection side of the device, a microscope objective is used to focus the light to optical components to split the scattered light from the fluorescent light, which are then detected by separate PMTs (photomultiplier tubes) and recorded.

PMTs are typically used because the fluorescent signals from samples are weak—these signals are weak for two reasons. First, the intensity of the fluoresced light is on the order of 1 million times less than the intensity of the excitation signal. Second, the fluorescent light is isotropic in direction, leading to a further weakening of the signal, as the light collected by the free-space collection scheme amounts to about 1–2% of the total light emitted. Beam shaping, mentioned earlier, is a very important step to employ in microchip-based devices, as it is relied on in conventional cytometry to obtain reliable and repeatable detection signals by tailoring the beam to the specific dimensional properties of the cells or particles of interest for interrogation.

Microchip-based devices have numerous advantages over the conventional-based systems. First, with a reduced channel size comes the benefit of a reduced requirement on the amount of sample fluid that is required, a drastic benefit when samples volume are rare or in extremely limited quantity, such as in prenatal diagnosis of genetic disorders (44), or when samples of the cerebrospinal fluid are obtained through a spinal puncture (44). Furthermore, large cost savings are realized from the reduced demands on expensive labeling dyes because of the reduction in sample fluids required for analysis while the small size of fluid handling permits smaller power requirements. Second, experimental run times are typically shorter than those of a conventional cytometer because of the reduced sample sizes (45). Third, owing

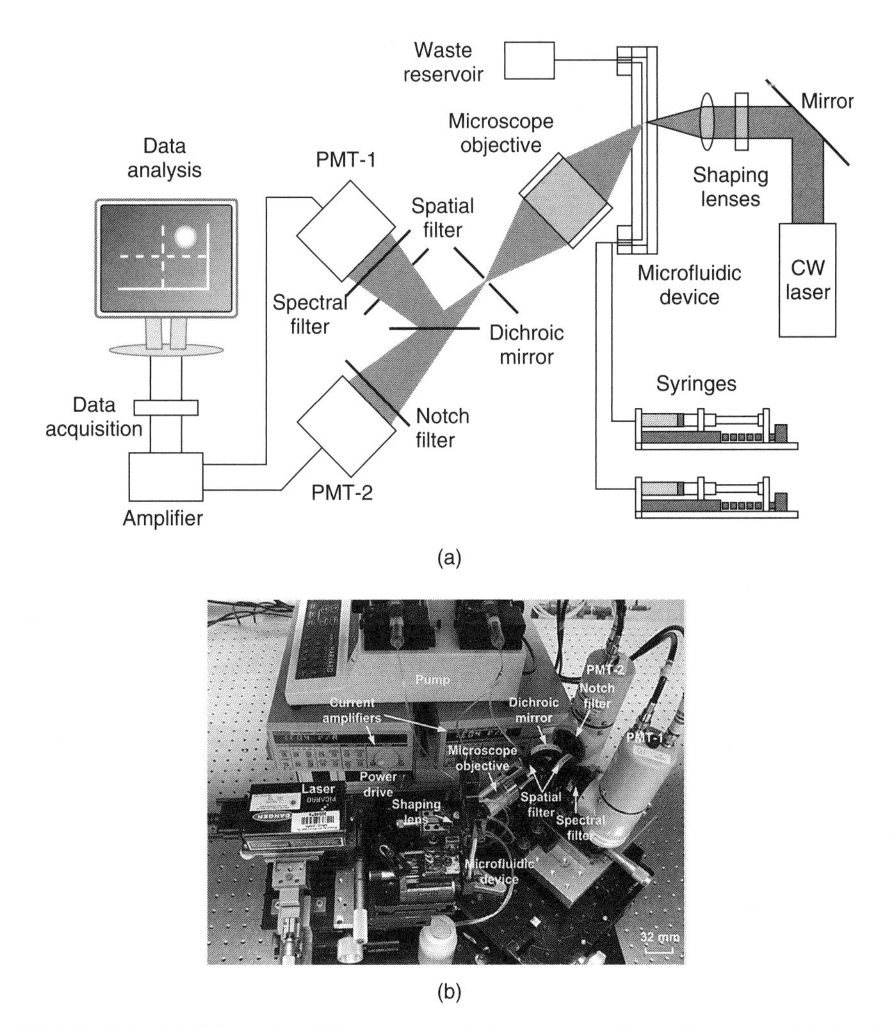

FIGURE 2.6 (a) Schematic of free-space optical system coupled with a microchip-based device showing the excitation optics and collection optics. (b) Bench top setup with the microchip flow cell in the middle with a vertical orientation. *Source:* Reprinted from Reference 28, with permission from Elsevier.

to the lithography-based fabrication procedures, these devices are easily produced in a mass-production type scheme that exploits economies of scale that will ease costs and increase quality assurance through repeatable and reliable measurements between devices (45, 46). Fourth, the chips are easily integrated with other on-chip fluidic handling procedures and devices, such as sorters (47–49), mixers, pumps and valves (50), and reactors, which will allow a steady evolution toward a fully lab-on-a-chip device (51). Furthermore, biohazard threats are minimized, as all fluid handling is done on-chip in a completely sealed and easily disposable platform while errors that may be introduced through human handling are eliminated.

2.3 MICROCHIP-BASED FLOW CYTOMETRY WITH INTEGRATED OPTICS

While a microchip-based flow cytometer can handle small samples at low cost, a free-space optics-based system is still bulky, expensive, and not portable. To overcome this problem while substantially reducing equipment costs, it is required that some sort of light-delivery mechanism is integrated onto the chip to perform excitation and detection functions. Integrating a light-delivery component onto the device creates automatic alignment to the channel during fabrication, as structures are designed with a CAD program—then converted to a photomask—allowing an optimal optical excitation once fabricated. Through this approach, it is possible to even achieve collection of the light on-chip for analysis. This eliminates the need for error prone, cumbersome, and difficult optical alignment that is required by conventional and microchip-based devices that require free-space optics. Another advantage is that the proximity to the sample is greatly enhanced—within a few tens of microns—allowing for great benefit from a collection type waveguide, as the solid angle of collection can be as great as 20% of the isotropically fluoresced light.

Integrating optical fibers onto a chip is accomplished by simply prefabricating a microtrench of a very specific size so as to allow the insertion of a fiber (50, 52–55). Etching the fiber before insertion (50, 54, 55) to achieve a very accurate diameter or use of an elastic of bulk material to hold the fiber (Fig. 2.7 (52, 53)) ensure that the fiber fits tightly in the device.

Planar ridge waveguides can also be fabricated on-chip (Fig. 2.8) by removing material on either side of a defined ridge of a certain width. One step further to integrating waveguide optics is to align the inserted fibers to the end of a fabricated waveguide integrated into the same plane of the device (55), allowing more complicated and sophisticated light handling on the chip via integrated waveguide structures such as beam splitters (56). Such a procedure does require a more complicated fabrication process; however, devices with integrated waveguides

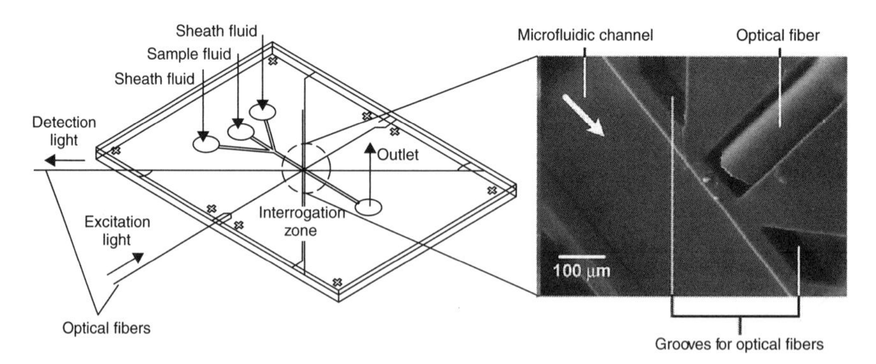

FIGURE 2.7 Schematic representation of a device with grooves in the substrate to guide optical fibers to precisely align to the microchannel on insertion. The device is fabricated from PDMS, allowing slight deformation of the material to allow a tight fit to hold the fiber. *Source:* Reprinted from Reference 53, with permission from Elsevier.

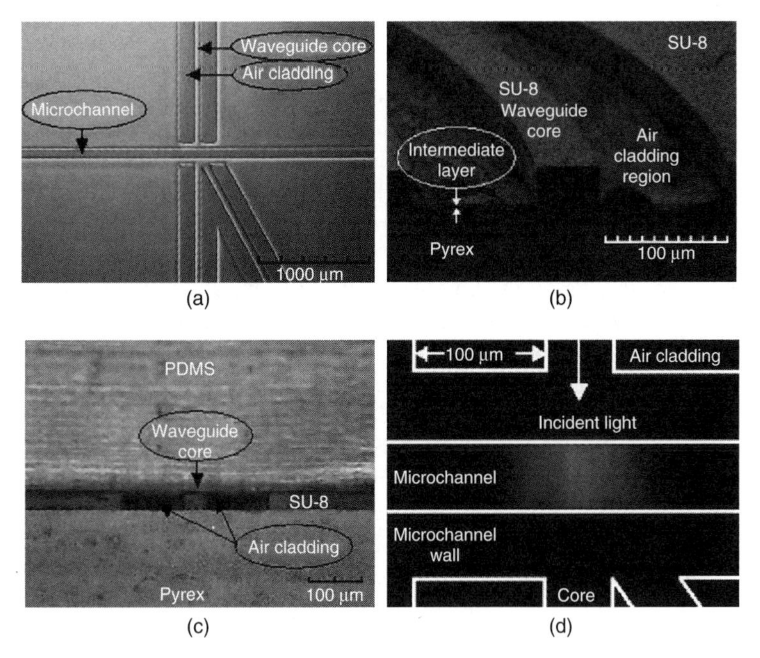

FIGURE 2.8 (a) Device that integrates a waveguide next to a microchannel (b) where the core of the waveguide has been formed by removing SU-8 material from either side of a ridge. (c) Upper and lower claddings are provided by lower index substrate and cover material. (d) Injection of the light is in a well-defined region in the microchannel. *Source:* Reprinted from Reference 58, with permission from John Wiley and Sons.

are widely fabricated with SU-8, an optically transparent yet rigid photoresist material. Through a simple and single photolithographic step, it is possible to form microchannels and waveguides in a one-shot processing method (57). Many devices use SU-8 as a core material to manufacture waveguides on the substrate (Fig. 2.8 (57–59)), while some others use PDMS as a core material (60), and a few others use PDMS to create microchannels that are then filled with a polymer of higher index to form a core guiding region (56).

Unfortunately, light brought to microfluidic channels through optical fibers and planar waveguides diverges significantly, with reduced power density through an unshaped beam, as it exits the fiber or planar waveguide and traverses the channel. In an attempt to increase the power density available for excitation, and thus the level of detection in optically integrated devices, microlenses have been integrated with the waveguides to focus light into the center of the channel with the primary goal of increasing the level of intensity of the excitation light to in turn boost detection signals (61–67). These lenses can be fabricated in the same step as the microchannel and waveguides, and, as processing is through a photolithographic based means (as discussed earlier), the lenses formed are 1D lenses that are cylindrical in shape. Single and compound lens systems have been explored with varying levels of improvement. Improvement to excitation intensity has been improved by 1.67 times with a single lens (66) and by seven times with a compound lens system (65).

Microlenses have also been used to shape the beam, as it traverses the microchannel and the sample stream, controlling the geometry of the beam that the cells see for excitation. Beam shaping, a ubiquitous practice in conventional flow cytometry, is the process used to carefully shape the beam at the interrogation region by controlling the beam waist parallel to the direction of fluid flow—defining the length (and time) of illumination for particles—and controlling the depth of focus, that is, the uniformity of the beam width and intensity across the microchannel. Tuning both these degrees of freedom in the beam geometry allows increased reliability of detection by ensuring that only one particle can be within the beam at one time and that every particle receives uniform illumination regardless of deviation in the sample stream. A microchip-based device with an integrated lens system (Fig. 2.9 (59)) has demonstrated a shaped beam in the microfluidic channel achieving a "bowtie-shaped" beam geometry that defined both the beam width and depth of focus across the particle flow. This device demonstrated the width of the formed beam to be adjustable through a range of 3.6–10 μm, while depth of focus control allowed regions of uniform intensity of 6.2 and 10.4 μm for formed beam widths of 3.6 and 10 μm, respectively.

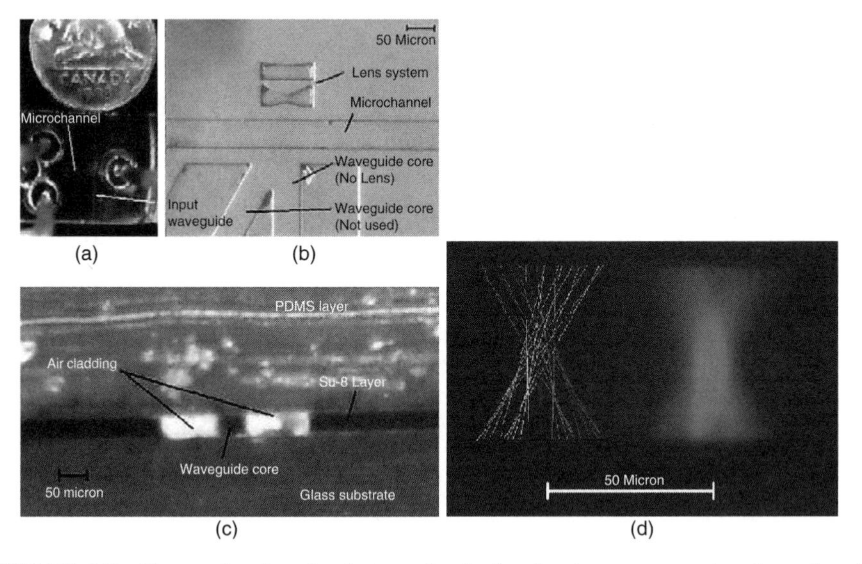

FIGURE 2.9 Picture showing the design of a device that integrates a microchannel and waveguides with a lens system in a single planar design showing the packaged device (a) a close up of the interrogation region (b) and a cross-section of the device (c). A picture of a beam designed to form a beam width of 10 μm in the microchannel with a superimposition of the simulated ray trace to show accurate predictions from simulations, (d). *Source:* Reprinted from Reference 59, with permission from OSA.

Collection of the fluorescent and scattered signals can be accomplished on-chip through integrated waveguides by simply integrating them across from the source in close proximity to the cell when it is under illumination. This configuration can be seen in the devices depicted in Figures 2.7–2.9. While it is simple to limit the proportion of the excitation light that can be coupled into the collection waveguides directly across the channel, it is still challenging to collect fluorescent and scattered signals via integrated waveguides at an acceptable signal-to-noise ratio (SNR). The use of PMTs and avalanche photo diodes are still widely required, which increase size, cost, and complexity of devices.

Simple placement of collection fibers next to the channel has been shown to allow similar collection efficiencies as can be obtained by expensive and elaborate confocal detection schemes (Fig. 2.10) by consideration of the geometry of the channel and waveguide (68, 69). Equation 2.1 gives the limited collection efficiency

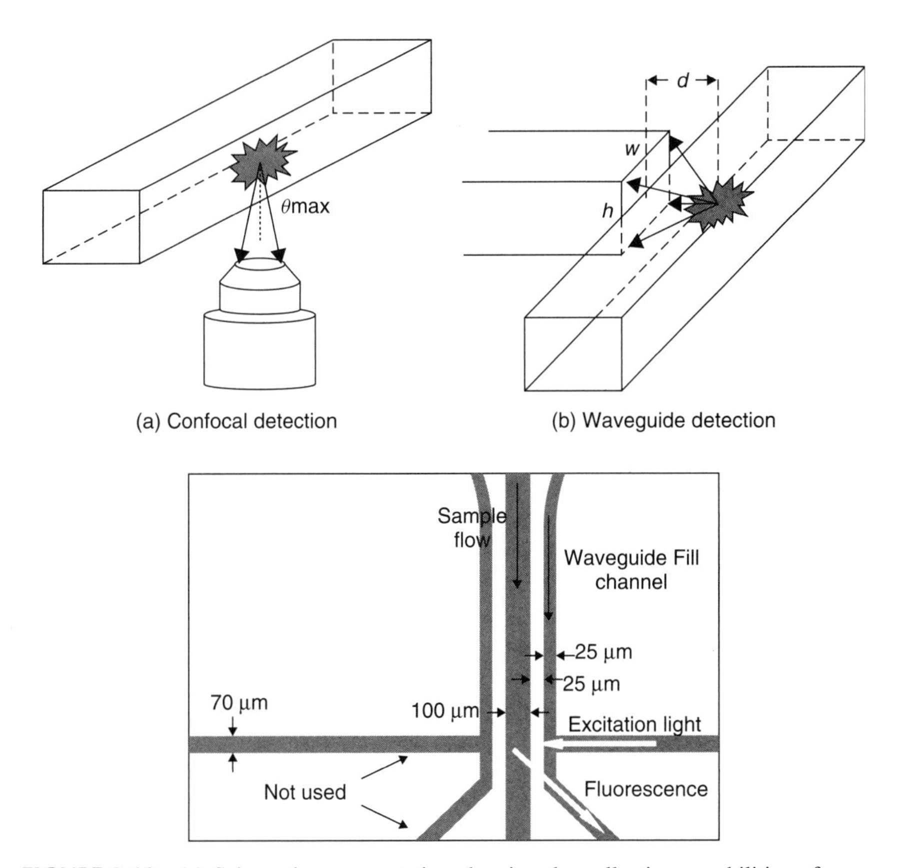

FIGURE 2.10 (a) Schematic representation showing the collection capabilities of a confocal system and an integrated waveguide and (b) a picture of a device where waveguide placement has been considered for proximity and SNR improvement. *Source:* Reproduced from Reference 68 by permission of The Royal Society of Chemistry.

(LCE) of a confocal system determined by the solid angle, Ω, or the largest cone of rays that is resolved by the objective subtended by the half-angle θ_{max}, also defined by the numerical aperture, NA. Equation 2.2 gives an approximation to the LCE based on the waveguide facet dimensions (given in Figure 2.10) while the maximum LCE (Eq. 2.3) is defined in situations, where d is very small and the LCE is thus limited by the NA of the waveguide and not the waveguide dimensions.

$$\mathrm{LCE} = \frac{\Omega}{4\pi} = \frac{1}{2}(1 - \cos\theta_{max}) \tag{2.1}$$

$$\mathrm{LCE} \approx \frac{wh}{4\pi d^2} \tag{2.2}$$

$$\mathrm{LCE}_{max} = \frac{\mathrm{NA}^2}{4n_s^2} \tag{2.3}$$

With microfabrication techniques, it is possible to eliminate the dependency of collection efficiency on the waveguide facet dimensions by transferring it to the NA of the waveguide, a simple quantity to control, as it depends on the core and cladding materials. For example, using waveguides manufactured out of SU-8 (devices shown in Figures 2.8 and 2.9), at a wavelength of 635 nm, SU-8 has an index of 1.59 (n_2) while air has an index of 1.00 (n_1). Calculating the NA of the resultant waveguide using Equation 2.4 (70) yields a value of 1.236, an extremely large value meaning that in a liquid medium ($n_a = 1.33$), the waveguide will collect light from a cone of rays subtended by a half-angle of $\theta_{max} = 68.1^\circ$. This value corresponds to about 12% of the light from an isotropic source, more than 10 times improvement over state of the art free-space detection techniques.

$$n_a \sin(\theta_{max}) = \sqrt{n_2^2 - n_1^2} = \mathrm{NA} \tag{2.4}$$

Along with the proximity of the waveguide, the placement of the waveguide can be used to increase the SNR in such devices. A microchip-based device (Fig. 2.10) showed that the placement of the collection waveguide 45° from the excitation waveguide (on the same side of the channel) drastically improved the SNR up to a value of 570 compared to 330 in conventional confocal schemes (68). With this collection geometry, the waveguide will collect back scattered light along with any fluorescent light demanding the need for a filtering scheme at the output of the waveguide. Filling the waveguide channels with a liquid allows the NA of the collection and excitation waveguide to be selected on an application-by-application basis or even allows for special waveguide function, such as filtering (69).

Microfabrication techniques also allow collection geometries and features that are difficult or not possible in conventional machines. With extensive pattering and replication capabilities, along with miniaturization of features, integrated optics on microchips can exploit the relatively large amounts of space to form complex structures and configurations. Beam splitters can be fabricated on a chip to divide a single source into multiple sources for multiple, sequential, uniform interrogations, a marked improvement over one signal per input fiber (56, 71, 72). An arrayed waveguide structure increasing the number of sequential interrogation points on the device and creating a regular interval of detection events (Fig. 2.11) allows increased detection algorithm functionality (56, 72).

One such algorithm to increase detection sensitivity, also shown in Figure 2.11, is the time-domain cross-correlation technique, which is able to retrieve buried information from a noisy signal. Considering the signal generated at the regularly space waveguides by particles with a uniform velocity, a correlation exists in the

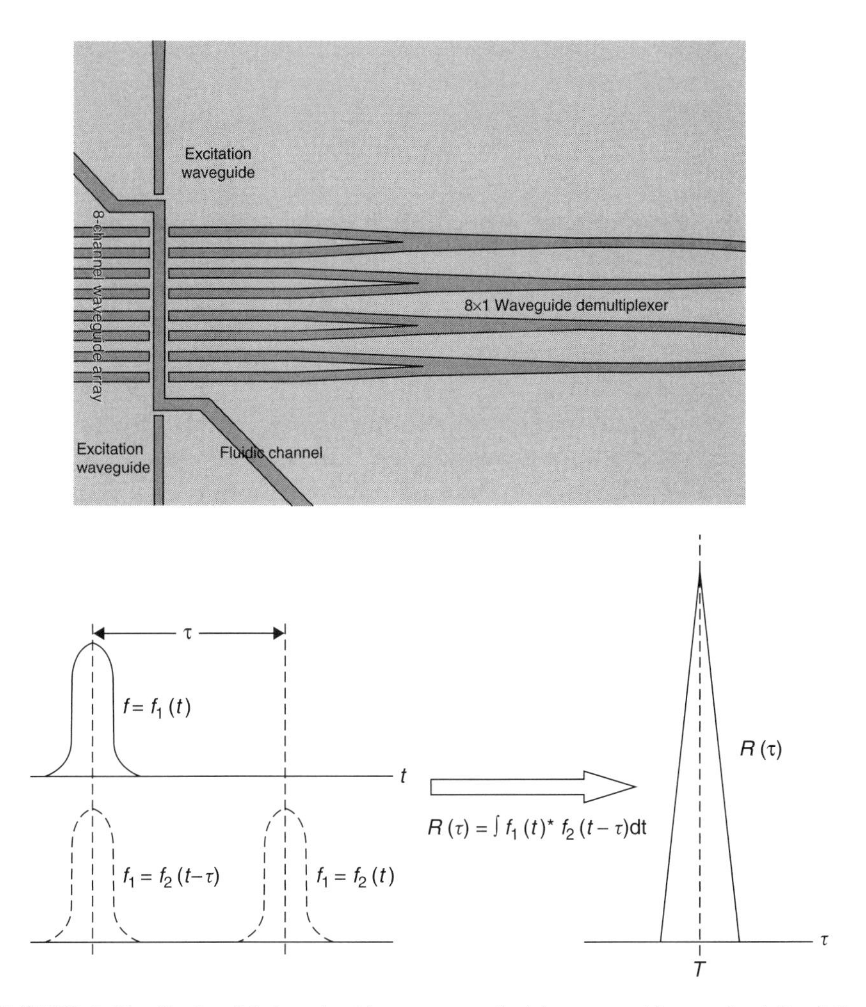

FIGURE 2.11 Device fabricated with an array of eight waveguides to the left of the channel and an 8 × 1 beam splitter to the right of the channel to demultiplex the signal for the time-domain cross-correlation technique. *Source:* Reprinted from Reference 56, with permission from IEEE. Copyright © 2005 IEEE.

output from the combination of the array of signals. With increasing number of waveguides, the algorithm is able to further separate the signal from the noise by random noise canceling out. Using this algorithm, it is possible to increase levels of SNR by many orders of magnitude, alleviating the necessity for very high power lasers or bulky detectors.

Integrated sources and detectors on the chip, two of the last external components to be integrated onto the chip, have also demonstrated manufacturing capabilities on-chip. A device consisting of a simple dye laser manufactured in the micro fluidic layer (SU-8) and coupled to integrated waveguides and a 1 × 5 beam splitter, was manufactured on a silicon substrate that incorporated multiple photodiodes for detection (73). Figure 2.12 shows the integrated device that only requires an external pump for the integrated laser and electrical connection to the photodiodes (73). This device shows the possibilities of a fully integrated device and the possibilities of miniaturization of an entire lab onto a portable and disposable chip.

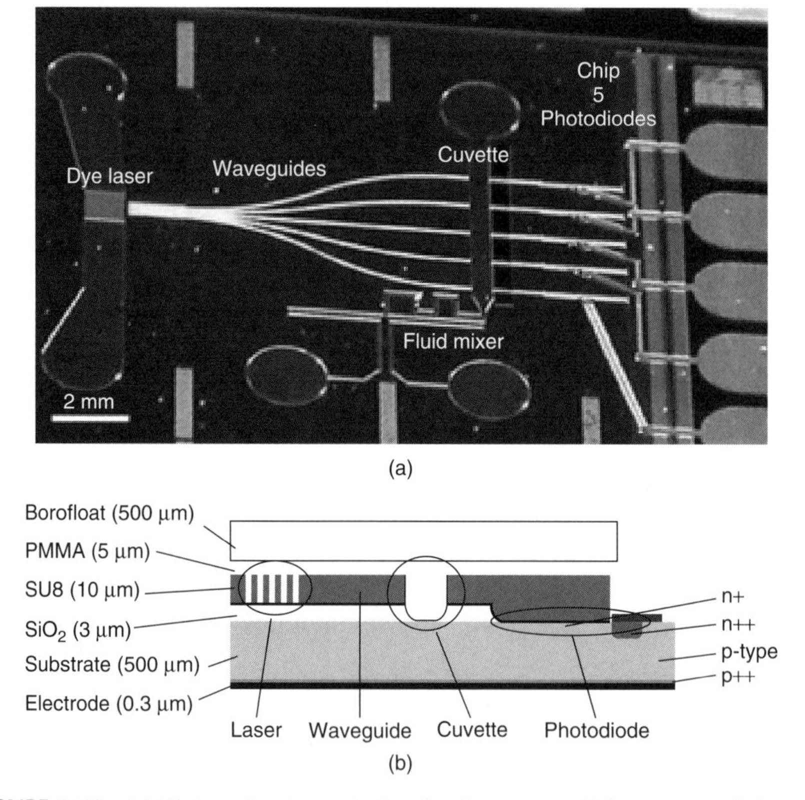

FIGURE 2.12 (a) Picture showing a device that integrates a light source and detectors with fluidic components and (b) a cross-sectional image showing the construction of such a device. *Source:* Reproduced from Reference 73 by permission of The Royal Society of Chemistry.

2.4 APPLICATIONS

Capabilities of a microchip-based flow cytometer to detect and characterize bacteria was first demonstrated in a simple detection and counting application of *Escherichia coli* bacterium (74). In this simple experiment, a microchip-based flow cytometer was demonstrated to be able to detect 94% of the labeled bacterium that were run through the device via electrokinetic flow. Samples of the detection signals obtained and the resulting histograms from the test (Fig. 2.13) show the scattered and fluorescent signals occurring simultaneously with varying intensities to indicate individual cells and their specific characteristics (74). The histogram plot shows a clear distribution of *E. coli* sizes, which reflects the sample population having a distribution through the life cycle, many midsized with few very large cells on the verge of division.

Relatively high throughput detections are also possible with a microchip-based flow cytometer. The detection efficiency of a microchip-based device is above 88% with a correlation (between scatter and fluorescent events) above 97% for 1.013 μm diameter latex beads while a throughput of 416 cells/s was reported (28). Comparing the performance of the device operating with microbeads to operation with real specimens yields experimental detection efficiencies with *E. coli* bacteria of approximately 90% while correlation of signals was 94.9% at a throughput of 350 cells/s (28). Figure 2.14 shows a sample of the detected signal, and the results of the detection algorithm that was used to determine a true detection as well as the histogram plots for the fluorescence and scattered signals of the FITC-labeled bacteria *E. coli* DH5α.

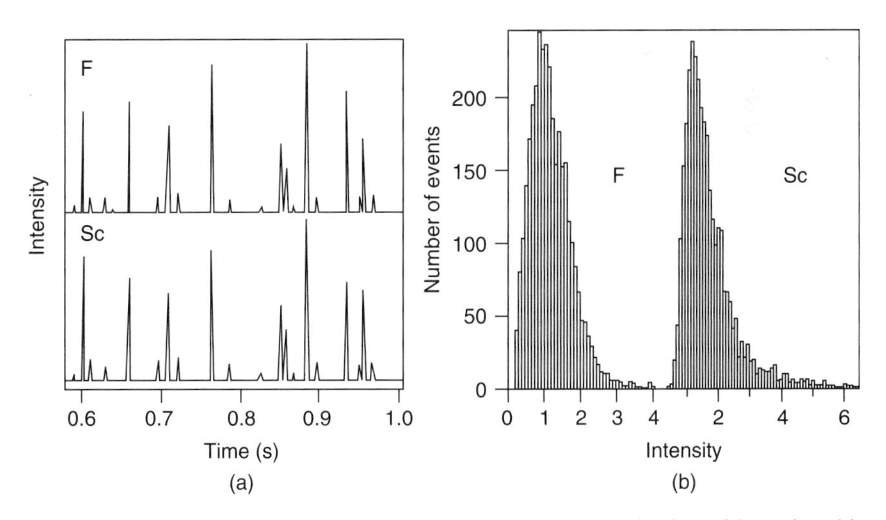

FIGURE 2.13 (a) Results from an *E. coli* bacteria counting application with a microchip-based flow cytometer device showing a sample of the obtained signal during processing samples with fluorescent (F) and scattered (Sc) signals and (b) a plot of the histogram were obtained. *Source:* Reprinted with permission from Reference 74. Copyright © 2001 American Chemical Society.

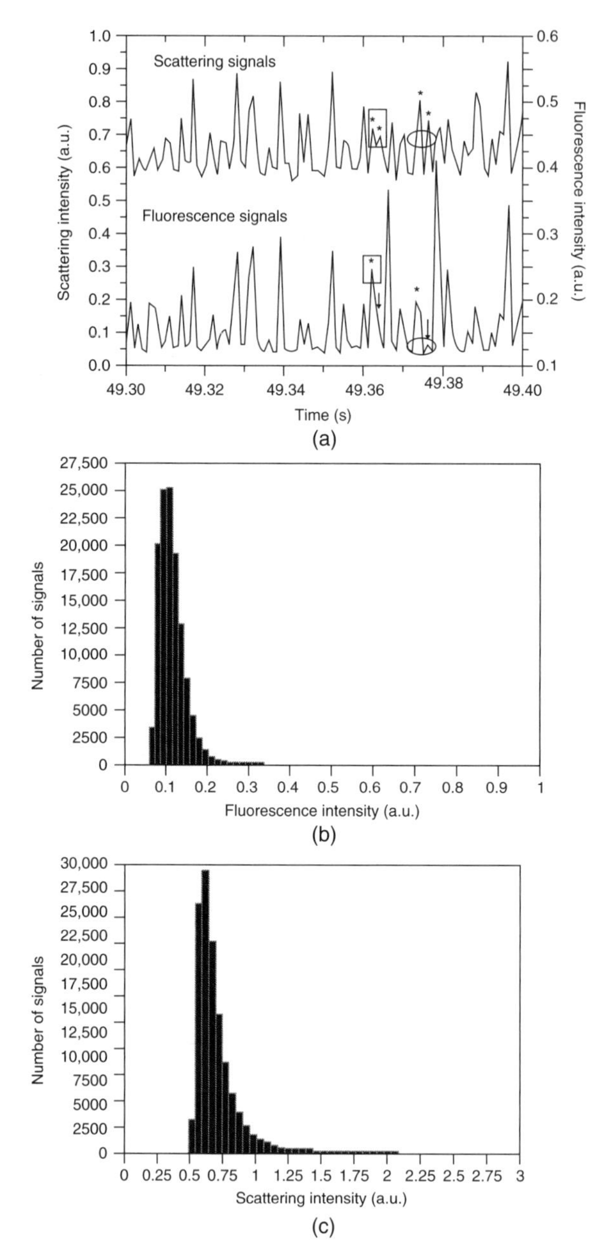

FIGURE 2.14 (a) a 0.1 s segment of the 300 s run of light scattering and fluorescence emission bursts for 5.0×10^6 cfu/ml FITC-labeled bacteria *E. coli* DH5α at an average particle throughput of 350 particles/s, the asterisk (*) indicates a event detected by peak algorithm, and the arrow represents a event missed by this algorithm. Histograms for (b) fluorescence emission and (c) light scattering intensities collected from FITC-labeled bacteria *E. coli* DH5α. *Source:* Reprinted from Reference 28, with permission from Elsevier.

A device that integrated optical waveguides with a simple lens was used in a scatter detection application for counting events (67). The fabricated device, shown in Figure 2.15, was constructed from SU-8 with the input waveguide coupled to a lens on the channel wall (top), and a collection waveguide to collect forward scattered light (0.5–5° from the optical axis) from the particle. Throughput was 2–25 beads/s, and a large scattered signal was collected from the top of the device—perpendicular to the plane of the device via free-space optics—in order to correlate forward scattered signals with actual events. The intensity of the forward scattered signal will yield useful information about the size of the particle; the greater the intensity the larger the particle. The lens was noted to focus the beam to a point near the sample flow with the benefit of increasing the intensity of the light available for excitation.

Analysis of the histograms in Figure 2.15 for the individual beads counting experiments reveal the CVs of the integrated device for the beads sized 2.8, 4.6, 5.8, and 9.1 μm were 26.6%, 27.4%, 28.2%, and 29.7%, respectively (67). The histograms indicate that the larger the bead size, the larger the average intensity from the bead's side scatter. The results were noted to be typical of other integrated optical devices; however, they are still far from the optimal values obtained by conventional machines of approximately 5% and 2% in the most ideal experimental circumstances. It was noted that improvements to device CV can be obtained by improving fluid focusing techniques through 3D focusing and steady pump control. By viewing the histogram, it is apparent that the devices still suffer from double detections and noise from impurities, which can be improved by simply tailoring the beam width to the particle size. In the work that generated this device, it was noted that the minimum distance between particles of interest should be twice the signal peak width. The low throughput was also noted to have room for improvement by reducing the width of the detection window.

To improve detection capabilities, a microchip-based device (Fig. 2.16) that permits 3D focusing on a planar microchip has been coupled with a free-space optical detection scheme that allows a flow cytometric function to be performed (75). Histograms of both populations of 8.32 μm fluorescent beads of different dye contents—approximately a 4 : 1 ratio for particle 2 to particle 1—mixed in a 1 : 1 ratio show improvements to the CV of the device because of less deviation of the particle in the flow stream (Fig. 2.16). The device was demonstrated to run at a throughput of 1730 particles/s. The CVs demonstrated are excellent −15.2% for particle 1 and 9.3% for particle 2—adequate when compared to a conventional cytometer. Furthermore, this device showed that the particle duration within the interrogation beam was nearly uniform and showed no statistical significance between the two particles despite the difference in fluorescent intensities.

Coupling flow cytometric fluorescence detection with a resistive pulse sensor (RPS)—a feature that detects a drop in the current along the channel as the cell of interest passes through a constriction in the channel—allows another degree of freedom in analysis. A microchip-based device coupled with free-space optics (Fig. 2.17) has been shown to have a performance level consistent with that of a conventional cytometer (76). Results from the device, also in Figure 2.17, obtained

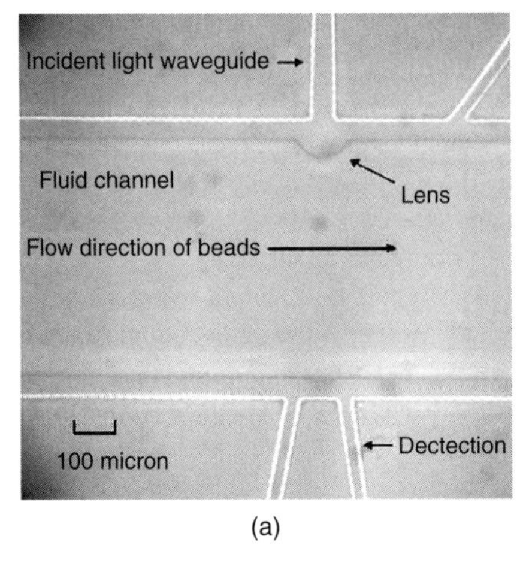

(a)

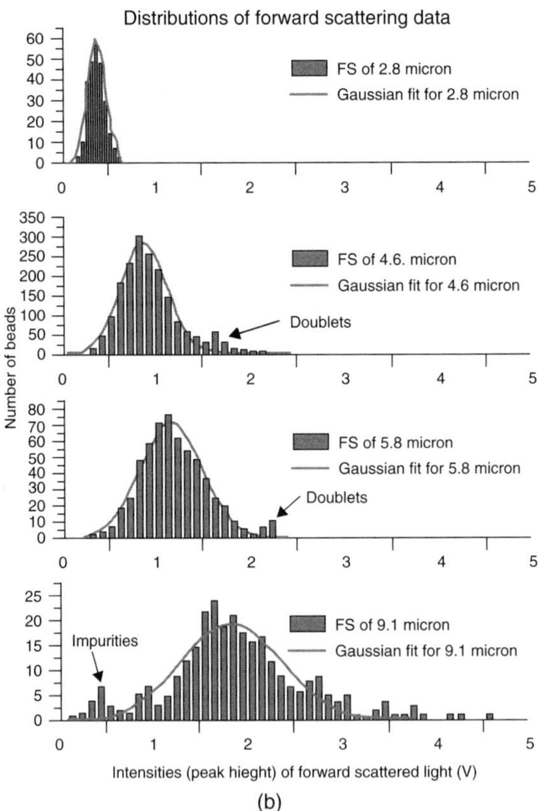

(b)

FIGURE 2.15 (a) Picture of a microchip-based device that incorporates waveguides for excitation and collection with a microlens for intensity improvement. (b) Scatter histograms showing the performance of the device at different bead sizes—note the Gaussian fit and the double detections for larger sized beads and introduction of impurities. *Source:* Reproduced from Reference 67 by permission of The Royal Society of Chemistry.

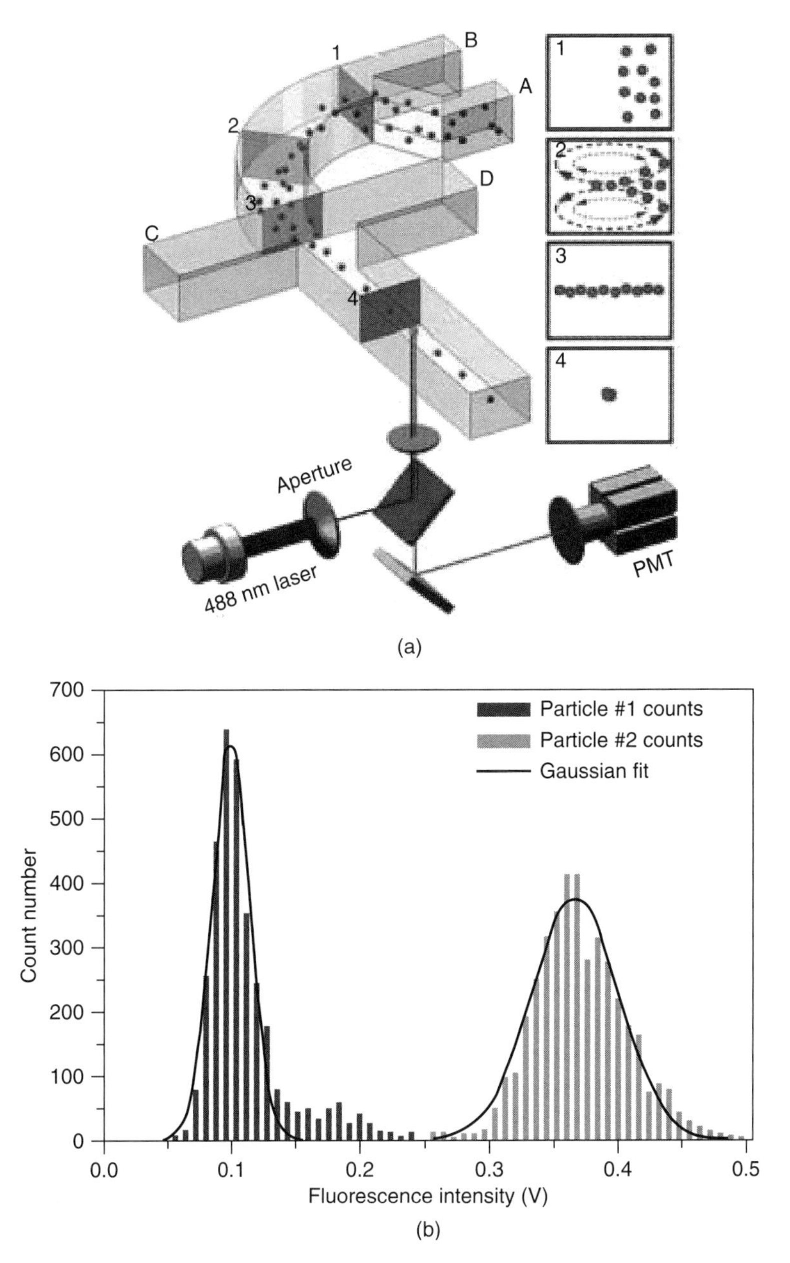

FIGURE 2.16 (a) Microchip-based device that incorporates 3D planar hydrodynamic focusing with a free-space optical detection scheme showing cross-sectional views of particle distribution in the channel at different points along the flow. (b) The histogram from two different fluorescent beads—both 8.32 μm in diameter with a 4 : 1 intensity ratio—demonstrate very good CVs. *Source:* Reproduced from Reference 75 by permission of The Royal Society of Chemistry.

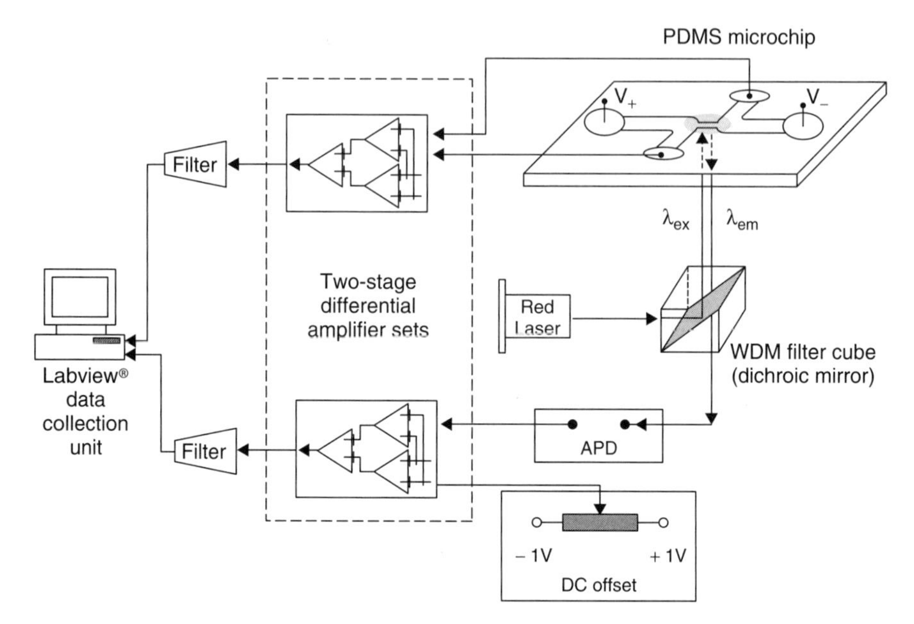

FIGURE 2.17 Schematic representation (bottom) of the PDMS-based microchip showing the integration with the free-space optical detection scheme and RPS counter. Multiple runs showed that the device is accurate when compared to a conventional machine. *Source:* Reproduced from Reference 76 by permission of The Royal Society of Chemistry.

through multiple runs determined the concentration of a mixture of similar sized beads—some fluorescent and some nonfluorescent—and a population of white blood cells where some were stained with a fluorescent dye (76). While the RPS counted all events of a bead passing the interrogation zone, the optics detected fluorescent particles, thus allowing the percentage of fluorescent beads in the mixture to be determined. When compared to the percentage obtained from established flow cytometry techniques, the microchip-based device was found to operate accurately within the confines of statistical deviations in the actual concentrations.

Microchip-based flow cytometers tend to easily integrate with a cell sorter, along with other fluid steps, as mentioned earlier. In fact, the process of flow cytometry is necessary to perform the decision criteria to sort a population of cells by a certain characteristic, thereby allowing very sophisticated sorting capabilities through the multiparameter capabilities of flow cytometry. Integration of such a complex fluidic handling system is easily fabricated on-chip by the addition of multiple fluid channels for cell sequestering, and the addition of separate electrical contacts with which to direct particles to a certain output via electrokinetic forces; all are easily integrated to the chip with very little sacrifice to complexity, cost, or size. Fu et al. showed a PDMS-based microchip flow cytometer with a free-space optical detection scheme that employed electroosmotic flow to sort between two colors of beads (48). Enrichment rates of beads in the output well (fractional increase in concentration of targeted specimens between output and input samples) reached 80–96-fold with run times that lasted 10 min to 3 h. Sorting of fluorescently labeled *E. coli* cells was also demonstrated and achieved enrichment levels of 30-fold. Improvements on this device—with respect to buffer compatibility, surface chemistry, and cell viability—were shown (47).

Flow cytometry is very useful when the application calls for detecting very rare events because of the high throughput and individual cell interrogation. Coupling a cell sorter with a cytometer allows rare cells to be removed from the natural population and sequestered with other rare cells allowing for more efficient analysis as the cells have been removed from the noise of the background population. However, with such rare occurrence of events in a large sample, an extremely long run time is needed in order to obtain a significant quantity of rare cells for analysis. A microchip-based device was demonstrated, which was capable of detecting the 1–10 rare cancerous blood cells that exist within a 10 ml volume of blood and achieving large enrichment rates without much sacrifice to run time (77). The device, Figure 2.18 by Micronics, manages to increase throughput dramatically by analyzing a vertical row of hydrodynamically focused cells (instead of individual cells), and sorting the row if even one cell gives a positive detection. This dramatically increases the concentration of the target cancer cells without slowing throughput to an infeasible rate. The device was demonstrated to be able to detect beads as small as 3 μm. Once sorted, the cells are held in a culturing chamber on-chip so that the chip can be removed and the concentrated rare cell population can be taken for special analysis elsewhere or further sorted to increase the sample concentration.

High speed applications of flow cytometry by a microchip-based device that incorporates 3D hydrodynamic focusing (Fig. 2.19) has demonstrated an ability to perform flow cytometry on microspheres at a rate of up to 17,000 particles/s (78). The fluorescence detection is comparable with a conventional devices and the throughput is similar to that of modest conventional machines.

Integrated optics coupled with a flow cytometric function and a sorting mechanism all on one chip has been demonstrated with integrated pumps and valves for fluid control. Figure 2.20 shows a picture of the device, a detail of the detection region, and details of the fluidic sorting region (79). The device was fabricated in PDMS to exploit the elasticity of the material to activate pumping mechanisms and to create a seal for valves on the sort channels. Microchannels were fabricated in the bulk material in order to guide the insertion of fibers to ensure automatic alignment for both excitation and detection signals. The device sorted labeled human lung cancer cells with a counting and sorting error rate of 1.5% and 2%, respectively (80).

Applications of detection and counting of bacteria in a natural environment are possible with a microchip-based device. Conventional cytometry has been used to remotely monitor environmental conditions of ocean water in a cytometer specially fabricated to operate within a buoy and transmit signals to users remotely (26, 27). A microchip-based flow cytometer was used to detect and count a mixture of *E. coli* and *Pseudomonas putida* bacterium from a suspension of river water (45). By comparing the count totals obtained from the microchip-based cytometer to a

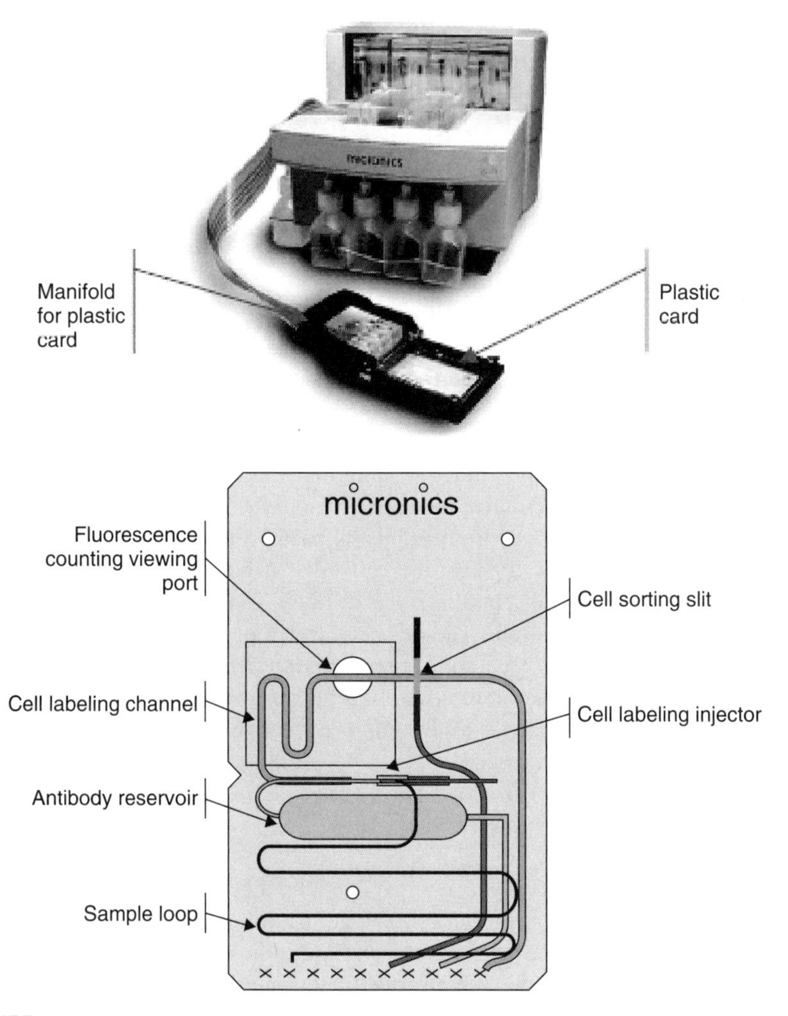

FIGURE 2.18 Micronics microFlow system adapted to sort a ribbon of cells for rare cancer cell detection and separation. Picture of the portable system (top) with interchangeable plug-and-play card (bottom). Enrichment rates of 1000-fold have been estimated. *Source:* Reprinted from Reference 77, with permission from Elsevier.

standard and widely adopted conventional total direct count (TDC), the accuracy of the microchip-based flow cytometer could be determined. A nearly one-to-one correlation was found between the two counting methods for simple runs of single *E. coli* and *P. putida* ($r^2 = 0.96$). An application to quantify bacterium in actual river water samples yielded ratios of 80% and 120% between the two counting methods, a figure that is still within error of the given counting methods. The study also concluded that owing to the high reproducibility that is afforded from microchip-based designs that the calibration of such devices could be done once via

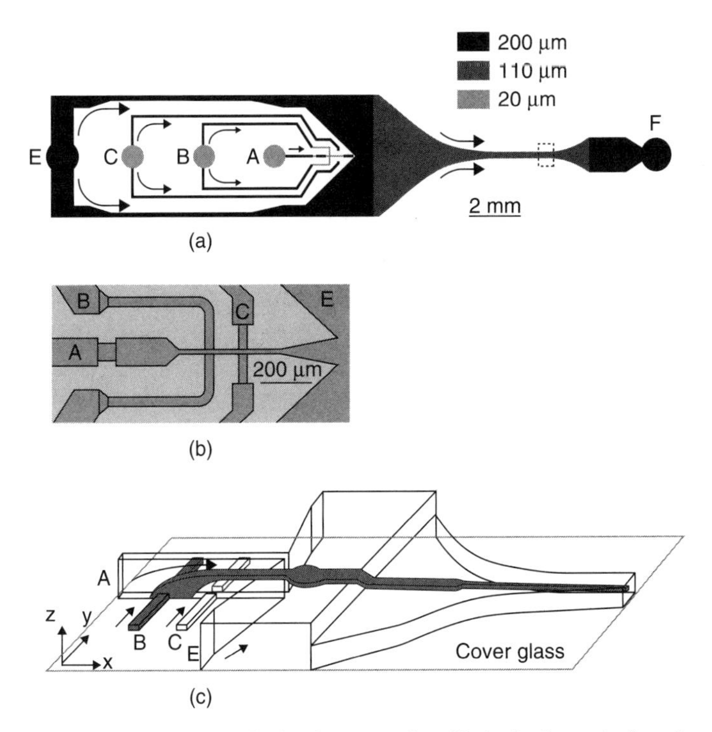

FIGURE 2.19 Microchip-based device incorporating 3D hydrodynamic focusing for high throughput flow cytometry. (a) shows a schematic vertical view of the chip with arrows indicating the fluid flow direction for the various fluid inputs (A and C are vertical sheath flows, B is sample flow, and E is horizontal sheath flow), (b) shows a picture of the 3D focusing fluid flow, and (c) shows a schematic diagram of the flow within the device. *Source:* Reprinted with permission from Reference 78. Copyright © 2006 American Chemical Society.

a fluorescent microbead solution before running with live samples. Conventional cytometers mix a known concentration of microspheres into the sample in order to accurately compare event detections to bacteria concentration. This leads to much simpler sample preparations as well as reliable and repeatable devices.

2.5 CONCLUSION

Flow cytometry is a very powerful analysis tool that has limitless capabilities when considered for applications involving characterizing a population of microparticles. However, owing to size, cost, and complexity restraints, conventional cytometers are limited in reach and scope when considered for applications outside a laboratory or large-scale facility. Microchip-based devices show much promise to take these powerful analysis capabilities out of the lab and into remote and rugged settings. Through lithography-based fabrication techniques that allow precise fabrication, miniaturization, large-scale integration, and mass-production capabilities,

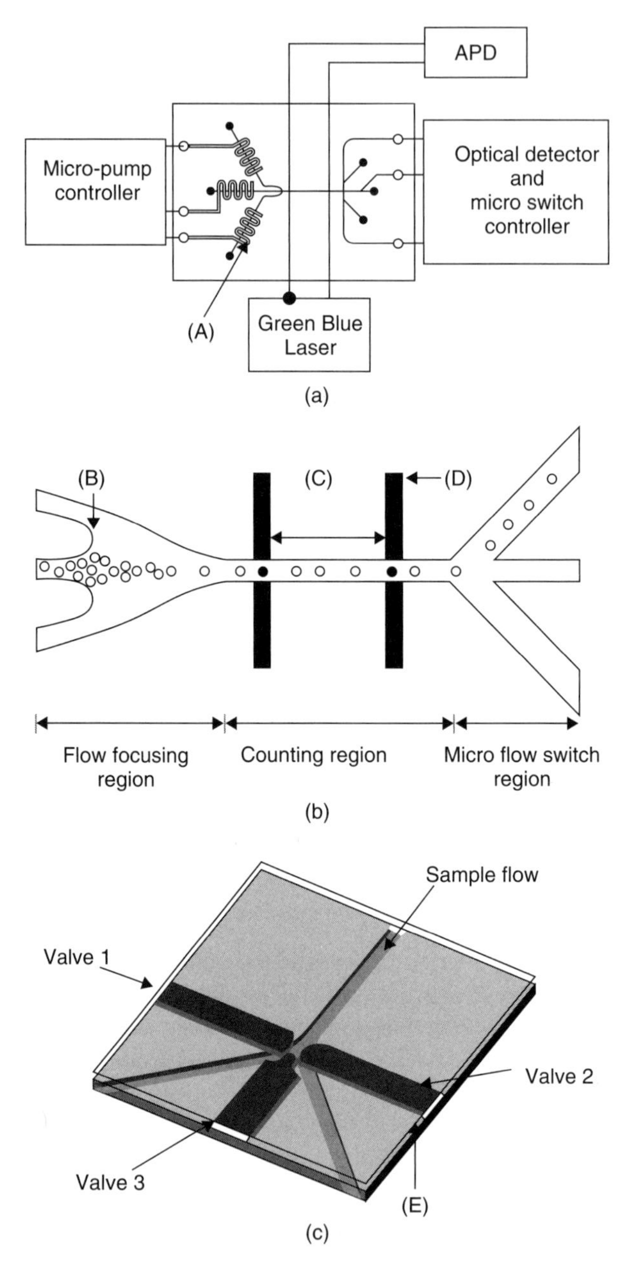

FIGURE 2.20 A microchip-based cell sorter with integrated fibers and micropumps and valves. (a) Schematic of the microchip-based device and the interface with external components and showing the integrated S-shaped micropumps (A). (b) Details of the microchannel near focusing region (B), interrogations fibers (C), (D) and the sorting structure. (c) The flow switch region showing the valves, (E), to direct sample flow based on sorting criteria. *Source:* Reproduced from Reference 79, with permission from the Japan Society of Applied Physics.

new applications for these microchip-based flow cytometers are being realized and created. Such devices are already showing detection capabilities comparable with conventional machines at comparable throughputs, while modest integration capabilities have been realized through integrated cell sorting and different fluid-handling techniques and basic integrated optics. With increased levels of on-chip integration such as fluid-handling components, excitation and collection optics, sources, and detectors, as well as thorough miniaturization of the device, microchip-based devices will gain independence from all external components to form a complete μTAS to allow more on-site and remote monitoring capabilities. Microchip-based cytometry is still in its infancy, however, and the level of integration on the devices is still quite primitive from a feasibility standpoint; the widespread adoption of such devices greatly depends on the level of functionality that can be adopted onto the chips. Even with eventual full-scale integration, the long-term goal of such devices should not be to replace conventional machines; on the contrary, they should serve to complement these machines wherever possible serving as a first-line detection/analysis technique for detecting anomalies and characterizing samples in traditional conventional applications, while allowing the new and novel applications opened up through integration and miniaturization capabilities.

REFERENCES

1. Sims CE, Allbritton NL. Analysis of single mammalian cells on-chip. Lab Chip 2007;7:423–440.
2. Life Technologies Corporation. 2011. Introduction to flow cytometry, Invitrogen. Available at http://probes.invitrogen.com/resources/education/tutorials/4Intro_Flow/player.html.
3. Perfetto SP, Chattopadhyay PK, Roederer M. Seventeen-colour flow cytometry: unravelling the immune system. Nat Rev Immunol 2004;4:648–655.
4. Baumgarth N, Roederer M. A practical approach to multicolor flow cytometry for immunophenotyping. J Immunol Methods 2000;243:77–97.
5. Gabriel H, Kindermann W. Flow-cytometry—principles and applications in exercise immunology. Sports Med 1995;20:302–320.
6. Garratty G, Arndt PA. Applications of flow cytofluorometry to red blood cell immunology. Cytometry 1999;38:259–267.
7. Bleesing JJH, Fleisher TA. Immunophenotyping. Semin Hematol 2001;38:100–110.
8. Giorgi J, Fahey JL, Smith DC, Hultin LE, Cheng HL, Mitsuyasu RT, Detels R. Early effect of HIV on CD4 lymphocytes *in vivo*. J Immunol 1987;138:3725–3730.
9. Stetler-Stevenson M, Braylan RC. Flow cytometric analysis of lymphomas and lymphoproliferative disorders. Semin Hematol 2001;38:111–123.
10. Bleesing JJH, Fleisher TA. Cell function-based flow cytometry. Semin Hematol 2001;38:169–178.
11. Darzynkiewicz Z, Bedner E, Smolewski P. Flow cytometry in analysis of cell cycle and apoptosis. Semin Hematol 2001;38:179–193.

12. Campana D, Behm FG. Immunophenotyping of leukemia. J Immunol Methods 2000;243:59–75.
13. Weir EG, Borowitz MJ. Flow cytometry in the diagnosis of acute leukemia. Semin Hematol 2001;38:124–138.
14. Steen HB, Boye E, Skarstad K, Bloom B, Godal T, Mustafa S. Applications of flow cytometry on bacteria: cellcycle kinetics, drug effects, and quantititation of antibody binding. Cytometry 1982;2:249–257.
15. Steen H, Boye E. Growth of *Escherichia coli* studied by dual-parameter flow cytometer. J Bacteriol 1981;145:1091–1094.
16. Alvarez-Barrientos A, Arroyo J, Canton R, Nombela C, Sanchez-Perez S. Applications of flow cytometry to clinical microbiology. Clin Microbiol Rev 2000;13: 167–195.
17. Harding CL, Lloyd DR, McFarlane CM, Al-Rubeai M. Using the microcyte flow cytometer to monitor cell number, viability, and apoptosis in mammalian cell culture. Biotechnol Prog 2000;16:800–802.
18. Nolan JP, Laurer S, Prossnitz ER, Sklar LA. Flow cytometry: a versatile tool for all phases of drug discovery. Drug Discov Today 1999;4:173–180.
19. Kuckuck FW, Edwards BS, Sklar LA. High throughput flow cytometry. Cytometry 2001;44:83–90.
20. De Rosa SC, Herzenberg LA, Herzenberg LA, Roederer M. 11-color, 13-parameter flow cytometry: identification of human naive T cells by phenotype, function, and T-cell receptor diversity. Nat Med 2001;7:245–248.
21. Roederer M, De Rosa S, Gerstein R, Anderson M, Bigos M, Stovel R, Nozaki T, Parks D, Herzenberg L, Herzenberg L. 8 color, 10-parameter flow cytometry to elucidate complex leukocyte heterogeneity. Cytometry 1997;29:328–339.
22. McSharry JJ. Uses of flow cytometry in virology. Clin Microbiol Rev 1994;7:576–604.
23. Ferris MM, McCabe MO, Doan LG, Rowlen KL. Rapid enumeration of respiratory viruses. Anal Chem 2002;74:1849–1856.
24. Davey HM, Kell DB. Flow cytometry and cell sorting heterogeneous microbial populations: the importance of singe-cell analyses. Microbiol Rev 1996;60:641–696.
25. Wedemeyer N, Potter T. Flow cytometry: an 'old' tool for novel applications in medical genetics. Clin Genet 2001;60:1–8.
26. Dubelaar GBJ, Gerritzen PL. CytoBuoy: a step towards using flow cytometry in operational oceanography. Sci Marina 2000;64:255–265.
27. Dubelaar GBJ, Gerritzen PL, Beeker AER, Jonker RR, Tangen K. Design and first results of CytoBuoy: a wireless flow cytometer ofr in situ analysis of marine and fresh waters. Cytometry 1999;37:247–254.
28. Mu C, Zhang F, Zhang Z, Lin M, Cao X. Highly efficient dual-channel and cytometric-detection of micron-sized particles in a microfluidic device. Sens Actuators B Chem 2011;151:402–409.
29. Rodriguez I, Spicar-Nihalic P, Kuyper CL, Fiorini GS, Chiu DT. Rapid prototyping of glass microchannels. Anal Chim Acta 2003;496:205–215.
30. Kikutani Y, Tokeshi M, Sato K, Kitamori T. Integrated chemical systems on microchips for analysis and assay. Potential future, mobile high-performance detection system for chemical weapons. Pure Appl Chem 2002;74:2299–2309.

31. Chandrasekaran A, Packirisamy M. Enhanced fluorescence-based bio-detection through selective integration of reflectors in microfluidic lab-on-a-chip. Sensor Rev 2008;28:33–38.
32. Fu JL, Fang Q, Zhang T, Jin XH, Fang ZL. Laser-induced fluorescence detection system for microfluidic chips based on an orthogonal optical arrangement. Anal Chem 2006;78:3827–3834.
33. Xia Y, Whitesides GM. Soft lithography. Annu Rev Mater Sci 1998;28:153–184.
34. Duffy DC, McDonald JC, Schueller OJA, Whitesides GM. Rapid prototyping of microfluidic systems in Poly(dimethylsiloxane). Anal Chem 1998;70:4974–4984.
35. Vallee R, editor. Microchip-based flow cytometry for Effective Detection and Count. Proceedings of the SPIE 2009, Photonics North; 2009 May 24–27; Quebec City. p. 7386: Mu C, Zhang Z, Lin M, Cao X, Zhang F.
36. Schrum DP, Culbertson CT, Jacobson SC, Ramsey JM. Microchip flow cytometry using electrokinetic focusing. Anal Chem 1999;71:4173–4177.
37. Fu LM, Yang RJ, Lin CH, Pan YJ, Lee GB. Electrokinetically driven micro flow cytometers with integrated fiber optics for on-line cell/particle detection. Anal Chim Acta 2004;507:163–169.
38. Lee G-B, Hung C-I, Ke B-J, Huang G-R, Hwei B-H, Lai H-F. Hydrodynamic focusing for a micromachined flow cytometer. J Fluids Eng 2001;123:672–679.
39. Sundararajan N, Pio MS, Lee LP, Berlin AA. Three-dimensional hydrodynamic focusing in polydimethylsiloxane (PDMS) microchannels. J Microelectromech Syst 2004;13:559–567.
40. Yang R, Feeback DL, Wang W. Microfabrication and test of a three-dimensional polymer hydro-focusing unit for flow cytometry applications. Sens Actuators A Phys 2005;118:259–267.
41. Simonnet C, Groisman A. Two-dimensional hydrodynamic focusing in a simple microfluidic device. Appl Phys Lett 2005;87:114104–1-3.
42. Chang C-C, Huang Z-X, Yang R-J. Three-dimensional hydrodynamic focusing in two-layer polydimethylsiloxane (PDMS) microchannels. J Micromech Microeng 2007;17:1479–1486.
43. Mao X, Waldeisen JR, Huang TJ. "Microfluidic drifting"—implementing three-dimensional hydrodynamic focusing with a single-layer planar microfluidic device. Lab Chip 2007;7:1260–1262.
44. Chung TD, Kim HC. Recent advances in miniaturized microfluidic flow cytometry for clinical use. Electrophoresis 2007;28:4511–4520.
45. Sakamoto C, Yamaguchi N, Nasu M. Rapid and simple quantification of bacterial cells by using a microfluidic device. Appl Environ Microbiol 2005;71:1117–1121.
46. Ramsey JM. The burgeoning power of the shrinking laboratory. Nat Biotechnol 1999;17:1061–1062.
47. Fu AY, Chou HP, Spence C, Arnold FH, Quake SR. An integrated microfabricated cell sorter. Anal Chem 2002;74:2451–2457.
48. Fu AY, Spence C, Scherer A, Arnold FH, Quake SR. A microfabricated fluorescence-activated cell sorter. Nat Biotechnol 1999;17:1109–1111.
49. Dittrich PS, Schwille P. An integrated microfluidic system for reaction, high-sensitivity detection, and sorting of fluorescent cells and particles. Anal Chem 2003;75:5767–5774.

50. Chang CM, Hsiung SK, Lee GB. A micromachine-based flow cytometer chip integrated with micro-pumps/valves for multi-wavelength detection applications. Mater Sci Forum 2006;505:637–642.
51. Dittrich PS, Tachikawa K, Manz A. Micro total analysis systems. Latest advancements and trends. Anal Chem 2006;78:3887–3907.
52. Chabinyc ML, Chiu DT, McDonald JC, Stroock AD, Christian JF, Karger AM, Whitesides GM. An integrated fluorescence detection system in poly(dimethylsiloxane) for microfluidic applications. Anal Chem 2001;73:4491–4498.
53. Tung YC, Zhang M, Lin CT, Kurabayashi K, Skerlos SJ. PDMS-based opto-fluidic micro flow cytometer with two-color, multi-angle fluorescence detection capability using PIN photodiodes. Sens Actuators B 2004;98:356–367.
54. Lee GB, Lin CH, Chang GL. Micro flow cytometers with buried SU-8/SOG optical waveguides. Sens Actuators A 2003;103:165–170.
55. Ruano-Lopez JM, Aguirregabiria M, Tijero M, Arroyo MT, Elizalde J, Berganzo J, Aranburu I, Blanco FJ, Mayora K. A new SU-8 process to integrate buried waveguides and sealed microchannel for a Lab-on-a-Chip. Sens Actuators B 2006;114:542–551.
56. Lien V, Zhao K, Berdichevsky Y, Lo YH. High-sensitivity cytometric detection using fluidic-photonic integrated circuits with array waveguides. IEEE J Sel Top Quantum Electron 2005;11:827–834.
57. Mogensen KB, El-Ali J, Wolff A, Kutter JP. Integration of polymer waveguides for optical detection in microfabricated chemical analysis systems. Appl Opt 2003;42:4072–4079.
58. Kowpak T, Watts B, Zhang Z, Zhu S, Xu CQ. Fabrication of photonic/microfluidic integrated devices using an epoxy photoresist. Macromol Mater Eng 2010;295:559–565.
59. Watts BR, Kowpak T, Zhang Z, Xu CQ, Zhu S. Formation and characterization of an ideal excitation beam geometry in an optofluidic device. Biomed Opt Expr 2010;1:848–860.
60. Fleger M, Neyer A. PDMS microfluidic chip with integrated waveguides for optical detection. Microelectron Eng 2006;83:1291–1293.
61. Watts BR, Kowpak T, Zhang Z, Xu CQ, Zhu S. A microfluidic-photonic-integrated device with enhanced excitation power density. Photonics West 2010, San Francisco, California, USA. Proceedings of SPIE 7555; p 7555-01–7555-04.
62. Ro KW, Lin K, Shim BC, Hahn JH. Integrated light collimating system for extended optical-path-length absorbance detection in microchip-based capillary electrophoresis. Anal Chem 2005;77:5160–5166.
63. Camou S, Fujita H, Fujii T. PDMS 2D optical lens integrated with microfluidic channels: principle and characterization. Lab Chip 2003;3:40–45.
64. Park S, Jeong Y, Kim J, Choi K, Kim KC, Chung DS, Chun K. Fabrication of poly(dimethylsiloxane) microlens for laser-induced fluorescence detection. Jpn J Appl Phys 2006;45:5614–5617.
65. Seo J, Lee LP, Actuators B. Disposable integrated microfluidics with self-aligned planar microlenses. Sens Actuators B 204;99:615–622.
66. Hsiung SK, Lin CH, Lee GB. A microfabricated capillary electrophoresis chip with multiple buried optical fibers and microfocusing lens for multiwavelength detection. Electrophoresis 2005;26:1122–1129.

67. Wang Z, El-Ali J, Engelund M, Gotsaed T, PerchNielsen IR, Mogensen KB, Snakenborg D, Kutter JP, Wolff A. Measurements of scattered light on a microchip flow cytometer with integrated polymer based optical elements. Lab Chip 2004;4:372–377.
68. Bliss CL, McMullin JN, Backhouse CJ. Rapid fabrication of a microfluidic device with integrated optical waveguides for DNA fragment analysis. Lab Chip 2007;7:1280–1287.
69. Bliss CL, McMullin JN, Backhouse CJ. Integrated wavelength-selective optical waveguides for microfluidic-based laser-induced fluorescence detection. Lab Chip 2008;8:143–151.
70. Yariv A, Yeh P. *Photonics: Optical Electronics in Modern Communications*. 6th ed. New York: Oxford University Press; 2007.
71. Mogensen KB, Kwok YC, Eijkel JCT, Petersen MJ, Manz A, Kutter JP. A microfluidic device with an integrated waveguide beam spitter for velocity measurements of flowing particles by fourier tanformation. Anal Chem 2003;75:4931–4936.
72. Lien V, Zhao K, Lo YH. Fluidic photonic integrated circuit for in-line detection. Appl Phys Lett 2005;87:194106.
73. Balslev S, Jorgensen AM, Bilenberg B, Mogensen KB, Snakenborg D, Gescheke O, Kutter JP, Kristensen A. Lab-on-a-chip with integrated optical transducers. Lab Chip 2006;6:213–217.
74. McClain MA, Culbertson CT, Jacobson SC, Ramsey JM. Flow cytometry of *Escherichia coli* on microfluidic devices. Anal Chem 2001;73:5334–5338.
75. Mao X, Lin SCS, Dong C, Huang TJ. Single-layer planar on-chip flow cytometer using microfluidic drifting based three-dimensional (3D) hydrodynamic focusing. Lab Chip 2009;9:1–8.
76. Wu X, Chon CC, Wang YN, Kang Y, Li D. Simultaneous particle counting and detecting on a chip. Lab Chip 2008;8:1943–1949.
77. Lancaster C, Kokoris M, Nabavi M, Clemmens J, Maloney P, Capadanno J, Gerdes J, Battrell CF. Rare cancer cell analyzer for whole blood applications: microcytometer cell counting and sorting subcircuits. Methods 2005;37:120–127.
78. Simonnet C, Groisman A. High-throughput and high-resolution flow cytometry in molded microfluidic devices. Anal Chem 2006;78:5653–5663.
79. Chang CM, Hsiung SK, Lee GB. Micro flow cytometer chip integrated with micro-pumps/micro-valves for multi-wavelength cell counting and sorting. Jpn J Appl Phys 2007;46:3126–3134.
80. Yang SY, Hsiung SK, Hung YC, Chang CM, Liao TL, Lee GB. A cell counting/sorting system incorporated with a microfabricated flow cytometer chip. Meas Sci Technol 2006;17:2001–2009.

CHAPTER THREE

Optofluidic Techniques for the Manipulation of Micro Particles: Principles and Applications to Bioanalyses

HONGLEI GUO
Microwave Photonics Research Laboratory, University of Ottawa, Ottawa, ON, Canada

GAOZHI XIAO
Institute for Microstructural Sciences, National Research Council Canada, Ottawa, ON, Canada

JIANPING YAO
Microwave Photonics Research Laboratory, University of Ottawa, Ottawa, ON, Canada

3.1 INTRODUCTION

The emerging needs required by the rapid pathogen detection, clinical diagnosis, and forensic science necessitate the development of lab-on-a-chip devices, which integrate laboratory functions and processes, and scale down the testing systems to a miniaturized chip format (1). Proper manipulation and handling of the micro-sized bioparticles is regarded as one of the most challenging problems to biologists,

Photonic Sensing: Principles and Applications for Safety and Security Monitoring, First Edition.
Edited by Gaozhi Xiao and Wojtek J. Bock.

medics, and colloidal physicists (2). In the last decade, several techniques have been reported to achieve this, including the pressure-driven flow (3), electrophoresis (4), and optoelectronic tweezer (5). In recent years, optofluidics, which is described as the combination of both optics and fluidics on a chip platform, has become a rapidly developing research area for the manipulation of microbioparticles owing to its advantages such as strong dependence on particle size, velocity and optical properties, extremely high optical trapping stability, and insensitivity to surface condition. Meanwhile, optofluidic techniques are benefited from the mature fabrication processes and components that are already developed by telecommunications industry (6).

The well-known optical tweezer (7), achieved by using a tightly focused light beam, is developed based on optofluidic techniques. It has been reported that an optical tweezer is capable of trapping microparticles in a nonintrusive way by the proper selection of lasers (8). The technique has already demonstrated the applications in a variety of biological research areas (9). In addition, by applying a spatial light modulator, the trapping laser can be split into many beams, making it possible to simultaneously trap and manipulate particles at different locations (10). However, the overall trapping strength of an optical tweezer is limited owing to its very short focal depth and large diffraction of light beams (11). To alleviate this limitation, other optofluidic techniques have been proposed and demonstrated.

In this chapter, we classify these optofluidic techniques into three categories, (i) fiber-based optofluidic technique; (ii) near-field optofluidic technique; and (iii) axial-type and cross-type optical chromatography. The fundamental mechanism and recent development of these optofluidic techniques are reviewed. Then, we present our work on a novel SU-8/PDMS (polydimethylsiloxane) optofluidic chip in which an on-chip lens structure is introduced to enhance the performance of microparticle manipulation. Finally, applications of optofluidic techniques to bioanalyses are discussed.

3.2 OPTOFLUIDIC TECHNIQUES FOR THE MANIPULATION OF PARTICLES

Optical radiation forces have been considered extremely effective in trapping and manipulating micrometer- and nanometer-sized particles since the first optical trapping demonstration was made by Ashkin (12). The radiation forces applied to a particle result from the momentum transfer between the photons and the particles and are classified into two categories: the gradient force and the scattering force. The gradient force drives the particle to the center of the light beam and eventually retains it at the focal point of the light beam, whereas the scattering force drives the particle in the propagation direction of the light beam. Details about these forces can be found in References 2 and 13. In this section, we present an overview on the state of the art of optofluidic techniques for manipulating particles.

3.2.1 Fiber-Based Optofluidic Techniques

The first optical fiber trapping system is reported in Reference 14. It is built on a microscope slide with a piece of sandwiched glass capillary tubes lying between the slide and a glass cover slip. The optical fibers are introduced through the two glass capillaries and input with a counter-propagating light beam. By properly aligning the two optical fibers in the axial direction and changing the separation of the fibers, the trapping of spherical particles with the diameter from 2 to 10 μm is achieved. Meanwhile, the insertion of the optical fibers into a sample cell at an angle is found to be helpful in increasing the trapping efficiency (15, 16).

Figure 3.1 illustrates the principle of using one optical fiber to manipulate the particle. In the vertical z direction, the vertical components of F_{ax} and F_{tr}, gravity, and buoyancy are in equilibrium. The total optical forces in the horizontal x direction are balanced with the viscous force generated by the liquid medium in the chamber. Thus the particle can be trapped at the equilibrium position. If the particle deviates in the $+x$ direction from the equilibrium position, the scattering force F_{ax} decreases as less optical radiation reaches the particle, induced by the increased distance between the optical fiber and the particle. Meanwhile, the gradient force F_{tr} increases owing to the greater excursion of the particle from the axis. In such a case, the components of optical forces in the x direction decrease and become negative, which pull the particle back to the equilibrium position. In contrast, if the particle deviates from the equilibrium position toward the $-x$ direction, both the scattering force F_{ax} and gradient force F_{tr} increase, which results in a greater total optical force in the $+x$ direction. Therefore, the particle will be pulled back to the equilibrium position.

As a strong gradient force is desirable in the single beam trapping, the implementation of a lensed optical fiber is proposed and investigated (16–19). The lensed optical fiber has a more confined optical field and long working distance in the output light beam (20). The optical manipulation of particles by using dual lensed optical fibers and single lensed optical fibers is shown in Figures 3.2 and 3.3, respectively.

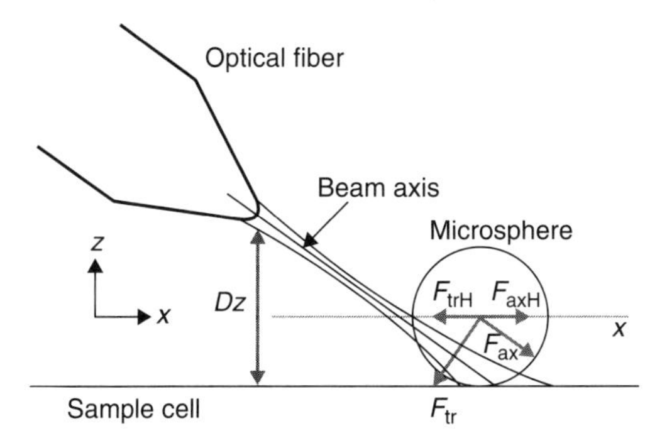

FIGURE 3.1 Illustration of optical fiber trapping system with a single beam. *Source:* Reprinted from Reference 16 with permission from IOP Publishing Ltd.

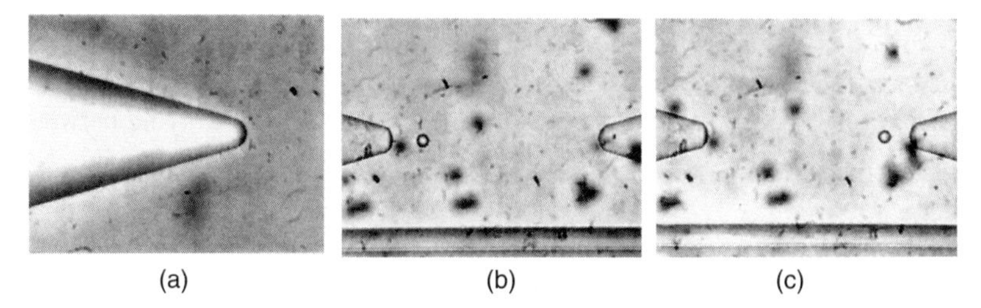

FIGURE 3.2 Optical manipulation of particles using dual lensed optical fiber. (a) a lensed optical fiber with a hemispherical microlens; (b) optical manipulation of a 10 μm polystyrene particle with a dual fiber trap; and (c) a new equilibrium position of the same particle with the use of a different optic power ratio among the two fibers. *Source:* Reprinted from Reference 17, with permission from American Institute of Physics.

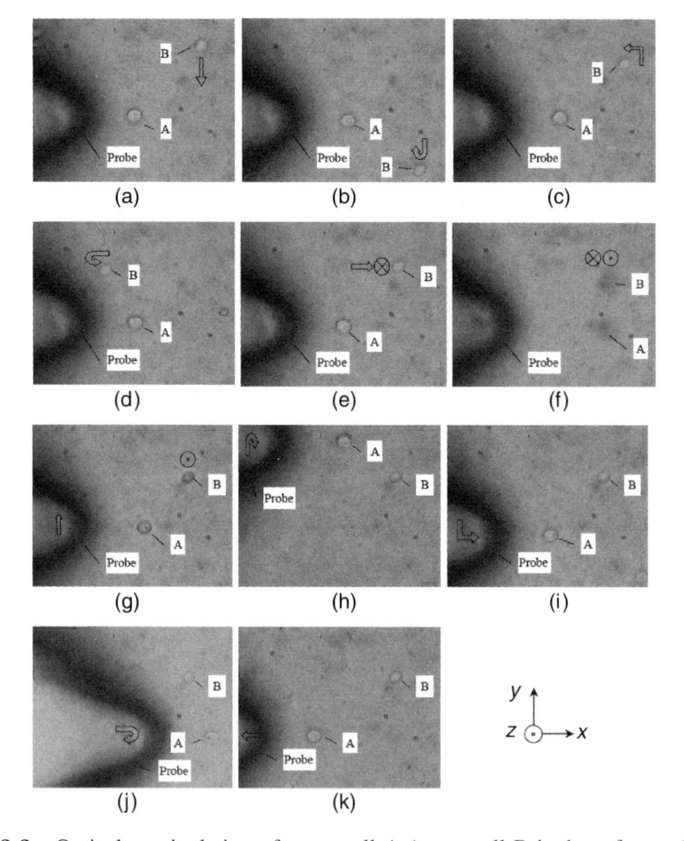

FIGURE 3.3 Optical manipulation of yeast cell A (yeast cell B is the reference). (a)–(g): The sample chamber moves with the nanostage in sequence in directions $-y \rightarrow +y \rightarrow -x \rightarrow +x \rightarrow -z \rightarrow +z$. (h)–(k): The lensed optical fiber moves in sequence in directions $+y \rightarrow -y \rightarrow +x \rightarrow -x$. *Source:* Reprinted from Reference 18 with permission from OSA.

The experimental results show that the advantages of using a lensed optical fiber for the single beam trapping are (i) that a lensed optical fiber is able to confine the output light to achieve a strong transverse optical field compared to a cleaved optical fiber, which makes it more effective in trapping and manipulating particle and (ii) that the focal point of the output light has a long distance from the fiber end in a lensed optical fiber compared with a cleaved optical fiber, which offers a longer working distance for trapping and manipulating a particle. Compared to other techniques, using a lensed optical fiber to trap and manipulate a particle is economical, is simple to operate, and requires relatively low optical power (16).

A pair of single-mode fibers (SMFs) or multimode fibers (MMFs) could be used for particle trapping, where two fibers form an optical trap with counter-propagating light beams (21). The larger core size of the MMF makes it possible to carry greater power and is simple in the alignment between the two fibers. Therefore, the MMF-based trapping system is relatively easy to set up. However, a complex intensity pattern at the transverse section of an MMF, as shown in Figure 3.4, makes the trapping very unpredictable. The applied optical forces are generated from an average optical field emitted from the two MMFs, and the value is dependent exactly on their positions. Considering that the optical forces are dependent on the particle size, the trapping behavior becomes even more complex. Sufficiently large particles will move between the peaks resulting from the transverse intensity pattern. As discussed in Reference 21, the trapped particles move independently of one another owing to the local optical field at their particular positions, which again proves the local nature of the trapping optical field from an MMF. In contrast, an SMF-based trapping system has only one stable equilibrium position (15, 22, 23). Although the alignment between two SMFs is harder to achieve, the trapping position and behavior are much more predicable and controllable. The trapping position can be fully controlled by changing the power of each SMF.

With the development of the lab-on-a-chip technology, a variety of biolaboratory functions and processes are proposed and demonstrated in a miniaturized chip format (24). Recently, a novel method for particle manipulation has been proposed and demonstrated by using dual beams in counter-propagating directions in an integrated optofluidic device as shown in Figure 3.5 (25).

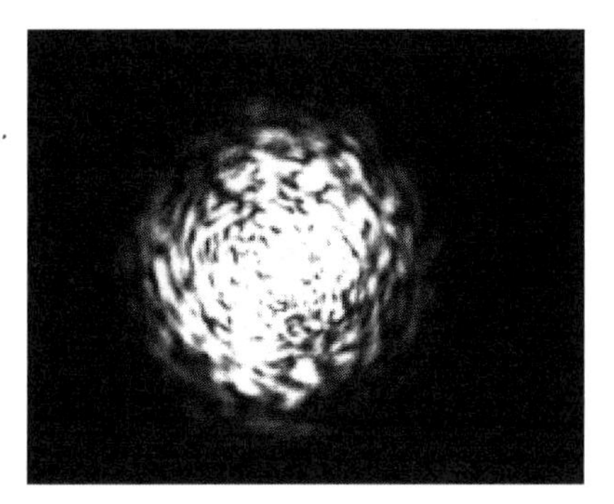

FIGURE 3.4 Intensity pattern at the transverse section of an MMF-based trapping system. *Source:* Reprinted from Reference 21 with permission from OSA.

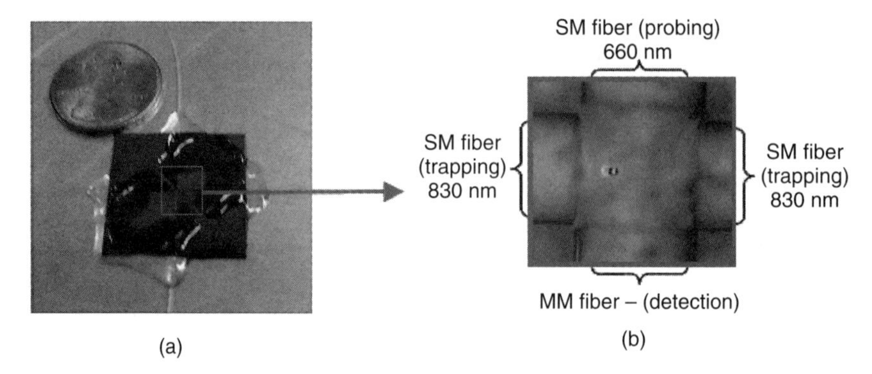

FIGURE 3.5 An integrated optofluidic device with buried fibers. (a) Overall view of the device. (b) A zoom-in view of the two SMFs for trapping, one SMF for probing, and one MMF for detection. *Source:* Reprinted from Reference 25 with permission from OSA.

This integrated optofluidic device consists of four angled V-grooves that form a symmetric diamond-shaped trapping area at the center. Two trapping SMFs are mounted into two opposing V-grooves, while a set of an SMF and an MMF is mounted into the other two opposing V-grooves, respectively, for the purpose of detecting the scattering light from the trapped particle. As discussed in References 21–23, the trapping position is determined by the point where the scattering forces from the two opposing SMFs are balanced with each other, and it can be adjusted by changing the optical power of each SMF. Besides the function of optical manipulation, the device is also able to detect the intensity of the scattering light.

The V-grooves used in the above optofluidic device are used for the easy insertion of fibers. They are fabricated with the steps as illustrated in Figure 3.6.

Step 1: the master mold with the V-groove pattern is fabricated by the standard lithography process. Step 2: a negative mold is fabricated by dispensing PDMS on the master mold. Step 3: dispense the PDMS on the negative mold and cure to create the final mold. Step 4: the final mold is fabricated by releasing the cured PDMS from the negative mold for the V-grooves (26). After the fibers are mounted

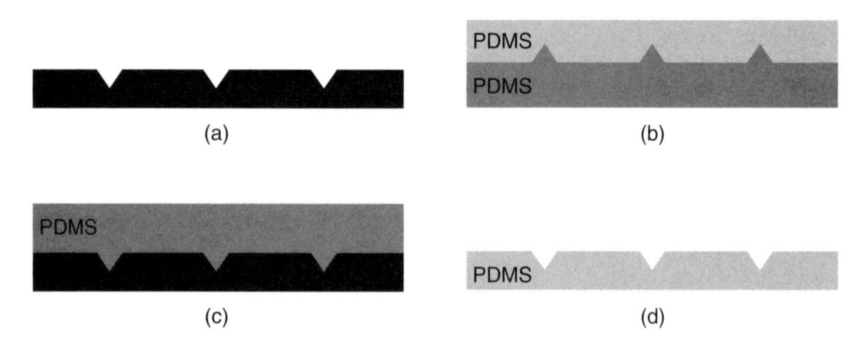

FIGURE 3.6 Overview of the V-groove fabrication steps. (a)–(d): From step 1 to step 4. *Source:* Reprinted from Reference 26 with permission from SPIE.

into the V-grooves, a microscope slide is used to cover the fibers and UV-curable adhesive is applicd to makc the seal.

A semiplanar structure, which is similar to the one in Figure 3.5b, is proposed and demonstrated in Reference 27, where an SMF is buried under a layer of polymeric material (NOA 61), and a microfluidic channel is fabricated by cutting across the whole device as shown in Figure 3.7 (27).

In this figure, an SMF fiber is suspended at 300 μm above a glass substrate, and then an UV-curable polymer is poured over them, which avoids the viscous spreading that occurs naturally under gravity during the spin coating or leveling of the liquid polymer. Once the polymer is cured after exposure to the UV light, a dicing saw is used to cut a groove with a width of 30 μm across the fiber and the polymer, thus forming a microfluidic channel as shown in Figure 3.7c. With the same principle as the one used in Figure 3.5, this chip is able to perform the particle manipulation with the simply built integrated chip.

Three-dimensional (3D) trapping of particles using two opposing optical fibers is first demonstrated in Reference 28. The use of multiple optical fibers for manipulating particles has also been reported in References 29 and 30. A single-fiber-based 3D light trapping of particles is discussed above by using a lensed optical fiber

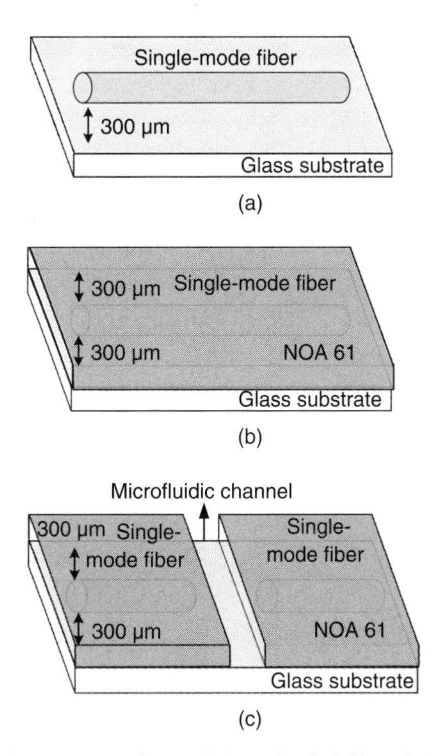

FIGURE 3.7 Fabrication process of a semiplanar buried fiber device (27). (a) An SMF is suspended at 300 μm above a glass substrate; (b) a layer of polymer is poured over (a); and (c) a microfluidic channel is formed by cutting the layer of polymer with a dicing saw.

(18). In addition, annular light distribution resulted from an etched conical-tapered optical fiber (31) is designed and created to achieve a purely optical 3D trap. The tapered region of an uncoated optical fiber consists of a hole filled with air inside and the surrounding medium acting as a cladding with a low refractive index. This structure is able to effectively confine the light in the annular silica region, hence generating an annular light distribution, which could be used to enhance the particle manipulation performance with a single fiber. An engineered fiber structure with microstructured end surface is also applied to build an optical 3D trap (32).

Another all-fiber probe with a core structure having an annular geometry is illustrated in Figure 3.8 (32).

As shown in Figure 3.8a, the optical fiber is specially cut at its endface. When the light is propagating along the fiber, it is bouncing between the core/cladding interface. At the exit, the light is deflected toward the focal point to achieve an annular light distribution. In this design, a highly focused light is generated by using total internal reflection instead of refraction in the annular core as shown in Figure 3.8b. The use of a multiple fiber bundle of these annular core fibers could enhance the optical trapping capability as shown in Figure 3.8c.

3.2.2 Near-Field Optofluidic Techniques

Different from the far-field methods, the near-field used for manipulating particles is actually the use of an optical evanescent field generated either from the cladding modes of waveguides or at the interface between two mediums under the total internal reflection condition (33). The principle is that the strength of an optical evanescent field decays exponentially with the distance from the source where the evanescent field is generated. The rapid decaying optical field can be regarded as a significant trapping region and used to manipulate particles. This concept is

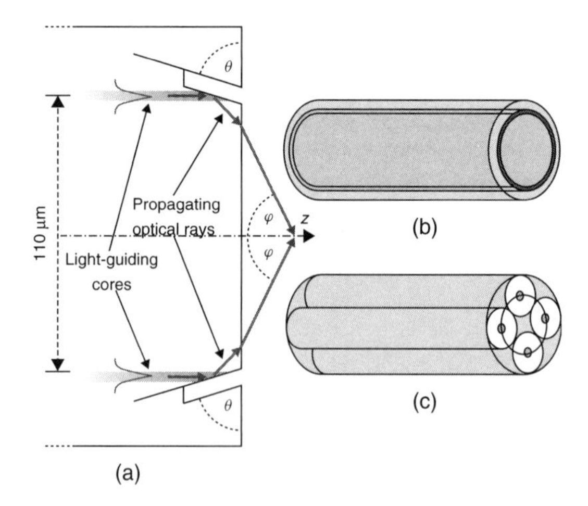

FIGURE 3.8 Geometry of the microstructured end surface. (a) Operation principle. (b) Single annular core fiber. (c) An optical fiber bundle consisting of four annular core fibers. *Source:* Reprinted from Reference 32 with permission from Nature Publishing Group.

first proposed with the use of a light-coupling prism (34) and also demonstrated with the use of a channeled waveguide (35). Figure 3.9 shows the experimental configurations of these two schemes.

Theoretical analyses of the above optical manipulation schemes are reported in References 36 and 37. As the optical evanescent field is not laterally localized, optical trapping of particles cannot be achieved using the above experimental configurations.

Similar to the use of a channeled waveguide for optical manipulation as discussed in Reference 35, waveguides made of ion exchange doped glasses are also used to generate optical evanescent fields (38). A novel approach for manipulating particles is demonstrated by using a Y-branched channel waveguide, especially to achieve the function of particle sorting (39) with the profile shown in Figure 3.10.

The Y-branched channel is 4 cm in length and fabricated by Cs^{+} ion exchange. An aluminum mask is used as a coating of the substrate, and the Y-branched stripe pattern is fabricated by the photolithography technique (39). The input waveguide has a width of 10 μm. The output waveguides have a width of 3 μm with a

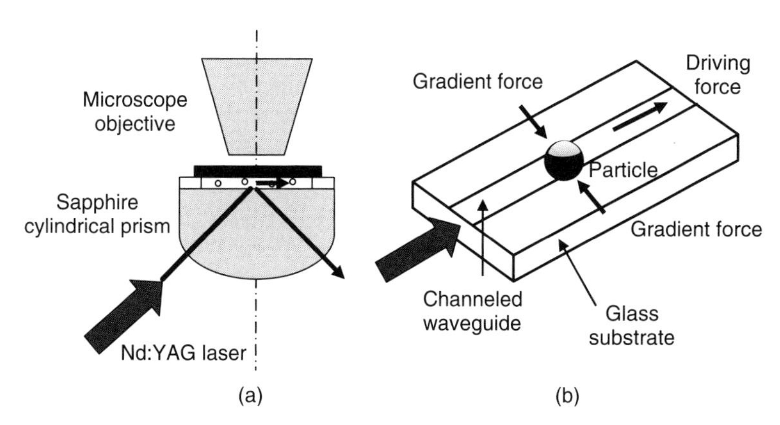

FIGURE 3.9 Experimental configurations for trapping particles using optical evanescent fields. (a) Total internal reflection by using a prism (34). (b) Evanescent field in a channeled waveguide (35).

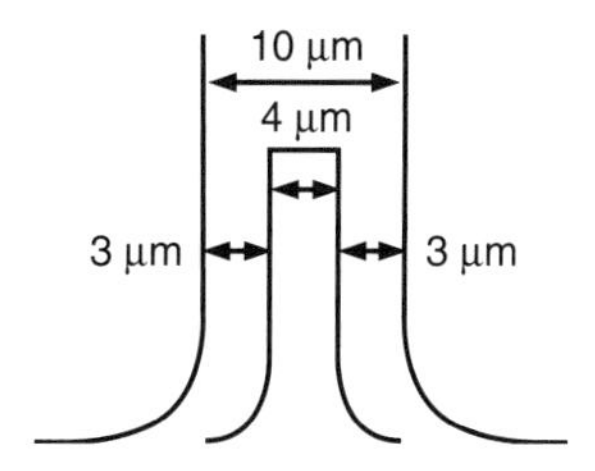

FIGURE 3.10 Illustration of the particle sorting based on a Y-branched channel waveguide (39).

separation of 4 μm at the junction region. The separation reaches 200 μm at the end with a branch bend radius of 8 μm. A fiber laser with the wavelength of 1066 nm is used as the light source and the light is coupled to the waveguide by butt-to-butt coupling. The optical fiber is mounted on a positioning stage in order to precisely control the position of the fiber along the waveguide input facet and to tune the power distribution in the two branches. From the experimental results, it is seen that the majority of particles are flowing through the branch that has a larger light power. The concept of this particle sorting technique can be potentially integrated into an on-chip system design with electro-optical switching and wavelength-dependent routing.

As mentioned above, an optical evanescent field is localized intrinsically and not laterally, and it can only work for optical transport rather than optical trapping. However, by using two counter-propagating light beams in a channeled waveguide, an optical trapping can be achieved (40). The experimental setup is shown in Figure 3.11.

Counter-propagating waves are first used for particle manipulation on a prism surface (41) and then in a configuration of dual optical fibers (21, 25). The use of optical waveguide fabricated with the same process as described in Reference 39 is a natural extension of the particle guiding/trapping setup. In the proposed design, the input light is equally split into two, which are coupled to a channeled waveguide from the two endfaces to make counter-propagating wave. The particles are first trapped in the evanescent field generated by the counter-propagating wave and then propelled back and forth along the waveguide depending on the difference of the light power coupled to each direction. Finally, particles will reach the equilibrium position. By moving the fibers at both ends of the waveguide, the fiber-to-waveguide coupling loss can be controlled. Thus the

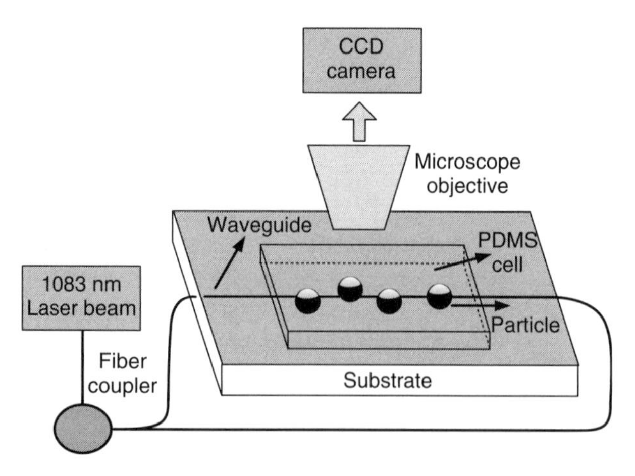

FIGURE 3.11 Experimental setup of particle manipulation by using the evanescent field in counter-propagating waves (40).

light power propagating in each direction is adjustable, which makes it possible to move the equilibrium position.

Silicon nitride waveguides are also used to generate an evanescent field for the optical manipulation of particles. A silicon nitride waveguide has a greater refractive index contrast ($\Delta n = 0.52$) between the core and the cladding, compared to waveguides formed by potassium ion exchange ($\Delta n = 0.01$) (35) and Cs^+ ion exchange ($\Delta n = 0.03$) (40, 41). The theoretical comparison of these technologies and the demonstration of optical manipulation using silicon nitride waveguides are discussed in Reference 42. With a similar experimental setup shown in Figure 3.11 but with a single light beam input, the optical propulsion of three different particles/cells is achieved. The theoretical analysis also shows that if the evanescent field decays faster in a channeled waveguide structure with a high refractive index contrast, optical forces excited from the evanescent field remain higher.

Optofluidic structures consisting of SU-8 polymer waveguides and PDMS microfluidic channels are widely used to perform the dynamic trapping and transport of particles in evanescent fields with a typical setup shown in Figure 3.12 (43).

In this type of structure, both optical waveguides and microfluidic channels are fabricated using standard photolithography (44) and soft lithography techniques (45). SU-8, an epoxy-based negative UV photoresist, is commonly used to fabricate the waveguide structure owing to its mechanical hardness and chemical resistance, as well as its high transparency in the wavelength range of 850–1100 nm. During the fabrication process, SU-8 is first poured on a substrate having the designed pattern. By using the standard photolithography technique, the pattern is replicated in the layer of SU-8 and is bonded to the substrate after the process of UV exposure, postexposure baking, and developing. Then, PDMS is cast into the SU-8-based pattern mold. Using the soft lithography technique, a layer of PDMS having the designed pattern transferred from the SU-8-based master mold is obtained. The PDMS channels, SU-8 waveguide and microscope slide are bonded together using a plasma-assisted technique by placing them in conformal contact (45). The waveguide structure consists of fused silica as the substrate ($n = 1.453$), exposed SU-8 as the core waveguide ($n = 1.554$), and water as the cladding ($n = 1.33$). The width and height of the SU-8 waveguide are selected to be 2.8 μm and 560 nm,

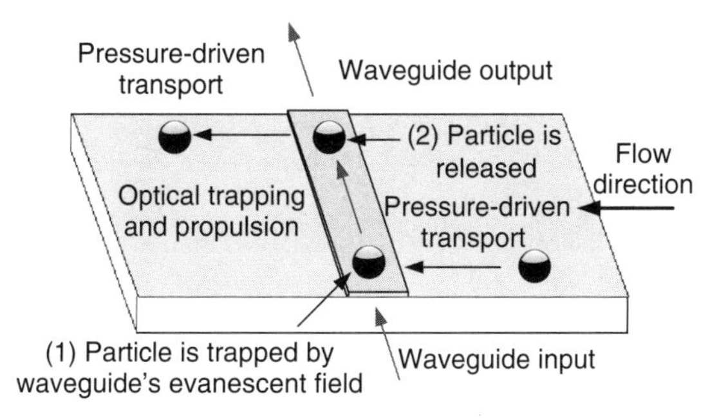

FIGURE 3.12 Optofluidic trapping and transport using SU-8 waveguide structure (43).

respectively, which is specifically designed and fabricated to achieve only the fundamental TM mode. Since a stronger discontinuity of the optical field that exists on the top surface of the waveguide will induce a stronger gradient force to trap and propel particles, the TM mode is of greater interest than the TE mode owing to its noncontinuous electrical field distribution at the interface of two mediums.

Similar to the use of an MMF in manipulating particles, the mode distribution in the waveguide also has a great influence on the optical manipulation performance when using an evanescent field. The use of multimode waveguide is investigated with the results shown in Figure 3.13 (46).

It is seen that optical trapping and transport of particles can also be achieved by using multimode waveguides, and the particles are trapped at the positions corresponding to the peaks of the mode distribution. Thus it is proven that the movement of particles is highly dependent on the mode distribution. To achieve a precise manipulation, the mode distribution in an SU-8 waveguide can be designed by tailoring the waveguide dimension (47).

Evanescent fields generated by total internal reflection methods can also be applied for optical manipulation of particles. One of these methods is using a high NA microscope objective. The demonstrations of optical trapping and sorting are reported in References 48 and 49, respectively. The concept is shown in Figure 3.14 (49).

In this concept, an off-axis collimated light beam is introduced into a high NA microscope objective. If the light is focused at an interface between two mediums of different refractive indices and the incident angle is greater than the critical angle

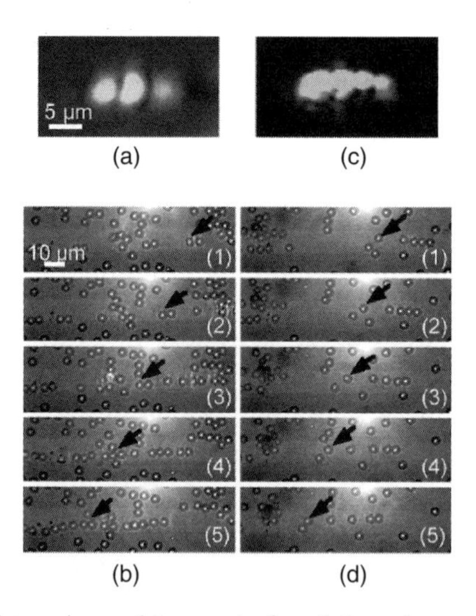

FIGURE 3.13 Optical trapping and transport of particles using multimode waveguide. (a) Near-field pattern of Ex02 mode. (b) Resultant movement of particles. (c) Near-field pattern of Ex03 mode. (d) Resultant movement of particles. *Source:* Reprinted from Reference 46 with permission from American Institute of Physics.

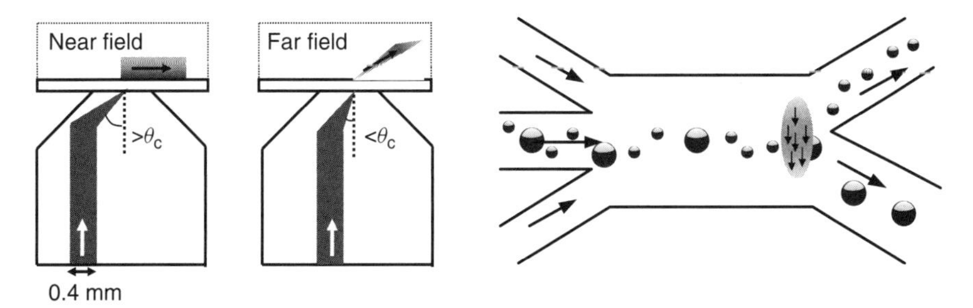

FIGURE 3.14 Concept of optical trapping and transport using evanescent field generated by a high NA microscope objective. (a) Near field and far field (48). (b) Passive particle sorting based on its size (49).

of the total internal reflection, an evanescent field is generated at the interface and its strength decays rapidly with the distance away from the interface. On the basis of a similar operation principle discussed in References 33 and 34, the generated evanescent field can be used to trap and manipulate particles.

Beside the micrometer-sized particles, an evanescent field is able to manipulate nanometer-sized particles and biomolecules in subwavelength slot waveguides (50, 51). The subwavelength slot waveguides have dimensions in nanometer scale with an illustration shown in Figure 3.15a. In addition, the fluid channel in this design is the gap between two silicon-based subwavelength waveguides. In this case, two

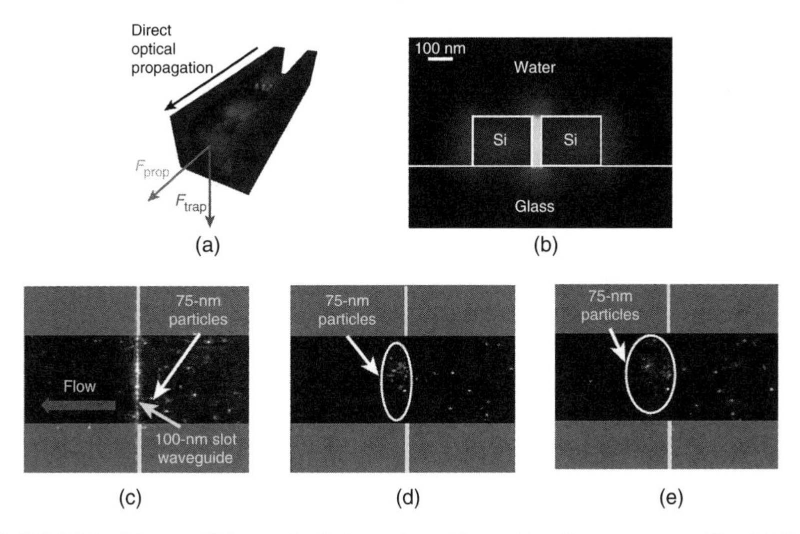

FIGURE 3.15 Nanoparticle manipulation using subwavelength nanowaveguide. (a) Schematic illustration of the slot waveguide. (b) Mode profile of the developed subwavelength slot waveguide. (c)–(e) Illustration of the trapping and transport of nanoparticles in a slot waveguide. *Source:* Reprinted from Reference 50 with permission from Nature Publishing Group.

evanescent fields from the two silicon-based waveguides are overlapped together with its mode profile shown in Figure 3.15b.

As can be seen from Figure 3.15a, nanoparticles are flowing inside the slot between the two subwavelength waveguides, which are controlled by two forces. One is F_{prop} representing the propagation forces introduced by the radiation pressure from fluid transport. The other is F_{trap} representing the trapping force in the slot region. Figure 3.15b shows the main trapping region with the overlapped high intensity modes generated by two evanescent fields inside the slot. In the conventional evanescent field methods, only a small portion of light interacts with particles and most of the light is confined within the solid waveguide core. In this structure, a majority of the light transmitting in the subwavelength waveguides is released from the core and generates a strong overlapped evanescent field in the slot, which is capable of continuously and efficiently confining particles. This slot waveguide structure is reported to be used to condense the accessible electromagnetic energy to the scale as small as 60 nm, and it allows us to overcome the fundamental diffraction problem (52). The experimental results of trapping nanoparticles in the proposed subwavelength slot waveguides are shown in Figure 3.15c–e. Nanoparticles with 75-nm diameters are flowing into the subwavelength slot, which has a width of 100 nm. Figure 3.15c shows that the particles are trapped in the slot when the lasers at both sides of the slots are turned on. These trapped nanoparticles get released and a “cloud” of nanoparticles forms when the lasers are turned off as shown in Figure 3.15d. Then, the released nanoparticles are carried down the channel by the fluid flow as illustrated in Figure 3.15e.

3.2.3 Optical Chromatography Techniques: Axial-Type and Cross-Type

Optical chromatography is an optical sorting technique that takes advantage of the differences in the equilibrium positions among individual particle types, which are usually determined by the size and refractive index of particles, and the fluid flow rate (13). This separation technique is first proposed and demonstrated in Reference 53, which is regarded as the axial-type optical chromatography. The operation principle is shown in Figure 3.16 (53).

In this optical sorting method, a weakly focused laser beam is introduced in a counter-propagating direction of the fluid flow. As shown in Figure 3.16a, a particle is introduced from the left-hand side to the right, and the laser beam is propagating from the right-hand side to the left. In Figure 3.16b, the gradient force drives the particle into the center line of the laser beam (7), and the particle flows along the channel and travels beyond the position of the laser beam waist owing to the initial velocity provided by the pressure-driven fluid flow. In Figure 3.16c, because of the scattering force, the particle turns around and is pushed toward the focal region of the laser beam with an acceleration movement. The highest velocity of the particle is achieved when the particle is pushed back at the position of the laser beam waist (focal point). As the particle moves away from the focal point, the particle moves against the fluid flow and the scattering force decelerates the particle as illustrated in Figure 3.16d. When equilibrium arises between the

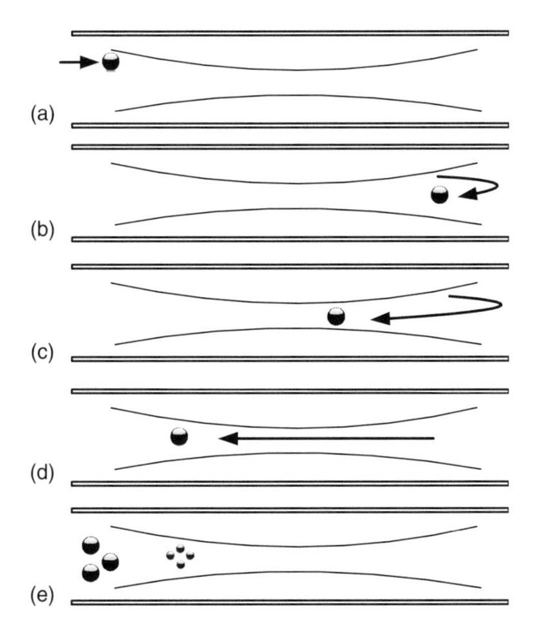

FIGURE 3.16 Schematic motion of particles (53). (a) Particle introduction; (b) particle driven toward the beam center by gradient force; (c) particle acceleration movement toward the beam center; (d) particle deceleration; (e) particles in equilibrium state.

competing optical radiation forces and fluid pressure driven force, the particle is trapped at a distance away from the focal point, which is termed as the *retention distance*. As mentioned earlier, theoretical analysis has shown that the retention distance is highly dependent on the particle properties, such as size and refractive index (22, 54). The experimental result, as shown in Figure 3.16e, is also indicative of the capability that this method could be used as a powerful and efficient tool for the separation of a diverse variety of biological particles.

The separation of particles has been demonstrated based on size and refractive index by using the axial-type optical chromatography technique (55, 56). This method has also been used to address the differences in closely related biological samples (57), environmental samples (58), and various other microsized particles (59, 60). The recent development of optofluidic techniques allows the axial-type optical chromatography technique not only to perform particle separation tasks but also to be exploited for applications in the concentration, cleanup of multicomponent particles (61), and performance enhancement in a PDMS microfluidic system (62).

In order to develop this method further, knowledge and understanding of the forces applied on the particles traveling through a specific optofluidic environment are required. Theoretical and numerical analyses of the optical radiation forces on particles have been well documented for nondynamic situations (63–65). The theoretical simulation of dynamic situations is achieved by the recent release of a commercial computational fluid dynamics package (66), which could numerically simulate the optical radiation forces and the hydrodynamic drag forces applied on particles flowing through a complex optical fluidic field.

Although the axial-type optical chromatography has been successfully applied as an efficient tool for sorting particles, it is not able to perform the continuous particle manipulation task and it also requires additional devices to deliver the separated particles away from the region. To overcome these limitations, cross-type optical chromatography technique is theoretically proposed (67) and experimentally demonstrated (68). In this method, the light beam perpendicularly propagates to the fluid flow direction. This orthogonal arrangement can manipulate particles in a continuous manner without the need of additional devices to deliver the target particles away. Theoretical analysis of the optical forces and its induced particle displacement are reported in Reference 69. We take the method a step further by introducing the beam forming technique with a lens structure built on the chip. In our design, the on-chip lens set is used to focus the light beam, and therefore, enhance the manipulation performance. In addition, manipulation of a particle by using two light beams is also achieved in Reference 70 on a hybrid SU-8/PDMS chip with this on-chip lens structure, which is described in the next section. Similar to other methods, the particle displacement induced by this method is highly dependent on the particle size and the refractive index; therefore, this method can be used for sorting particles. The separation performances, such as the resolution of particle separation (71) and the optical mobility (72), have been fully investigated.

Besides the three main categories of optofluidic particle/cell maneuvering techniques discussed above, other techniques, such as interference field (73), hollow-core waveguide (74), and liquid-core waveguide (75), have also been demonstrated recently.

3.3 ENHANCING OPTICAL MANIPULATION WITH A MONOLITHICALLY INTEGRATED ON-CHIP STRUCTURE

On the basis of previously discussed cross-type optical chromatography method (71), we have demonstrated a novel microfluidic chip with an improved particle manipulation performance (70). The improved performance results from the reduced light beam waist radius with the use of a monolithically integrated on-chip lens set. To understand its operation principle, a theoretical model is set up to investigate the effect of the light beam waist radius on the manipulation performance. The schematic illustration of this model is shown in Figure 3.17.

An analytical expression of the particle displacement induced by an ideal cross-type optically driven propulsion is obtained by calculating the particle velocity (70). Theoretical results show that the displacement is counter-proportional to the light beam waist radius at the microfluidic channel. This conclusion is important since we can increase the displacement by reducing the light beam waist radius rather than increasing the input light power.

The key significance of our work is the incorporation of a monolithically integrated on-chip lens structure into a microfluidic chip, which can be used to reduce

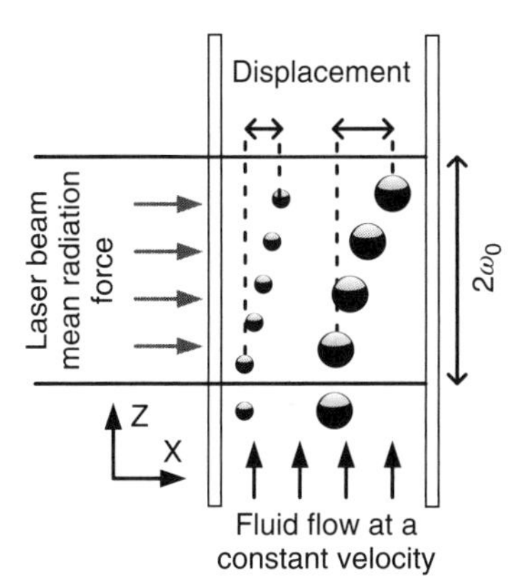

FIGURE 3.17 Schematic illustration of the cross-type optical chromatography. *Source:* Reprinted from Reference 70 with permission from IEEE. Copyright © 2010 IEEE.

the light beam waist radius, and therefore, to an improved performance in the optofluidic transport and optical manipulation.

The proposed on-chip lens structure is shown in Figure 3.18.

To achieve a specific light beam waist radius, two on-chip lenses (such as On-Chip Lens 1 and On-Chip Lens 2) as shown in Figure 3.18 are used as an on-chip lens set (such as On-Chip Lens Set 2) with the lens dimension settings for the On-Chip Lens Set 2 listed in Table 3.1. On the right side of the microfluidic channel, three sets of lenses with different lens dimensions are fabricated and placed along the channel. This means that three different light beam waist radii can be achieved by the on-chip lens sets. The "symmetrical lenses architecture" labeled in Figure 3.18 stands for the design of a symmetric architecture by incorporating on-chip lens sets on the other side of the microfluidic channel. This design is to achieve the optical manipulation by using two counter-propagating light beams.

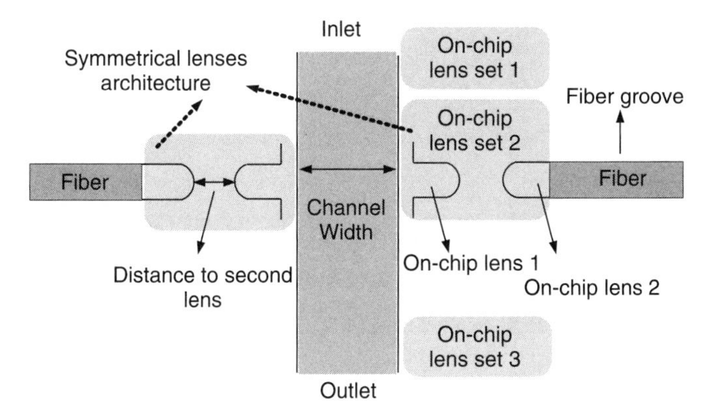

FIGURE 3.18 Schematic illustration of the proposed on-chip lens set. *Source:* Reprinted from Reference 70 with permission from IEEE. Copyright © 2010 IEEE.

TABLE 3.1 Dimensions of the On-Chip Lens Set 2 Structure[a] (70)

Lens 1	Curvature	-300^b
	Aperture Radius	500
Distance to second lens		300
Lens 2	Curvature	190
	Aperture Radius	500

[a] The unit is micrometer (μm).
[b] Negative curvature means a concave lens, while positive curvature means a convex lens.

An example of an overall field view of the fiber groove and the lens set structures in both bright and dark fields is shown in Figure 3.19. And the illustration of reduced light beam waist radii of the three lens sets is shown in Figure 3.20.

The light beam waist radius in the microfluidic channel is estimated by comparing its size with the known channel width, which is 100 μm defined by the photomask. From Figure 3.20b, the light beam waist radii are measured as ~75, ~50, and ~100 μm for the On-Chip Lens Set 1, 2, and 3, respectively.

The structure is implemented in a hybrid SU-8/PDMS optofluidic chip as illustrated in Figure 3.21.

All microfluidic chips can be generally classified into two categories based on the materials used for the fabrication of the chips, that is, inorganic-material-based architectures and polymeric-material (such as PDMS)-based architectures. Inorganic-material-based architectures can be fabricated by using the thin-film deposition method or the anisotropic etching technique, which usually requires two or four photolithographic steps, and lasts up to a couple of weeks (76). The fabrication process can be simplified by using the polymeric materials (44), such as the SU-8 (negative photoresist) and PDMS. The use of polymeric materials can

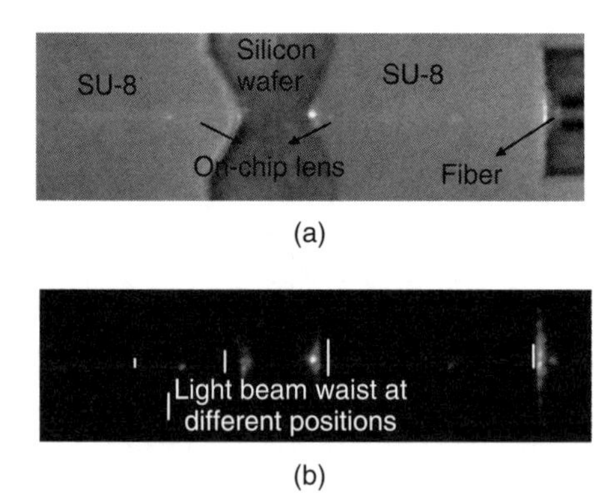

FIGURE 3.19 Microscope images of one on-chip lens set. (a) Bright field image. (b) Dark field image with the incident light beam of He–Ne laser (632.8 nm, 20 mw). *Source:* Reprinted from Reference 70 with permission from IEEE. Copyright © 2010 IEEE.

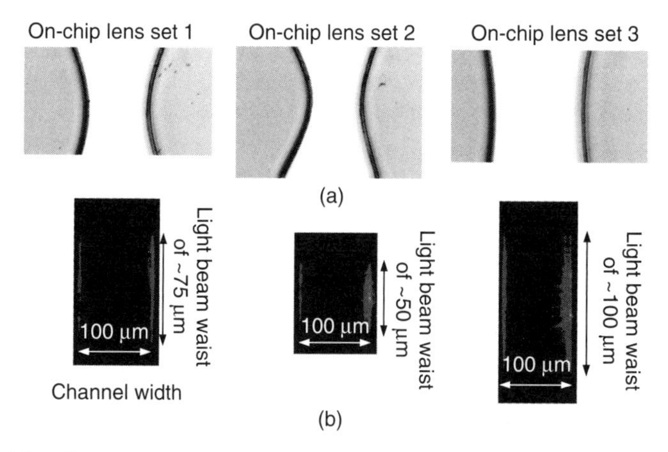

FIGURE 3.20 Illustration of the three on-chip lens structures. (a) On-chip lens structure. (b) Light beam waist in the microfluidic channel. (a) and (b) are not in the same scale. *Source:* Reprinted from Reference 70 with permission from IEEE. Copyright © 2010 IEEE.

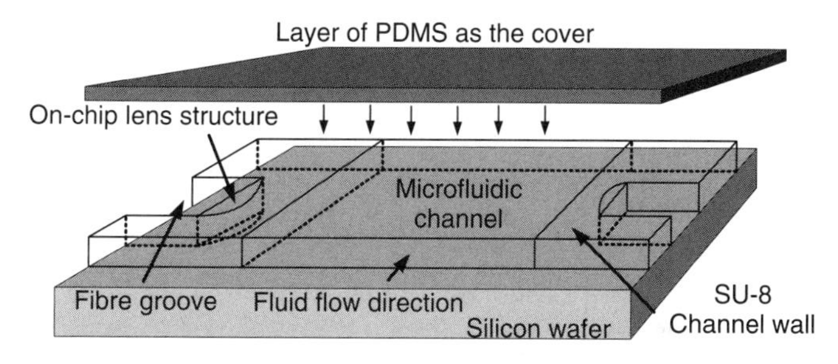

FIGURE 3.21 Schematic representation of the proposed SU-8/PDMS hybrid microfluidic chip with an on-chip lens set. *Source:* Reprinted from Reference 70 with permission from IEEE. Copyright © 2010 IEEE.

significantly reduce the complexity in the chip fabrication process and shorten the fabrication cycle to around two days, which is required to finish the steps of master mold fabrication and the PDMS molding (77).

Incorporation of optical elements, such as an on-chip lens, into the optofluidic platform will generally complicate the fabrication process. For example, peeling off the PDMS layer from the master mold and bonding it to the substrate is highly probable in distorting the optical elements owing to the micro features of the optical elements. In developing the "lab-on-a-chip" or μTAS system, the ability to fabricate device/system in a fast manner with low complexity is highly desirable.

In our design, because the on-chip lens set, the microfluidic channel and the fiber grooves are all defined in the same SU-8 layer, only one mask is required in the fabrication process. The proposed SU-8/PDMS hybrid microfluidic chip can be made in a single fabrication step. This significantly reduces the processing time

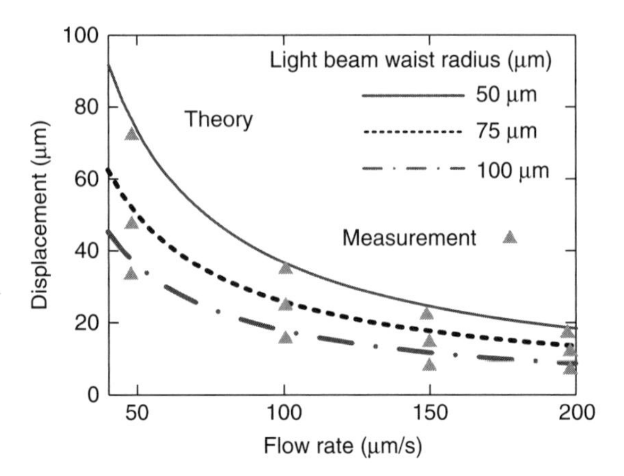

FIGURE 3.22 Measured and simulated displacements as a function of the flow rate and the light beam waist radius. The comparison is made for a microparticle with a 3-μm radius. *Source:* Reprinted from Reference 70 with permission from IEEE. Copyright © 2010 IEEE.

compared with other existing methods where a long processing time of a few days is usually needed. In addition, the fabrication complexity is also significantly reduced.

Figure 3.22 shows a comparison between the theoretical and experimental results using the home-fabricated optofluidic chips. The particle displacements are evaluated as a function of the light beam waist radius and the flow rate. It is seen that the measured displacements agree well with the theoretical predictions in all cases, and the displacement is increased by reducing the light beam waist radius. This conclusion accords well with our theoretical prediction and provides a potential solution to enhance the performance of optical manipulation of particles by defining the optofluidic architecture rather than increasing the laser power.

The optical manipulation of a particle using two counter-propagating light beams in the home-fabricated optofluidic chip is also carried out and the results are shown in Figure 3.23.

There are six steps from Figure 3.23a–f. Figure 3.23a shows the area of the control region. At the beginning, the two lasers are off. After the particle flows into the control region at position A, laser 2 is switched on and input light beam 2 drives the particle to the left as shown in Figure 3.23b. After the particle has been driven with a displacement to the left direction, laser 2 is off and the particle continues to flow along the microfluidic channel as shown in Figure 3.23c. Then, laser 1 is turned on and input light beam 1 pushes the particle to the right, as shown in Figure 3.23d. When the particle reaches a position as shown in Figure 3.23e, laser 1 is turned off. The particle then leaves the control region at position B, as shown in Figure 3.23f.

It is seen that the output position and the moving path inside the control region can be defined by the two lasers. The architecture for the manipulation of the particle shown in Figure 3.23 can be considered as an optical manipulation unit.

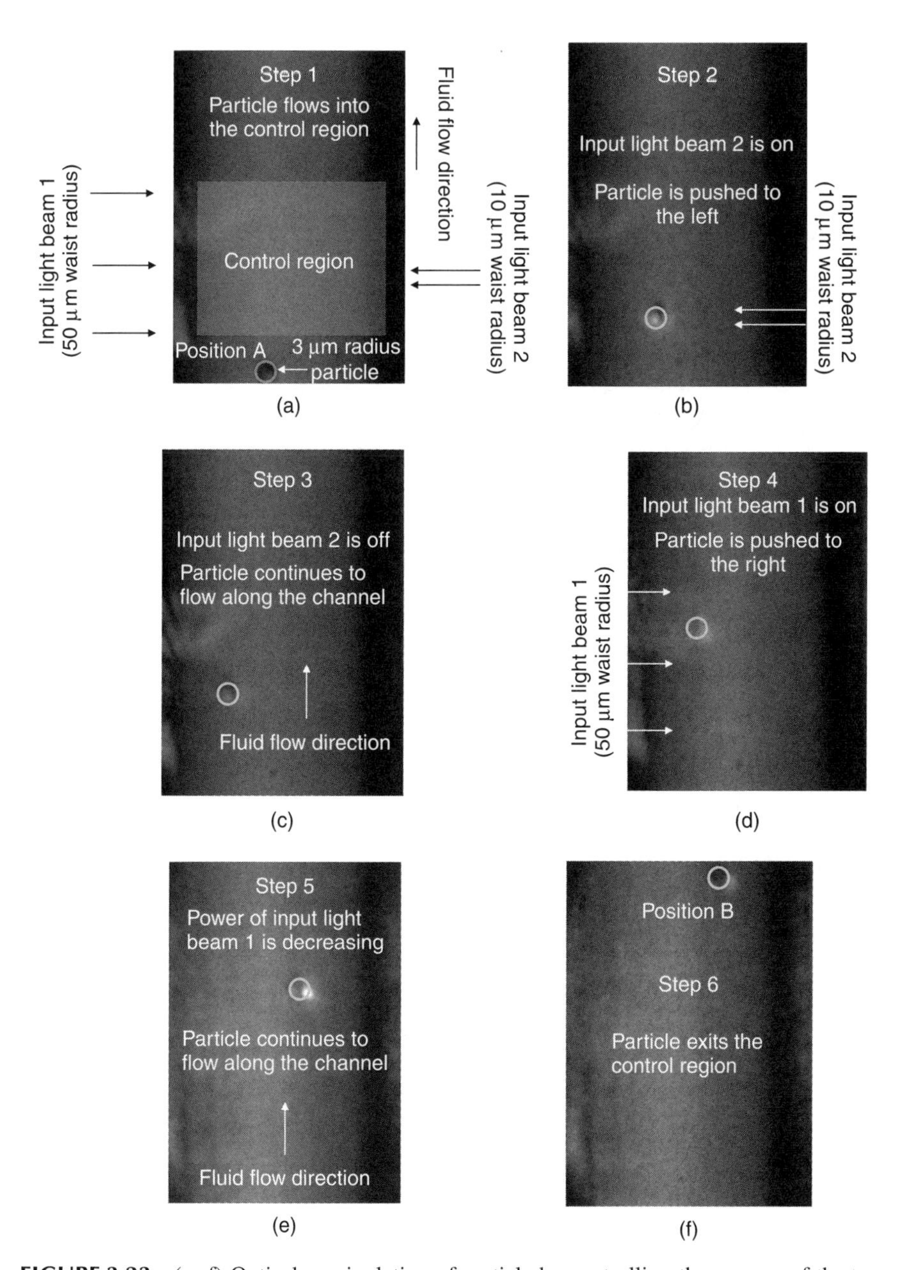

FIGURE 3.23 (a–f) Optical manipulation of particle by controlling the powers of the two light beams. *Source:* Reprinted from Reference 70 with permission from IEEE. Copyright © 2010 IEEE.

By placing a number of these units along the microfluidic channel, the transport of a particle to a specific position can be accomplished. With the modified waist radius of the input light beams, the optofluidic transport performance can be maximized in each unit. This will improve the performance of the optical manipulation.

3.4 APPLICATIONS

Optical manipulation is to use laser light to move small particles at the microscopic scale, allowing characterization of small objects such as individual cells, biomolecules, or even nanoparticles. In general, all optical manipulation applications can be divided into two categories, the precise motion control and the accurate force measurement (78). Importantly, the manipulation and force sensing is noninvasive and sterile, because no foreign objects are introduced and are in touch with the delicate samples, which makes the optical manipulation ideal for the applications in biology and materials science (8, 9). Recently, sorting of different types of particles has also been achieved by using optofluidic techniques as described in References 49 and 79.

Precise motion control of particles is very useful. The first optical manipulation application is to use optical tweezers to manipulate bacteria and viruses (80). The state-of-the-art techniques could hold and translate the chosen sample. Thus it is able to manipulate a single molecule to interact with a bead, which is commonly used to study single DNA (81) and enzymes (82). Besides the applications in biological analysis, optical manipulation can also be used to develop further novel optofluidic devices, such as an optically driven fluid pump (83) and an all-optical switch (84).

In the optical driven fluid pump, two beads with 5 μm diameter are trapped within a microfluidic channel and driven to counter-rotate by using a circularly polarized laser beam as shown in Figure 3.24. The results show that the rotating speed of the two beads could reach up to 10 Hz, resulting in a flow rate up to 200 $\mu m^3/s$. The fluid flow direction could also be reversed by changing the sense of the rotation of the two beads.

The all-optical switch uses the same structure as shown in Figure 3.7c, where two SMFs are used to introduce light beams to manipulate the particle. However, in this application, the particle is placed at the middle position between the two SMFs and manipulated by an optical tweezers rather than the two SMFs. The particle could be positioned to block a portion of the light transmitting from one SMF to the other. Thus the movable particle controlled by an optical tweezers can be used as an optical switch (84).

In the optical manipulation, optical traps confine the particle at the trap center and the particle rarely moves. When applying a foreign force to the particle, the value of the force has a linear relationship with respect to its displacement from the center of the trap; if the displacement keeps small, this is dominated by the Hooke's Law (85). It provides a solution for the quantitative scientific measurement of forces in biological analysis. The first force measurement is to use an optical

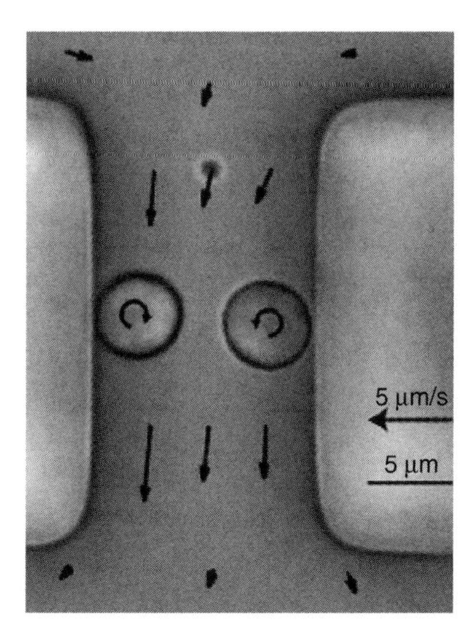

FIGURE 3.24 Illustration of the fluid flow pumped by two rotating beads driven by optical tweezers. *Source:* Reprinted from Reference 83 with permission from The Royal Society of Chemistry.

tweezers to hold a bead and control a kinesin molecule to attach to the bead. Then, the bead is moved to a microtubule that is fixed on a glass slide. Then, the motion between the bead and the kinesin molecule is analyzed when the molecule moves along the microtubules (86).

In the biological analysis, it is also of great interest to find out the elastic behavior of the tethered molecules and measure the extension of both the molecules (87), where two DNA-microspheres of radii r_1 and r_2 are held by two optical traps with stiffness k_1 and k_2, and tethered by a molecule of dsDNA of stiffness k_{DNA}. By applying a given tension, the extension of the molecule ε, the displacements of the two microspheres from the center x_1 and x_2, and the distance R_{12} could be identified. A schematic illustration of this process is shown in Figure 3.25 (87).

Recently, commercial optical trapping systems have been released in the market and are widely applied in the biology and materials science with the capabilities of optical manipulation and force measurement. An overview and basic specifications of these systems are provided in Reference 78. Figure 3.26 shows a commercial product by Carl Zeiss. Its model is PALM MicroTweezers IV, which has a maximum of 10 traps and a force measurement range from 5 to 250 pN.

In the biological analysis, especially in the sample preparation stage, it is desirable to separate or sort particles or cells. Optofluidic techniques have been demonstrated for this purpose (88). Optical trapping could be achieved by a number of schemes, including optical tweezers (79) and an optical lattice (89). The use of an optical lattice is based on the varying affinity of different particles to the periodic three-dimension structure of optical lattice, where latex and silica particles could be successfully sorted as shown in Figure 3.27.

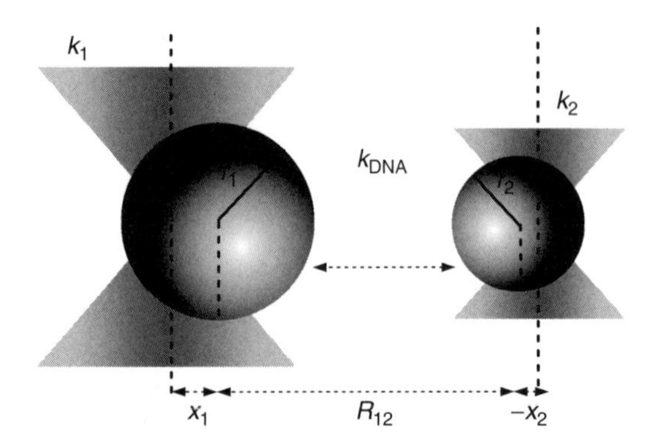

FIGURE 3.25 Illustration of the tethered extension measurement (87).

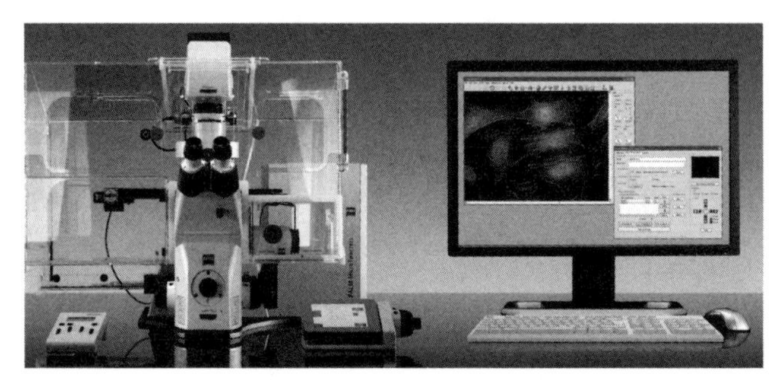

FIGURE 3.26 Commercial optical manipulation system by Carl Zeiss. *Source:* Reprinted with permission from Carl Zeiss MicroImaging.

Coupling an optical sorter with an online analyzer permits the device to identify the type of particles and isolate the rare ones from the natural populations. Figure 3.28 shows the layout of a microfluidic device that integrates both analysis stage and sorting stage in one chip (90).

An integrated fiber-based device with a similar structure is proposed and developed as shown in Figure 3.29 (91). In this device, optical waveguides are buried in the chip and attached with fiber for simple operation.

3.5 CONCLUSION

In this chapter, the state-of-the-art optofluidic techniques for the manipulation of particles have been categorized into three classes: fiber-based technique, near-field-based technique, and optical chromatography technique. A review of the recent development of each class was provided. Then, we discussed a novel SU-8/PDMS optofluidic chip, which incorporated on-chip lens structures to control the light beam waist radius. Both theoretical and experimental results have shown that the optical manipulation performance could be enhanced by the proposed on-chip lens

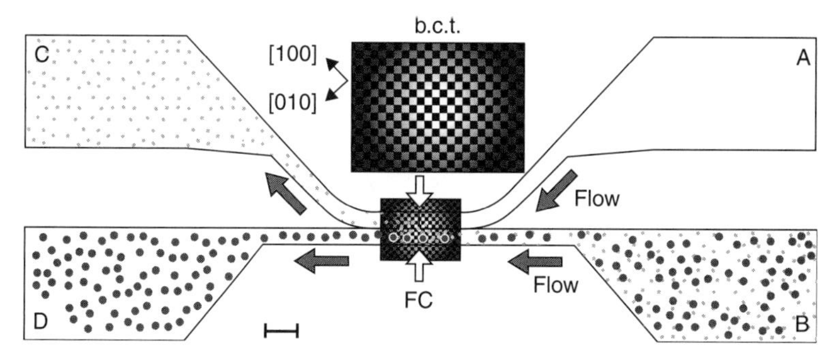

FIGURE 3.27 Illustration of optical sorting based on an optical lattice. *Source:* Reprinted from Reference 89 with permission from Nature Publishing Group.

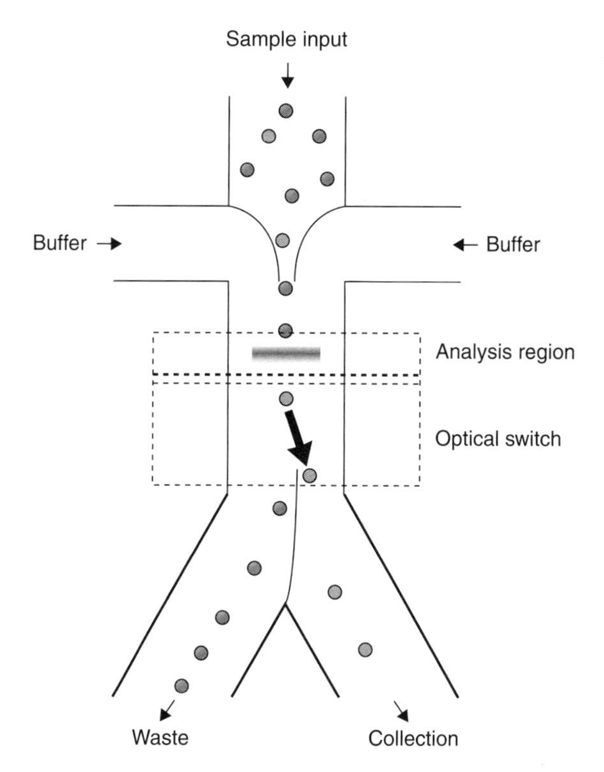

FIGURE 3.28 Layout of the microfluidic device with analysis region and sorting region. *Source:* Reprinted from Reference 90 with permission from Nature Publishing Group.

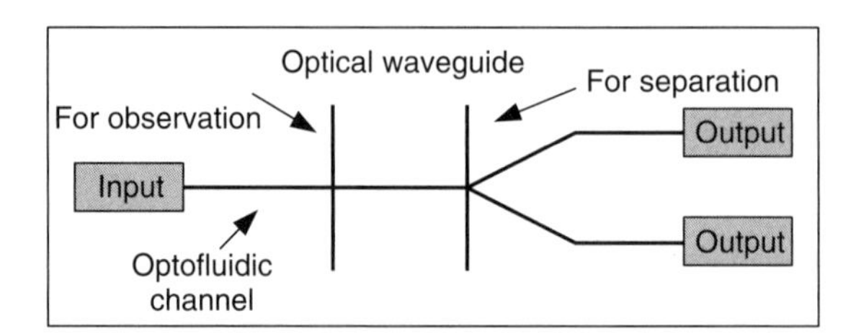

FIGURE 3.29 Illustration of a fiber-based device with analysis region and sorting region (91).

structures, because the light beam waist radius was reduced by these lens structures. Optical manipulation of particles by using this chip was also demonstrated. At the end, various applications of optofluidic manipulation techniques in the bioanalyses were discussed.

ACKNOWLEDGMENTS

This work was supported in part by the Centre for Photonics Fabrication Research (CPFR) and the National Research Council (NRC) Canada.

REFERENCES

1. Godin J, Chen CH, Cho SH, Qiao W, Tsai F, Lo YH. Microfluidics and photonics for Bio-System-on-a-Chip: A review of advancements in technology towards a microfluidic flow cytometry chip. J Biophoton 2008;1:355–376.
2. Bartlett P. *Colloid Science: Principles, Methods and Applications*. Oxford: Wiley-Blackwell Publishing Ltd; 2005.
3. Unger MA, Chou HP, Thonen T, Schemer A, Quake SR. Monolithic microfabricated valves and pumps by multilayer soft lithography. Science 2000;288:113–116.
4. Bousse L, Cohen C, Nikiforov T, Chow A, Koph-Sill AR, Dubrow R, Parce JW. Electrokinetically controlled microfluidic analysis systems. Annu Rev Biophys Biomol Struct 2000;29:155–181.
5. Dholakia K. Gene regulation: kissing chromosomes. Nature 2005;4:579–580.
6. Erickon D, Yang AHJ. Optofluidic trapping and transport using parnar photonic devices. In: Fainman Y, Psaltis D, editors. *Optofluidics: Fundamentals, Devices, and Applications*. Oxford: Wiley Publishing Ltd; 2009.
7. Ashkin A. Forces of a single-beam gradient laser trap on a dielectric sphere in the ray optics regime. Biophys J 1992;61:569–582.
8. Neuman KC, Chadd EH, Liou GF, Bergman K, Block SM. Characterization of photodamage to *Escherichia coli* in optical traps. Biophys J 2008;77:2856–2863.
9. Kuyper CL, Chiu DT. Optical trapping: a versatile technique for biomanipulation. Appl Spectrosc 2002;56: 300A–312A.

10. Liesener J, Reicherter M, Haist T, Tiziani H. Multi-functional optical tweezers using computer-generated holograms. Opt Commun 2000;185:77–82.
11. Neuman L, Block S. Optical trapping. Rev Sci Instrum 2004;75:2787–2809.
12. Ashkin A. Acceleration and trapping of particles by radiation pressure. Phys Rev Lett 1970;24:156–159.
13. Dholakia K, MacDonald M, Zemanek P, Cizmar T. Cellular and colloidal separation using optical forces. Methods Cell Biol 2007;82:467–495. Elsevier Inc.
14. Constable A, Kim J, Merivis J, Zarinetchi F, Prentiss M. Demonstration of a fiber-optical light-force trap. Opt Lett 1993;18:1867–1869.
15. Taguchi K, Ueno H, Hiramatsu T, Ikeda M. Optical trapping of dielectric particle and biological cell using optical fibre. Electron Lett 1997;33:413–414.
16. Taguchi K, Watanabe N. Single-beam optical fiber trap. J Phys Conf Ser 2007;61:1137–1141. DOI: 10.1088/1742-6596/61/1/225.
17. Lyons E, Sonek G. Confinement and bistability in a tapered hemispherically lensed optical fiber trap. Appl Phys Lett 1995;66:1584–1586.
18. Hu Z, Wang J, Liang J. Manipulation and arrangement of biological and dielectric particles by a lensed fiber probe. Opt Express 2004;12:4123–4128.
19. Knoer G, Ratnapala A, Nieminen T, Vale N, Heckenberg N, Rubinsztein-Dunlop H. Optical force field mapping in microdevices. Lab Chip 2006;6:1545–1547.
20. Barnard C, Lit J. Single-mode fiber microlens with controllable spot size. Appl Opt 1991;30:1958–1962.
21. Rudd D, Lopez-Mariscal C, Summers M, Shahvisi A, Gutierrez-Vega J, McGloin D. Fiber based optical trapping of aerosols. Opt Express 2008;16:14550–14560.
22. Sidick E, Collins S, Knoesen A. Trapping forces in a multiple-beam fiber-optic trap. Appl Opt 1997;36:6423–6433.
23. Gussgard R, Lindmo T, Brevil I. Calculation of the trapping force in a strongly focused laser beam. J Opt Soc Am B 1992;9:1922–1930.
24. Craighead H. Future lab-on-a-chip technologies for interrogating individual molecules. Nature 2006;442:387–393.
25. Jensen-McMullin C, Lee H, Lyons E. Demonstration of trapping, motion control, sensing and fluorescence detection of polystyrene beads in a multi-fiber optical trap. Opt Express 2005;13:2634–2642.
26. Jensen-McMullin C, Au A, Quinsaat J, Lyons E, Lee H. Fiber-optic-based optical trapping and detection for lab-on-a-chip (LOC) application. Proc SPIE 2002;4622:188–194.
27. Domachuk P, Cronin-Golomb M, Eggleton B, Mutzenich S, Rosengarten G, Mitchell A. Application of optical trapping to beam manipulation in optofluidics. Opt Express 2005;13:7265–7275.
28. Constable A, Kim J, Mervis J, Zarinetchi F, Prentiss M. Demonstration of a fiber-optical light-force trap. Opt Lett 1993;18:1867–1869.
29. Collins S, Baskin R, Howitt D. Microinstrument gradient-force optical trap. Appl Opt 1999;38:6068–6074.
30. Gauthier R, Frangioudakis A. Optical levitation particle delivery system for a dual beam fiber optic trap. Appl Opt 2000;39:26–33.
31. Taylor R, Hnatovsky C. Particle trapping in 3-D using a single fiber probe with an annular light distribution. Opt Express 2003;11:2775–2782.

32. Liberale C, Minzioni P, Bragheri F, Angelis F, Fabrizio E, Cristiani I. Miniaturized all-fibre probe for three-dimensional optical trapping and manipulation. Nat Photonics 2007;1:723–727.
33. Dholakia K, Reece P, Gu M. Optical micromanipulation. Chem Soc Rev 2008;37:42–55.
34. Kawata S, Sugiura T. Movement of micrometer-sized particles in the evanescent field of a laser beam. Opt Lett 1992;17:772–774.
35. Kawata S, Tani T. Optically driven mie particles in an evanescent field along a channelled waveguide. Opt Lett 1996;21:1768–1770.
36. Almaas E, Brevik I. Radiation forces on a micrometer-sized sphere in an evanescent field. J Opt Soc Am B 1995;12:2429–2438.
37. Lester M, Nieto-Vesperinas M. Optical forces on microparticles in an evanescent laser field. Opt Lett 1999;24:936–938.
38. Grujic K, Helleso O, Wilkinson J, Hole J. Optical propulsion of microspheres along a channel waveguide produced by Cs+ ion-exchange in glass. Opt Commun 2004;239:227–235.
39. Grujic K, Helleso O, Hole J, Wilkinson J. Sorting of polystyrene microspheres using a Y-branched optical waveguide. Opt Express 2005;13:1–7.
40. Grujic K, Helleso O. Dielectric microsphere manipulation and chain assembly by counter-propagating waves in a channel waveguide. Opt Express 2007;15:6470–6477.
41. Garces-Chavez V, Dholakia K, Spalding G. Extended-area optically induced organization of microparticles on a surface. Appl Phys Lett 2005;86:031106–1–031106–3.
42. Gaugiran S, Getin S, Fedeli J, Colas G, Fuchs A, Chatelain F, Derouard J. Optical manipulation of microparticles and cells on silicon nitride waveguides. Opt Express 2005;13:6956–6963.
43. Schmidt B, Yang A, Erickson D, Lipson M. Optofluidic trapping and transport on solid core waveguides within a microfluidic device. Opt Express 2007;15:14322–14334.
44. Eldada L, Shacklette L. Advances in polymer integrated optics. IEEE J Sel Top Quantum Electron 2000;6:54–68.
45. Duffy D, McDonald J, Schueller O, Whitesides G. Rapid prototyping of microfluidic systems in Poly(dimethylsiloxane). Anal Chem 1998;70:4974–4984.
46. Tanaka T, Yamamoto S. Optically induced propulsion of small particles in an evanescent field of higher propagation mode in a multimode channeled waveguide. Appl Phys Lett 2000;77:3131–3133.
47. Beche B, Pelletier N, Gaviot E, Zyss J. Single-mode TE_{00}—TM_{00} optical waveguides on SU-8 polymer. Opt Commun 2004;230:91–94.
48. Gu M, Haumonte J-B, Micheau Y, Chon J, Gan X. Laser trapping and manipulation under focused evanescent wave illumination. Appl Phys Lett 2004;84:4236–4238.
49. Marchington R, Mazilu M, Kuriakose S, Carces-Chavez V, Reece P, Krauss T, Gu M, Dholakia K. Optical deflection and sorting of microparticles in a near-field optical geometry. Opt Express 2008;16:3712–3726.
50. Yang A, Moore S, Schmidt B, Klug M, Lipson M, Erickson D. Optical manipulation of nanoparticles and biomolecules in sub-wavelength slot waveguides. Nature 2009;457:71–75.
51. Almeida V, Xu Q, Barrios C, Lipson M. Guiding and confining light in void nanostructure. Opt Lett 2004;29:1209–1211.

52. Born M, Wolf E. *Principles of Optics*. Oxford: Pergamon; 2003.
53. Imasaka T, Kawabata Y, Kaneta T, Ishidzu Y. Optical chromatography. Anal Chem 1995;67:1763–1765.
54. Kaneta T, Ishidzu Y, Mishima N, Imasaka T. Theory of optical chromatography. Anal Chem 1997;14:2701–2710.
55. Imasaka T. Optical chromatography. A new tool for separation of particles. Anal Mag 1998;26:M53–M55.
56. Hart S, Terray A, Arnold J. Particle separation and collection using an optical chromatographic filter. Appl Phys Lett 2007;91:171121.
57. Hart S, Terray A, Leski T, Arnold J, Stroud R. Discovery of a significant optical chromatographic difference between spores of *bacillus anthracis* and its close relative Bacillus thuringiensis. Anal Chem 2006;78:3221–3225.
58. Hart S, Terray A, Kuhm K, Arnold J, Leski T. Optical chromatography for biological separations. Proceedings of SPIE, Denver, CO, USA, vol. 5514; 2004. p 35–47.
59. Hart S, Terray A. Refractive-index-driven separation of colloidal polymer particles using optical chromatography. Appl Phys Lett 2003;83:5316–5318.
60. Makihara J, Kaneta T, Imasaka T. Optical chromatography: size determination by eluting particles. Talanta 1999;48:551–557.
61. Terray A, Arnold J, Sundbeck S, Leski T, Hart S. Preparative separations using optical chromatography. Proceedings of SPIE, San Diego, CA, USA, vol. 6644, 66441U; 2007.
62. Terray A, Arnold J, Hart S. Enhanced optical chromatography in a PDMS microfluidic system. Opt Express 2005;13:10406–10415.
63. Gauthier R. Computation of the optical trapping force using an FDTD based technique. Opt Express 2005;13:3707–3718.
64. Bonessi D, Bonin K, Walker T. Optical forces on particles of arbitrary shape and size. J Opt A: Pure Appl Opt 2007;9:S228–S234.
65. Nieminen T, Loke V, Stilgoe A, Knoner G, Branczyk A, Heckengerg N, Rubinsztein-Dunlop H. Optical tweezers computational toolbox. J Opt A Pure Appl Opt 2007;9: S196–S203.
66. Terray A, Ladouceur H, Hammond M, Hart S. Numerical simulation of an optical chromatographic separator. Opt Express 2009;17:2024–2032.
67. Kim S, Kim J, Kim S. Theoretical development of *in situ* optical particle separator: cross-type optical chromatography. Appl Opt 2006;45:6919–6924.
68. Kim S, Yoon S, Sung H, Kim S. Cross-type optical particle separation in a microchannel. Anal Chem 2008;80:2628–2630.
69. Kim S, Kim S. Radiation forces on spheres in loosely focused Gaussian beam: ray-optics regime. Appl Opt 2006;23:897–903.
70. Guo H, Zhao P, Xiao G, Zhang Z, Yao J. Optical manipulation of microparticles in an SU-8/PDMS hybrid microfluidic chip incorporating a monolithically integrated on-chip lens set. IEEE J Sel Top Quantum Electron 2010;16:919–926.
71. Kim S, Yoon S, Sung H, Kim S. Resolution of cross-type optical particle separation. Anal Chem 2008;80:6023–6028.
72. Kim S, Jung E, Sung H, Kim S. Optical mobility in cross-type optical particle separation. Appl Phys Lett 2008;93:044103.

73. Zemanek P, Karasek V, Sasso A. Optical forces acting on Rayleigh particle placed into interference field. Opt Commun 2004;240:401–415.
74. Measor P, Kuhn S, Lunt E, Phillips B, Hawkins A, Schmidt H. Hollow-core waveguide characterization by optically induced particle transport. Opt Lett 2008;33:672–674.
75. Mandal S, Erickson D. Optofluidic transport in liquid core waveguiding structures. Appl Phys Lett 2007;90:184103.
76. Mogensen K, Kwok Y, Eijkel J, Peterson N, Manz A, Mutter J. A microfluidic device with an integrated waveguide beam splitter for velocity measurements of flowing particles by Fourier transformation. Anal Chem 2003;75:4931–4936.
77. Chang-Yen D, Eich R, Gale B. A monolithic PDMS waveguide system fabricated using soft lithography tcchniques. IEEE/OSA J Lightwave Technol 2005;23:2088–2093.
78. Piggee C. Optical tweezers: not just for physicists any more. Anal Chem 2009;81:16–19.
79. Ozkan M, Wang M, Ozkan C, Flynn R, Birkbeck A, Esener S. Optical manipulation of objects and biological cells in microfluidic devices. Biomed Microdevices 2003;5:61–67.
80. Ashkin A. Optical trapping and manipulation of viruses and bacteria. Science 1987;235:1517–1520.
81. Chu S. Laser manipulation of atoms and particles. Science 1991;253:861–866.
82. Harada Y, Ohara O, Takatsuki A, Itoh H, Shimamoto N, Kinosita K. Direct observation of DNA rotation during transcription by *Escherichia coli* RNA polymerase. Nature 2001;409:113–115.
83. Leach J, Mushfique H, Leonardo R, Padgett M, Copper J. An optically driven pump for micrfluidics. Lab Chip 2006;6:735–739.
84. Domachuk P, Cronin M, Eggleton B, Mutzenich S, Rosengarten G, Mitchell A. Application of optical trapping to beam manipulation in optofluidics. Opt Express 2005;13:7265–7275.
85. Svoboda K, Schmidt C, Schnapp B, Block S. Direct observation of kinesin stepping by optical trapping interferometry. Nature 1993;365:721–727.
86. Svobods K, Block S. Force and velocity measured for single kinesin molecules. Cell 1994;77:773–784.
87. Moffitt J, Chemla Y, Lzhaky D, Bustamante C. Differential detection of dual traps improves the spatial resolution of optical tweezers. Proc Natl Acad Sci U S A 2006;103:9006–9011.
88. Applegate R Jr., Squier J, Vestad T, Oakey J, Marr D. Optical trapping, manipulation, and sorting of cells and colloids in microfluidic systems with diode laser bars. Opt Express 2004;12:4390–4398.
89. MacDonald M, Spalding G, Dholakia K. Microfluidic sorting in an optical lattice. Nature 2003;426:421–424.
90. Wang M, Tu E, Raymond D, Yang J, Zhang H, Hagen N, Dees B, Mercer E, Forster A, Kariv I, Marchand P, Butler W. Microfluidic sorting of mammalian cells by optical force switching. Nat Biotechnol 2005;23:83–87.
91. Kirei H, Oroszi L, Valkai S, Ormos P. An all optical microfluidic sorter. Acta Biol Hung 2007;58:139–148.

CHAPTER FOUR

Optical Fiber Sensors and Their Applications for Explosive Detection

JIANJUN MA and WOJTEK J. BOCK
Département d'informatique et d'ingénierie, Université du Québec en Outaouais, Gatineau, Québec, Canada

4.1 INTRODUCTION

After 9/11, monitoring and protection of critical infrastructures such as airports, bridges, buildings, railways, utilities, and water reservoirs, as well as concentrations of people clustered inside or around them, became a priority of governments around the world. At the same time, awareness increased the need to enhance surveillance of borders, perimeters, and transportation systems. All these concerns led to significant government-sponsored industrial and academic research and development. One of the most intensive efforts to counterpotential terrorist threats has been directed to explosive detection. In this connection, a significant document issued by the US Transportation Security Administration and the US Congress gives priority attention to improving the ability of screening checkpoints to detect explosives and other hazardous materials on passengers, in their luggage, and in all other cargo transported by air (1).

Explosive detection for air transportation security purposes focuses on two major areas: cargo security (2) and passenger screening (3). There are several hundreds of explosive materials officially listed under the United States Code (4). We are concerned here with three types. The first type consists of nitrate organic compounds

Photonic Sensing: Principles and Applications for Safety and Security Monitoring, First Edition.
Edited by Gaozhi Xiao and Wojtek J. Bock.

such as 2,4,6-trinitrotoluene (TNT) or 2,4-dinitrotoluene (2,4-DNT). In practice, mixtures of high explosives are often seen, with TNT as one of the major components because of its low cost. The second type contains inorganic nitrates, chlorates, or perchlorates. The third type includes unstable peroxide groups such as triaceton triperoxide (TATP). Nitrated organic explosives are usually used for military purposes. However, a significant number of terrorists have used "easy to obtain–easy to make" improvised explosives, including sodium nitrate or perchlorate, or more recently, TATP. Figure 4.1 illustrates the molecular architectures of several of the most widely used nitroaromatic explosive materials.

To identify different explosives, various principles have been applied, often involving more than one area of research. The first well-known explosive-detection principle is purely animal based, with trained dogs as the most salient example. Dogs can be trained to track several common explosive materials using their noses that are very sensitive to scents. A similar example being studied is the use of

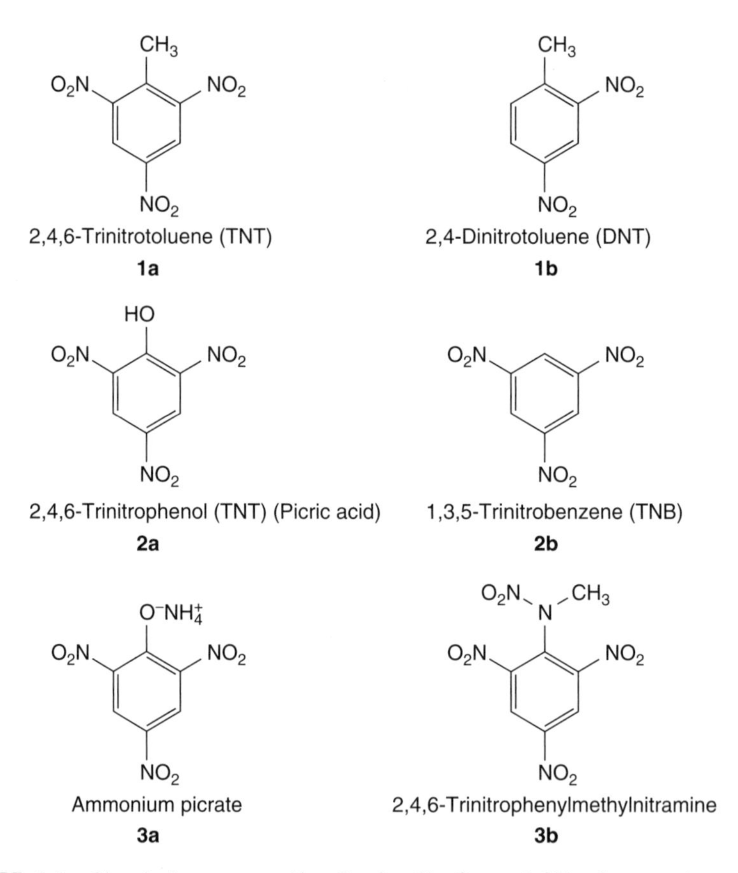

FIGURE 4.1 Chemical structures (**1a, 1b, 2a, 2b, 3a**, and **3b**) of some nitroaromatics explosive materials.

honeybees, which can be trained to map mine fields. The locations of mines are identified by the appearance of a large population of honeybees. Optical technology has been adopted to track the density distribution of the honeybees (5).

The second principle is based on the variation of density, with X-ray machines as the representative technology. Operators can identify explosive items using computed axial tomography, supported by an explosives threat library and color coding on their monitor screens. Special machines have also been designed to bombard suspected explosives with neutrons. Chemical compositions are then determined by the γ-radiation decay signatures. It has been reported that, with this technique, concealed explosives can be sniffed out with a standoff distance of 3 m within 400 s (6). Some explosives such as RDX have nitrogen in a crystalline form, whose interaction with radio frequency (RF) energy causes nuclear quadruple resonance (NQR). NQR detectors are commercially available with a false alarm rate as low as 3% (7).

The third principle is based on spectroscopy. This encompasses perhaps the most abundant list of technologies, including ion-mobility spectrometry (IMS), Raman scattering spectroscopy, chemiluminescence sensing, and fluorescence detection. IMS was designed to detect very low concentrations of samples based on the differential migration of gas-phase ions passing through a homogeneous electric field. It has become an excellent tool for detecting extremely low concentrations of explosive vapor. However, compared with the much easier vapor sampling techniques offered by some other detectors, IMS requires a more challenging particle sampling technique.

Raman scattering spectroscopy identifies explosives by detecting changes in the state of vibrational energy under laser excitation. The unique fingerprint of an explosive molecule is revealed by the vibration of its chemical bond, as it is related not only to the composition of materials but also to the molecular structure of each constituent. Even a trace amount of explosive residue can provide measurable Raman-scattered light. It has been demonstrated that a field-prototype system can detect and identify submilligram quantities of explosives at distances of up to 30 m (8).

Chemiluminescence is defined as the characteristic radiation of light generated from exothermic chemical reaction. The phenomenon occurs when highly explosive compounds containing nitric oxide (either NO_2 or NO_3) react with ozone. Gas chromatography (GC) is often combined with chemiluminescence sensing to separate molecules in order to provide good selectivity. The combination of GC and mass spectroscopy (MS) is well suited for detection of explosive substances, while thermal energy analysis (TEA), specifically designed for detecting nitro- and nitroso compounds, is more efficient when coupled with GC (9).

However, all these technologies have several drawbacks in common: not only are they expensive but their large size and need for additional accessories limit their portability and utility for field use. In addition, dogs need extensive training and continuous care, while IMS, GC-MS, and GC-TEA are destructive to the sample. In contrast, fluorescence-quenching-based detection of TNT and DNT explosives in the vapor phase is nondestructive and, as we shall see later, is much easier to adapt to create a portable device for field use and even for mass deployment. The key element of such a device is a specific organic conjugated polymer that offers a tremendously amplified quenching effect when in contact with vapor-phase analytes. Such amplifying fluorescent polymers (AFPs) consist of fluorescing chromophores linked together by polymer chains. In the absence of TNT, AFPs fluoresce (emit visible light) when excited. Binding of explosive molecules such as TNT to the AFP quenches or "turns off" the fluorescence. Unlike the traditional concept of fluorescence quenching, where binding of a single molecule of explosive quenches only the chromophore to which the explosive molecule is bound, with AFP, binding of a single molecule of explosive turns off the fluorescence of many chromophores, resulting in a dramatic amplification of the quenching. A deeper insight into the underlying mechanism shows that this fluorescent quenching is based on the photoinduced electron transfer from the excited polymer donor to the explosive acceptors. The unoccupied low energy π^* orbitals in nitroaromatic explosives can accept an electron from the excited polymers, leading to nonbonding electrostatic interactions (10) between the electron-rich polymer and the electron-deficient DNT and TNT molecules.

The origin of AFP lies in the research conducted by Dr. Swager and coworkers at MIT. In 1998, they reported the detection of nitroaromatic explosives using fluorescent quenching of peniptycene-derived conjugated polymer (11). The rigid peniptycene group prevents π-stacking of the polymer backbones, reduces inter-polymer interactions, and generates cavities large enough to allow diffusion of organic molecules into the film. More electron-rich polymers and larger cavities will produce a greater fluorescence response to nitroaromatic explosives. The AFP-based method has proved to be simple, sensitive, and rapid. At present, it is the most popular technology based on the vapor phase. The quenching percentage depends on many factors, including the vapor pressure, the exergonicity of electron transfer, the strength of the binding interaction between the AFP and the explosives particles, and the ability of analytes to diffuse through polymer films (11).

Following Dr. Swager's invention, Nomadics Inc. successfully incorporated his AFP into their Fido product line (12) and achieved a limit of detection as low as 1 fg (1×10^{-15} g), a sensitivity 10 times higher than it would be expected from the most sensitive IMS-based technology. Their products were adopted by US troops in Iraq for detection of improvised explosive devices (IEDs).

In principle, AFPs can be cast onto virtually any surface to form a thin layer of film. Fido products use a glass capillary with polymer coated onto their internal walls. Other typical coating substrates include planar glass surfaces and the curved sidewalls of optical fibers. Spin coating and simple dipping can create uniform films on both these surfaces.

Fido products have demonstrated excellent performance in explosive-detection operations, particularly in applications where simultaneous remote, nondestructive, and *in situ* operation is desired. Remote detection is important not only for safety reasons but also to ensure that the method is unobtrusive during operation—a primary concern both for privacy reasons and to avoid alerting the explosive carrier to the presence of a detection system. The combination of remote sensing and *in situ* assay enormously reduces costs by cutting down on the number of operational personnel needed and eliminating time-consuming laboratory-based professional sampling procedures.

Among the available technologies, fiber-optic sensing is uniquely suited to meet the demands of explosive detection, as it offers the desirable features of low cost, low attenuation, flexibility, and lightweight, as well as the capability of remote, *in situ*, and unobtrusive operation. Existing fiber-optic network technologies provide a ready means for distributing large numbers of fiber sensors at various critical spots and monitoring them from a single remote site. The focus of this chapter is on fiber-optic explosive detection incorporating the use of AFPs and describing some detailed work done in our group.

4.2 A BRIEF REVIEW OF EXISTING FIBER-OPTIC-BASED EXPLOSIVE DETECTORS

Since the appearance of fiber-optic sensors in the mid-1970s, more than 30 years have passed. Hundreds of fiber-optic sensor principles and their derivative designs have been proposed and explored for a variety of applications, covering parameter demands from the physical, chemical, biological, medical, and clinical sectors. In this chapter, we review five of the most important examples.

First, we must stress that optical fiber is not intrinsically a sensing medium. Fibers per se can only serve as optical signal carriers, with the sensing sections provided by other free-space- or micro-optics-based components. However, a modified fiber can be used for both sensing and signal delivery purposes. As for fiber types, multimode and single-mode fibers are two typical candidates. Owing to the influence of the fiber-optic communications sector, the predominance of single-mode fiber as the major signal carrier has led to the availability of very low cost fibers. In the fiber-sensing world, researchers generally prefer single-mode fibers as sensing media. For chemical or biological sensing, however, multimode fibers, especially large-core multimode fibers, are often chosen for signal delivery purposes.

At present, only a small number of explosive detectors are based on optical fibers. Within this category, several devices exploit the evanescent-wave (EW) tails of the guided modes for sample assay. One of the most successful examples of this concept, shown in Figure 4.2, is an off-the-shelf product line of Research International (13–15) originating in research done at the Naval Research Lab (NRL) (16, 17). Figure 4.2a and b shows two different principles for strengthening the excitation power carried by EW fields, which is then used to excite the sample species. Figure 4.2a depicts the well-known fiber taper developed by NRL researchers. Unlike regular fibers with their cylindrical architecture, a tapered fiber segment not only increases the EW-power level transferred from higher-order modes but also converts the higher-order modes carrying the incoming signal light to lower-order ones. Lower-order modes are much better confined within the fiber core and thus less affected by various sources of interference such as fiber bends and defects on the cladding surface. Figure 4.2b represents a different design used in Research

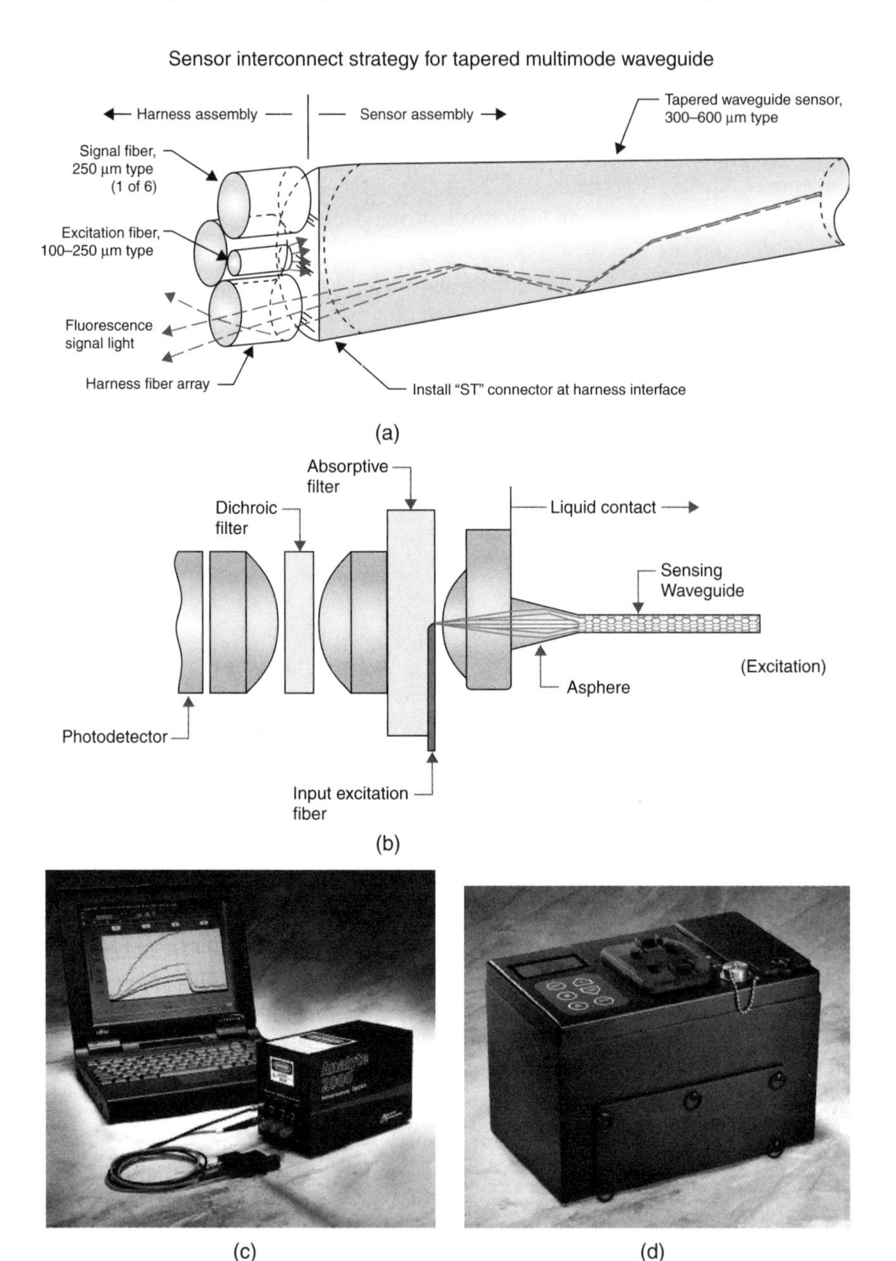

FIGURE 4.2 EW-based fiber-optic explosive detectors developed by NRL and manufactured by Research International. (a) Tapered large-core fiber for sensing tailed with signal-collection fibers and one excitation fiber; this is the original design from NRL. (b) Modified taper by Research International that maximizes the excitation power level in the cladding side of the sensing waveguide. Both excitation and collection fibers have been replaced by free-space optics. (c) The Analyte 2000. (d) The Raptor, automated and portable version designed for quicker assay in a battlefield environment.

International products (13, 14), employing a regular cylinder waveguide for sampling, tailed with a short tapered section called a *concentrator*. This increases the excitation power level even further while maintaining the same mode-conversion feature as the original design as long as the cylindrical section is short. Figure 4.2c shows the Analyte 2000 product model with four waveguides, which can analyze four samples simultaneously, while Figure 4.2d illustrates the Raptor model, which is the automated and portable version designed for a battlefield environment (13).

The second type of fiber-optic explosive detector was developed by Dr. Walt's group (18) and has also been commercialized. Although an earlier version of the detector used imaging fibers for sample immobilization and fluorescent signal transmission (19), here only the system based on large-core fiber is presented. As shown in Figure 4.3a, this system uses 1.52 m of fiber with an NA (numerical aperture) of 0.22 and a core diameter of 300 μm to form a six-around-one bundle. The tip of each fiber is functionalized with different sensory material or material combinations, including AFPs from Dr. Swager's group at MIT. The idea behind this concept is that each sensory material responds differently to vapor exposure so that the fluorescence response patterns are unique for each analyte. The instrument is computationally "trained" to discriminate target response patterns from nontarget patterns and background environments. As shown in Figure 4.3b, the optical system uses beam splitters F2 and F3 for excitation light and fluorescent signal delivery. Each fiber serves both to input excitation light from the LEDs and to transmit the fluorescent signal. A CCD camera is used to identify the responses from each individual sensory material. Figure 4.3c shows the field-deployable system based on this principle and Figure 4.3d highlights the compact size of the sensing head (sniffer). This system has a capability of blind detection of 2,4-DNT to the level of 120 parts per billion (ppb).

The third example is also an EW-based fiber-optic detector (20), illustrated in Figure 4.4. Unlike its traditional EW-based single-fiber counterpart that uses one fiber for both excitation light delivery and fluorescent signal collection, fiber X in Figure 4.4 serves only to deliver the excitation light. The explosive-sensitive fluorescent PAH (Polycyclic Aromatic Hydrocarbons) compound is bonded with the distal end of the fiber sidewall through specific chemical attachment (i.e., covalent or ionic bonding of derivatized PAH, encapsulation or entrapment of a PAH). The excitation light interacting with this compound coating comes from the EWs of the fiber core modes. Fluorescent light emanating from this coating is captured by the free-space photodetector AG directly rather than by a fiber.

In this system, an integrating sphere with inner reflective surface AA is adopted to enhance the efficiency of fluorescence collection by redirecting the emission rays traveling in other directions to the acceptance cone of detector AG. The distal end of the fiber is encapsulated within a glass tube AC. A large volume of air sample AK, drawn by a pump, enters the glass tube. When the sample goes past the compound AE, explosive analytes contained in the vapor are captured and fluorescence quenching will be observed. The quenching percentage depends on the amount of explosive captured, which in turn is strongly associated with the appropriate inner diameter of the glass tube. The amount of captured analyte can

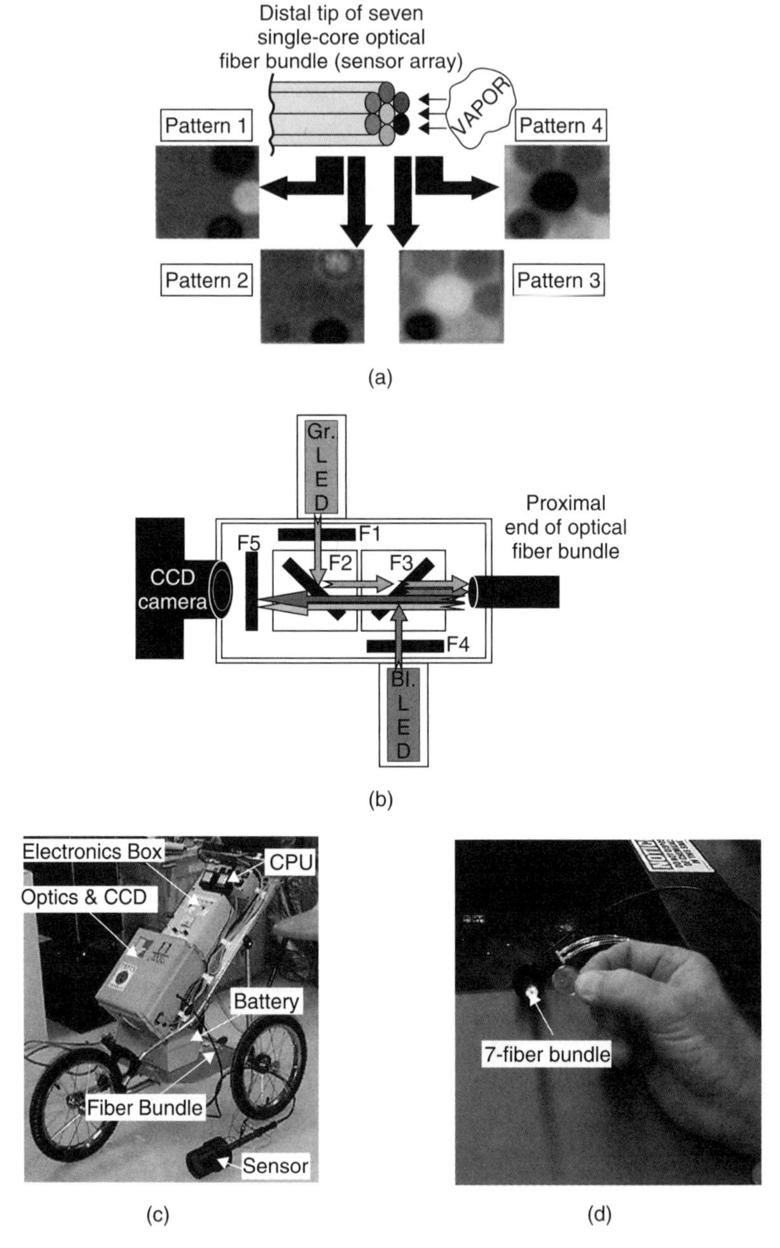

FIGURE 4.3 Field-deployable system developed by Dr. Walt's group. (a) Six-around-one fiber bundle with fiber tips coated with different sensory materials to generate unique patterns for explosives. (b) Operating principle of the system. (c) A system ready for field test. (d) Detector head smaller than a dime in size.

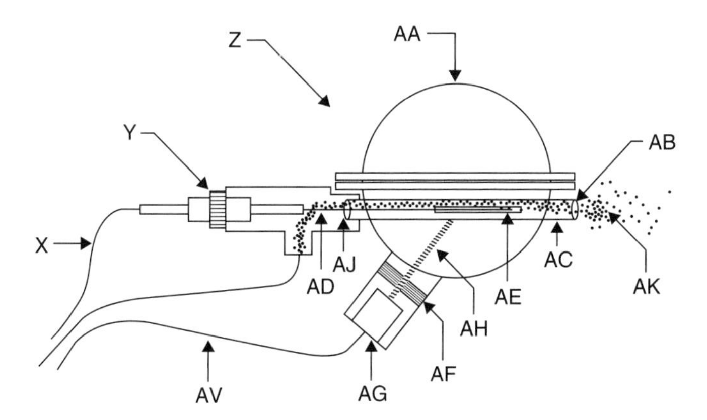

FIGURE 4.4 EW-based fiber-optic high explosive detector incorporated with integrating sphere, from US Patent No. 5,17,261. Key elements in this figure: AA, integrating sphere; AB, air entry; AC, glass tube; AD, optical fiber; AE, chemical sensor; AF, band-pass filter; AG, photodiode; AH, fluorescent emission; AJ, exit aperture; AK, explosive molecules; AV, electrical output; X, fiber; Y, fiber-optic connector; Z, sensor assembly.

be further increased by improving the preconcentration system, which is discussed in detail in Section 4.7.

The fourth interesting example is again an EW-based fiber-optic system, which is built with 200-μm-core multimode fiber. As shown in Figure 4.5, unlike the system from Research International, where only very short-fiber waveguides are used, this fiber-optic explosive detector employs a long fiber for sensing, excitation, and signal transmission purposes (21). The middle segment of the fiber is decladded, coiled, and coated with sensory materials based on triphenylene ketals (22). These receptors bind TNT molecules through charge transfer, causing attenuation of the transmitted light and a color change from transparent to red. This system can be used for online and *in situ* TNT detection down to the sub-ppb level.

The fifth example is a product from Nomadics Inc. (12). Strictly speaking, their core optical element for signal transmission and AFP coating is a capillary tube rather than a conventional optical fiber. However, this tube actually functions as a hollow-core circular waveguide. Figure 4.6a illustrates the operating principle of their Fido model. The AFP is coated onto the inner wall of the capillary tube via the spin-coating process. Excitation light is sent directly from an external source, which is positioned perpendicular to the tube via a free-space optics arrangement. The advantage of this design is that the AFP film receives the maximum possible excitation light, which in turn generates the maximum fluorescent signal level. Part of this signal, entering the glass via EW-field coupling, will propagate from the excited AFP site to the detector. The fixed ratio of the coupling process means that this design will yield a maximal level of detectable signal similar to that achieved by the sensors shown in Figures 4.2 and 4.5. Additional advantages of this design are that minimal stray excitation light arrives at the detector and that there is no need for extra housing to hold and evacuate the explosive vapor.

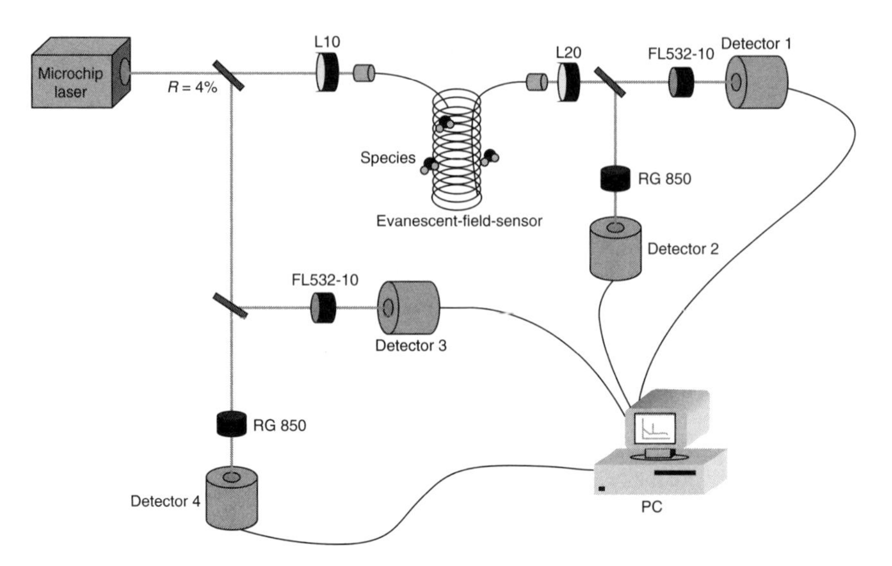

FIGURE 4.5 An EW-based fiber-optic explosive detector employing large-core fiber for excitation and signal light transmission, as well as sensing purposes.

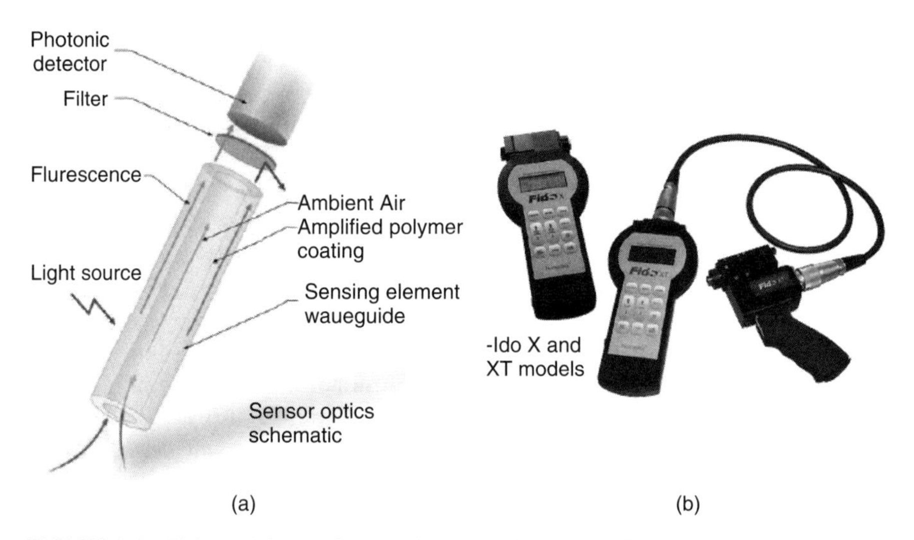

FIGURE 4.6 Fido model manufactured by Nomadics Inc. (a) Operating principle based on a capillary waveguide. (b) View of the hand-held Fido model.

4.3 HIGH PERFORMANCE FIBER-OPTIC EXPLOSIVE DETECTOR BASED ON THE AFP THIN FILM

4.3.1 Optimizing Fiber-Optic Explosive Detector Architecture

For the detection of quenched fluorescence light introduced by AFP thin film, only a few of the hundreds of existing fiber-optic sensing architectures are suitable for the purpose, having an appropriate combination of features such as low cost, compact construction, robustness, and high sensitivity. Among them, one- or two-fiber architecture is often preferred, with a tendency for large-core multimode fibers to be used as the sensing and signal transmission media. Substantiating evidence of the limited number of possible architectures is found in the literature and the range of available products (23–26). The one-fiber configuration has the best light-collection efficiency and the smallest sensing head. However, it requires an additional fiber coupler or a beam splitter and a collimating-focusing assembly, which causes loss of signal during transmission. The alignment of these elements is difficult and the overall system cost is high. Figures 4.2b and 4.3b show typical examples of this architecture. A two-fiber version system is easy to make, but it suffers from poor signal-collection efficiency because of the small overlapping area determined by the acceptance cones of the exits of the two fibers. Solutions to this issue (23, 24) mainly focus on liquid samples.

An alternative optimized design we proposed is shown in Figure 4.7 (27), which simultaneously solves a number of technical challenges, including the efficient use of excitation power, collection of the EW-form fluorescent signal, suppression of stray excitation light, and quality control of the formation of a very thin film layer on the fiber core-cladding surface. As discussed later, the proposed architecture also enables the end-face-TIR (total internal reflection) of a specific mode group, which boosts the initial thin-film fluorescence (TFF) level up to 22%.

This sensor is simple, lightweight, and robust and has a small sensing head, offering both flexibility and cost-effective system construction. In this design, two

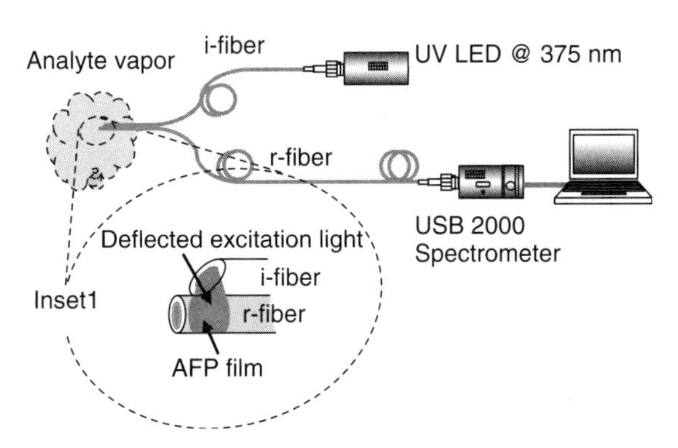

FIGURE 4.7 One of the optimized fiber-optic explosive detectors with combined performance enhancement.

fibers are arranged side by side, with one fiber (the illuminating fiber or i-fiber) serving for excitation light delivery and the other, for fluorescent signal collection (the receiving fiber or r-fiber). Both fibers are BFL37-800 large-core multimode fiber with core/cladding/jacket/NA of 800 μm/830 μm/1400 μm/0.37. The cladding and jacket of the end segments of both fibers are stripped off. In addition, as illustrated in Inset 1, the end face of the i-fiber is angle-polished to 45° to partially deflect the incoming excitation light toward the sidewall of the r-fiber. The r-fiber is coated with AFP film (27, 28) through a simple dipping process. Before dipcoating, the fiber core-cladding surface is pretreated with specific chemicals to increase the quality of the film. The pretreatment process is discussed in Section 4.4.

In contrast to the procedure followed with a conventional two-fiber sensor, the fluorescent signal generated from the film is not fed into the r-fiber core from the fiber end face. Instead, what we call higher-order guided modes (As an aside, note that this is a recognized although not accurate concept. Tunneling modes, and even some radiation modes should be also included, a point that is discussed later in this chapter.) in the r-fiber receive this signal through their EW tails, which reside in an extremely thin layer very close to the fiber core-cladding surface. The thickness of this layer is in the approximate range of from several hundred nanometers to one wavelength. This arrangement allows the guided modes to respond to the events occurring in this layer, which is occupied by the AFP film in our sensor. Evanescent fluorescence (EF) is the name given to the form of fluorescent signal referred to here. The incoming EF signal is directly fed into a USB 2000 spectrometer (Ocean Optics Inc.) without any filters in between. As will be seen later, the spectrum will show a stray excitation light spectrum. However, this light is caused by fiber defects only and will not affect the fluorescent signal under detection. The computer receives data from the spectrometer, displays all signal spectra and performs data processing. The light source used is a USB-LS-450 LED (Ocean Optics Inc.), coupled into the i-fiber.

4.3.2 Experimental Demonstration of Fluorescent Quenching Detection and Discussion

The AFP polymers, used for TNT explosive detection in our project, are provided by Dr. Wang and his team at Carleton University, Ottawa, Canada (27). As shown in Figure 4.8, these polymers are synthesized by palladium-catalyzed Suzuki cross-coupling condensation of dibromoterephthalates with bisboronic acid ester monomers, which are derived from alkylated fluorine and *N*-hexyldiphenylamine. The obtained polymers are completely soluble in certain organic solvents such as toluene, THF, DMF, and chloroform.

Decayed TNT or DNT in powder form was used to generate the explosive vapor. The DNT powder was sealed inside a glass container with a small pinhole allowing for insertion of the fiber probe. At room temperature, the vapor released from the powder will remain trapped inside the bottle until the vapor concentration reaches its dynamic equilibrium status.

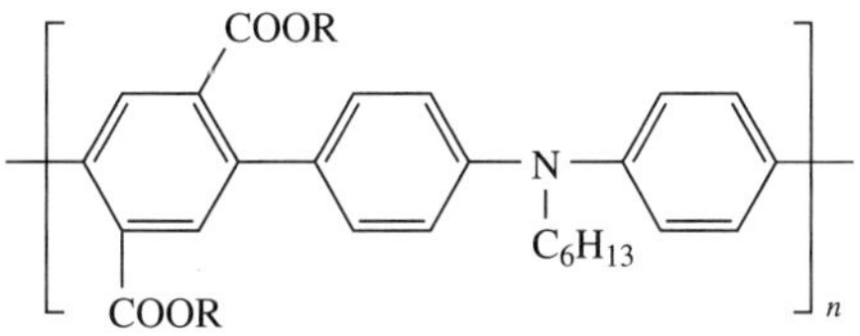

COOR
N
C_6H_{13}
COOR
n

4a or **4b**

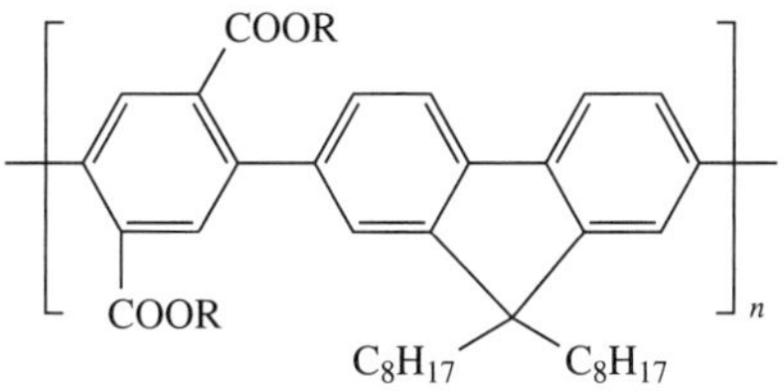

COOR
COOR
n
C_8H_{17} C_8H_{17}

5a or **5b**

a = $-CH_3$

H_3C
CH_3
CH_3
CH_3
H_3C

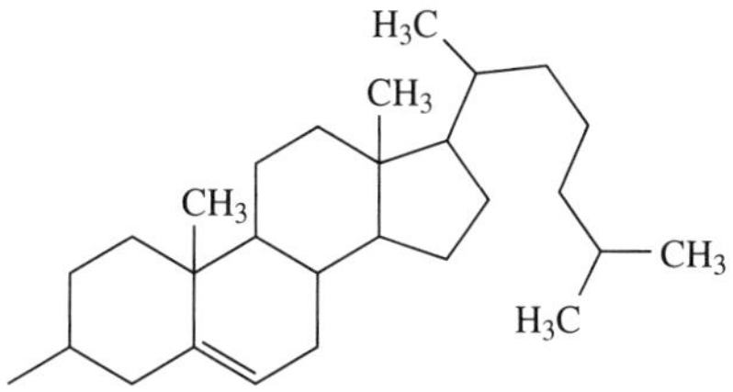

b =

FIGURE 4.8 Chemical structures of polymers (**5a, 5b, 4a**, and **4b**).

Measurement of quenching has a stringent requirement for a stable fluorescent signal level. To achieve a result with high confidence, the following two steps are critical:

1. Before the sensor is exposed to the vapor, the initial fluorescent signal level has to be monitored for a length of time comparable to the duration of the entire measurement, which is usually about a couple of minutes. Comparison of the maximum fluctuation of the signal to the quenching percentage allows any signal drop caused by fluctuation to be excluded.
2. The entire sensor system must be rigid, allowing absolutely no movement between components and the fiber itself. In addition, the sensing head should enter the bottle through the predrilled hole without touching its edge.

Having met these requirements, during the actual measurement we record both the initial EF signal spectrum before exposure and the time of this recording as the first reference I_{R_1}. We also record the signal at the moment when exposure of the sensor to the analyte vapor begins. This signal is used as the second reference I_{R_2}, which ideally should be identical or at least very close to the first reference I_{R_1}. This step will increase confidence in the result by identifying any possible effect of the internal volume of the bottle on the EF spectrum, which could be an issue if the area surrounding the sensing head changes during the experiment. To ensure that the two references were in fact closely identical, the experiment was performed in a dark laboratory room. After recording I_{R_2}, we collected signal drops for several typical time intervals and compared them with the reference I_{R_2}. Polymer **5a** was used in this measurement.

The results summarized in Figure 4.9 and Table 4.1 lead to the following conclusions:

- The proposed sensor incorporating this sensory polymer film is able to quickly respond to the preconcentrated analyte vapor, as indicated by a dramatic signal drop of 26% within 10 s following exposure.
- The minimum 26% signal drop shown in Table 4.1 demonstrates the superior capability of the proposed sensor to identify TNT explosives. This capability

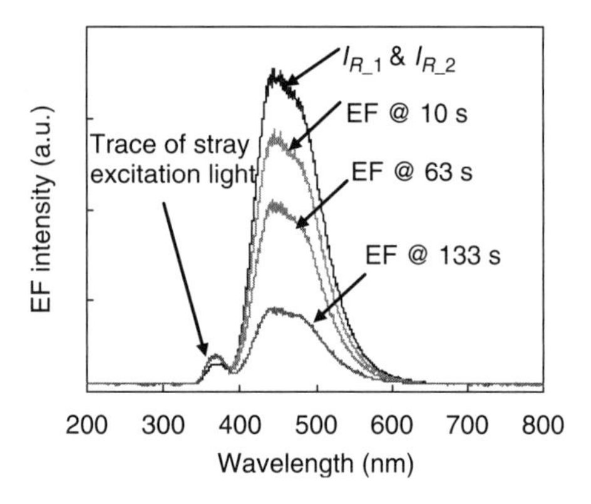

FIGURE 4.9 EF spectra before and after exposure to analyte vapor. I_{R_1} and I_{R_2} are references for all other quenched EF signals.

TABLE 4.1 Experimentally Observed EF Signal Drop Caused by Analyte Vapor Quenching

EF Fluctuation	2% within 180 s		
Time, s	10	63	133
Signal drop caused by quenching, %	26	74	87

is due to the fact that the excitation power from the UV LED is launched to the thin film at a level that balances the conflicting demands of generating a sufficiently high EF signal and avoiding material degradation caused by high power as indicated by fluctuation of the EF signal.

- Figures 4.7 and 4.9 clearly demonstrate that even without a filter to remove stray light, the proposed sensor still generates only a trace level of stray excitation light, owing to the fact that the deflected light enters the fiber core at an angle that allows no guided-mode excitation. This trace level is solely connected with the defects existing along the light path of the excitation light, as discussed in Reference 29.
- The quenching percentage is not determined exclusively by the polymer molecular architecture. There is also a high association with the quality of thin film on the core-cladding surface. Very thin film is the key to achieving a high quenching efficiency and the result of 87% EF quenching suggests that our dipping technique generates excellent film quality. Determining the film thickness remains a challenging task because traditional tools such as an ellipsometer or stylus profiler are not appropriate for the curved sidewall of the fiber core. However, a rough estimation was performed using an atomic force microscope, showing that the optimum film thickness is approximately 90 nm.
- Integration time is another factor to consider when evaluating the performance of the sensor design. Given an identical type and quality of film, the same excitation power level, the same area under illumination, and the same spectrometer, better sensor design gives a shorter integration time, suggesting higher fluorescent signal-collection efficiency and a higher overall signal level. These pulses can facilitate the design of a signal processing system, while minimizing the possibility of false alarms and the overall cost. Although the results in Figure 4.9 are obtained under only 400-ms integration time, the high signal-to-noise ratio in the spectrum reveals the excellent capability of this sensor.
- The excitation light source used in this investigation was a USB-LS-450 UV LED centered at 375 nm. We measured the overall output power after the 600-μm-core fiber at 67.5 μW by an HP 8152A optical-average power meter. Deflecting this power level at the angled fiber end without additional treatment can provide only a small portion of the necessary illuminating level. Compared with another of our designs (29) that calls for a 630 μW Argon laser power for excitation, the architecture proposed here shows better use of excitation power. Moreover, the solution proposed in Reference 29 needs 300 ms integration time to achieve a clean spectrum, demands a larger sensing head and most particularly, must have an additional complex fixture to establish a rigid sensing head. In contrast, the elegant design of the proposed architecture described in this chapter has a more competitive potential for prototyping purposes.

4.3.3 Unique Advantage of the Optimized Detector—Dramatically Increased Fluorescence Collection through the End-Face-TIR Process

The proposed sensor design (27) in Figure 4.7 may appear similar to the design of a conventional two-fiber sensor (23, 24). However, the fundamental difference is that its overall signal is in EF form and the signal is not fed mainly to guided modes. In terms of ray optics, EF power from the analyte sample, rather than being channeled into either meridional or long-pitched helical guided rays is instead carried by helical rays belonging to the category of tunneling rays (30), which are superior to guided rays in the collection of EF power. In terms of mode theory, these rays are associated with tunneling modes, which operate just below cutoff. While tunneling modes are leaky in nature, the associated loss is so weak that these rays persist in a fiber even after kilometers of travel and can in fact be treated as guided modes (30).

Our recent work (31) gives several important results for rays/modes contributing to the overall detectable fluorescent signal level. For a liquid sample such as Rhodamine 6G (R6G) and the sensing architecture shown in Figure 4.10, we found that a surprising 91% of the fluorescent signal detected at the fiber end is contributed by rays capable of TIR from the end face. With strong experimental backup, we also indicated that a large number of longer-pitched helical tunneling rays plus a few very steep guided rays are responsible for this enormous percentage. However, for this phenomenon to be observed, the sample must be positioned at the sidewall of the fiber core in such a way that there is no cladding between the sample and the end face. In terms of wave optics, the modes associated with these rays have high percentages of their mode fields extending into the fiber cladding, allowing efficient coupling of fluorescent power for detection. This is especially true of the longer-pitched tunneling modes, as they operate just below cutoff and have much wider light acceptance cones than would be expected from very steep guided modes. In

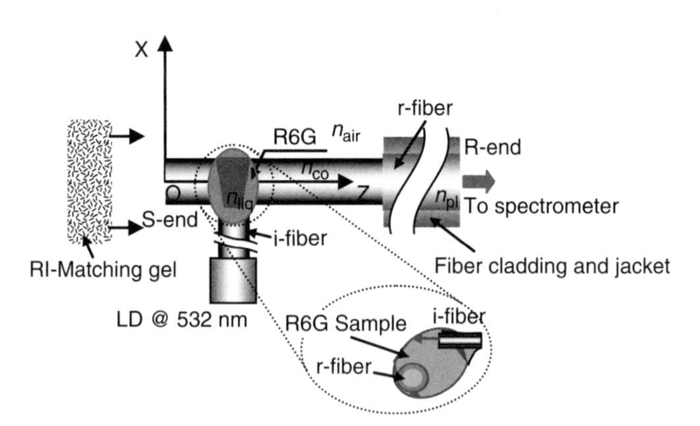

FIGURE 4.10 Sensing architecture to verify the existence and dramatic impact of end-face-TIR on overall detectable fluorescent signal level (31).

particular, our further theoretical investigation contradicts our experimental observation that the fluorescent signal carried by end-face-TIR-capable rays will not be directly detectable because of the presence of cladding with a high refractive index (RI), such as polymer cladding, which will block all these end-face-TIR-capable modes. However, we believe that the mechanism of nonpropagating-mode excitation in fact takes over and carries the signal power to the detecting end. Although nonpropagating modes will not survive in a very long fiber, in the case of a fiber sensor up to several meters in length, fluorescent power carried by end-face-TIR-capable modes can still be detected without a significant loss.

While the above conclusions are based on a liquid-sample study, they are applicable to any other form of sample as long as the sensor follows the requirement given above. However, solid sensory polymer thin film behaves differently from a liquid sample. The thin film in the work discussed here has two salient features: first, as it is synthesized using polymer materials, it commonly has a higher RI than the fiber core, resulting in a leakage of TFF power, which obviously does not occur with the liquid sample; and second, the overall received excitation power will be less than that received from the liquid sample. The reshaping process described in Reference 32 clearly indicates that the overall fluorescent power level can be significantly enhanced by reshaping the liquid sample droplet because this process maximizes the excitation power received near the sidewall of the fiber core. As a result, the EW fields of these end-face-TIR-capable modes are fully used to absorb the emitted fluorescent power. A thin-film sample will not benefit from this process. In fact, the use of excitation power is far from being optimized for the case of thin-film polymer. Another factor contributing to the low coupling efficiency is that the thickness of the film is around 90 nm, implying that the EW fields (which can reach as deep as several hundred nanometers or a wavelength) in the cladding are severely underfilled, so that the signal collecting capability of the end-face-TIR-capable modes is wasted to a high degree.

We have experimentally examined how these factors affect the fluorescent power level coupled to the end-face-TIR-capable modes for the thin-film sample. In the setup shown in Figure 4.7, the receiving fiber is coated with approximately 90 nm of AFP thin film and the spectrometer continuously monitors the TFF level during the entire experiment. Once a sufficiently stable initial TFF signal is achieved, the RI matching gel shown in Figure 4.10 is introduced to the fiber end by following the same steps as in our earlier work (31) in order to eliminate the end-face-TIR. Figure 4.11 shows the strength of end-face-TIR for the thin-film sample. Curve ② indicates a 22% drop in TFF power when end-face-TIR is eliminated. This figure is significantly smaller than what we observed with the R6G liquid sample following the reshaping process (31, 32), which demonstrated a 91% drop in fluorescent power. This enormous difference reveals that the contribution of end-face-TIR-capable modes to the overall TFF power is far less than would be expected from the liquid sample. In fact, this is evidence that the end-face-TIR-capable modes are not efficiently excited owing to the underfilled EW fields discussed above. In other words, we see that 78% of overall TFF power

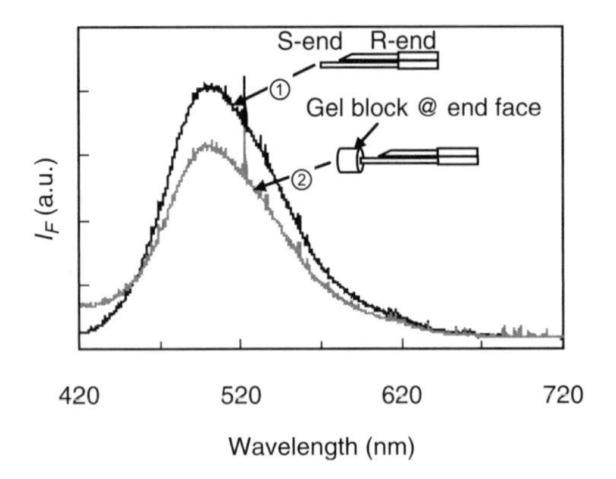

FIGURE 4.11 Effects of end-face-TIR on overall TFF power collection. Curve ①: initial TFF spectrum with contribution of end-face-TIR-capable modes; Curve ②: TFF spectrum showing 22% power drop when end-face-TIR is eliminated from the fiber end-face with gel block.

comes from modes incapable of end-face-TIR, or more precisely, from a large number of higher-order guided modes near cutoff.

We now take a closer look at the modes contributing to this 78% of overall TFF power. To facilitate our analysis, we adopt the RIs of a specific multimode fiber used in our study, which has RIs of $n_{co} = 1.46, n_{air} = 1$, and $n_{pl} = 1.41$. The critical angle of each individual segment is determined by $\alpha_c = \sin^{-1}(n_{cl}/n_{co})$. Obviously, we have $\alpha_{c\_air} < \alpha_{c\_pl}$. The very steep guided rays, associated with very high order guided modes, are those with incident angles close to α_{c_air} and possess the highest EW-power collection capability within the category of guided rays.

In terms of ray optics, the rays that may contribute to the overall power detected at the fiber end can be categorized according to angles θ_z and α by Ma et al. 31:

$$\text{Guided rays:} \quad 0 < \theta_z < 90^\circ - \alpha_c, \tag{4.1}$$

$$\text{Tunneling rays:} \quad 90^\circ - \alpha_c < \theta_z < 90^\circ, \text{ and } \alpha_c < \alpha < 90^\circ \tag{4.2}$$

where θ_z represents the angle of incidence on the fiber end face, and α is the incident angle at the core-cladding interface.

For the fiber under study, all tunneling rays obey $47^\circ < \theta_z < 90^\circ$ for $\alpha_{c\_air} = 43^\circ < \alpha_{c\_pl}$, suggesting that they all experience TIR at the fiber end face. Thus, the rays contributing to the 78% of TFF power follow the following conditions:

$$\text{Guided rays:} \quad 0 \ll \theta_z < 43^\circ, \tag{4.3}$$

$$\text{Tunneling rays:} \quad \text{No eligible rays.} \tag{4.4}$$

Equations 4.3 and 4.4 reveal that in the case of the thin-film sample, most of the higher-order guided modes, excluding those very steep end-face-TIR-capable ones, account for 78% of overall collected TFF power. This figure is reduced to 9% when a reshaped liquid sample such as R6G is involved (31). Again, it is the severely underfilled EW fields of these modes, extending deeply into the cladding owing to their operation near and just below cutoff (30, 31), that account for this enormous difference.

4.4 GENERATING HIGH QUALITY POLYMER FILM—PRETREATMENT WITH ADHESION PROMOTER

High quality of the polymer film is critical to the performance of an AFP-based explosive sensor. However, generating a uniform polymer film strongly attached to the substrate can be challenging in the sense that this cannot be done simply by providing a clean substrate. In general, before the polymer coating process, a proper pretreatment of the substrate with another specific chemical material, an adhesion promoter, is required to assist in bridging, or establishing a chemical bond between the organic polymer and the inorganic substrate. Among available adhesion promoters, organofunctional silanes are excellent for the purpose. Their underlying principle is shown in Figure 4.12. They are used as adhesives or sealant materials because in addition to enabling good adhesion, they have excellent mechanical properties and storage stability (33). In the preparation of our fiber probe, we used 3-aminopropyltrimethoxysilane, $H_2N(CH_2)3Si(OCH_3)_3$, as the adhesion promoter. Purchased from Fluka, it has a concentration of 97%. A simple dipping process is used to prepare the probe, following the sequence of promoter, water, and finally the AFP solution. Dipping into water is an important step to provide moisture, which causes the silane's alkoxy groups to hydrolyze and react with the inorganic substrate, forming a bond covalently to the inorganic substrate.

Using the setup shown in Figure 4.7, we verified the effect of the adhesion promoter using two-fiber probes coated with polymer (PL**4a** shown in Figure 4.8, which has a mass concentration of 0.03%). One probe was treated with the adhesion promoter and one was not. The results in Table 4.2 clearly indicate that treating the fiber probe with adhesion promoter greatly increases the quenching percentage. In fact, the difference between these two probes is visible even from an inspection

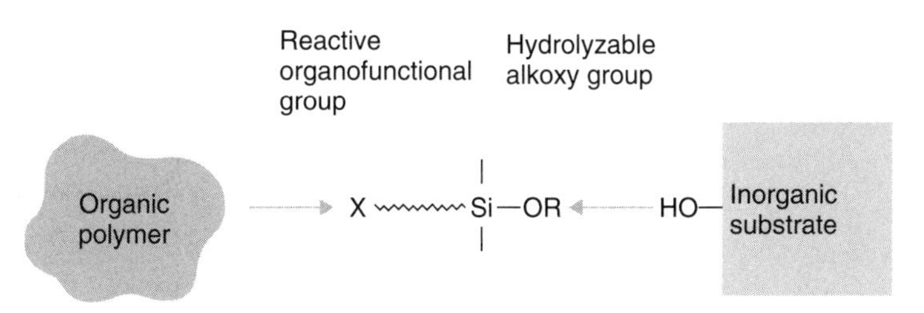

FIGURE 4.12 Organofunctional silanes as a bridge between organic polymers and inorganic materials (33).

TABLE 4.2 Experimental Verification of the Effect of Treating or Not Treating PL4a-Coated Fiber Probes with Adhesion Promoter

	Quenching Percentage, %		
PL4, a coated Fiber Probe	10 s	20 s	30 s
Promoter treated	13	26	38
Not treated	8	17	26

The sample probes were prepared from PL**4a** solution exposed to ambient light for over 12 h.

TABLE 4.3 Effect of Ambient Light on the Lifetime of AFP Polymer 4a

		Quenching Percentage, %		
Concentration, mass percentage	Exposure Time to Ambient Light, h	10 s	20 s	30 s
0.03%	~12	13	26	38
	<0.5	30	56	68

of the coated area with the naked eye. The entire coated area of the treated fiber probe appears almost transparent while irregular variations in thickness are obvious for the untreated probe. This visible irregularity implies that there are in fact huge variations in thickness, causing an enormous remaining background fluorescence and a poor signal-to-noise ratio.

4.5 EFFECT OF PHOTODEGRADATION ON AFP POLYMER

All quenching-related sensing devices suffer from photodegradation when exposed to ambient or excitation light for a sufficiently long time. The latter type of degradation can be minimized using pulse-form excitation light with low intensity. Ambient light is easily blocked by sealing the sensing segment inside the housing, which is in fact considered as part of the preconcentration system discussed in the next section.

In our experiments, we observed the dramatic effect of ambient light on AFP polymer. Table 4.3 shows the results of quenching measurements when polymer **4a** is exposed to ambient light for less than half an hour and for more than 12 h. After about 12 h of ambient light exposure, the quenching effect is dramatically reduced to about half of that effect when the polymer is exposed to ambient light for only half an hour. Our further study indicates that the quenching effect completely disappears after the polymer is continuously exposed to ambient light for a few days.

4.6 OPTIMIZING POLYMER CONCENTRATION FOR OPTIMIZED AFP-FILM THICKNESS

Forming an optimum AFP film on the fiber sidewall means juggling several different factors as indicated in Section 4.3.2. Perhaps the most important consideration is the compromise between thickness and integration time, which must often be made, as reflected in Table 4.4. A shorter integration time with a strong initial fluorescence and a better signal-to-noise ratio comes at the cost of a much lower quenching percentage, implying the existence of an underlayer of the AFP film that is never in contact with the explosive vapor and remains as a permanent background. A thicker film is also likely to lead to an issue of reversibility as vapor molecules might be trapped within this underlayer and never be removed. A very thin AFP layer, however, introduces a very low initial signal level and a poor signal-to-noise ratio, and therefore requires a complicated signal processing system in order to accurately retrieve the information from the target sample. In our experience, an integration time of 5 s gives the best results, all factors considered.

Freshly prepared AFP polymer **4a** was used for all sample fibers, which are tested when the films are fresh.

4.7 EXPLOSIVE VAPOR PRECONCENTRATION AND DELIVERY

AFP is an extremely sensitive material for explosive detection. However, as it is intended for use with low concentration vapor down to the level of parts per trillion (ppt), directly detecting sparsely distributed particles with a millimeter-size fiber-sensor head is a difficult proposition. On the other hand, such a low concentration, plus the sticky property of the explosive vapor mentioned below, makes for a detection threshold below the limit of the most commercially available analytical instruments (34). These considerations suggest that an AFP-based sensing device cannot effectively do its job without the assistance of an additional assembly called a *preconcentrator*. The preconcentrator helps by collecting a large volume of surrounding air, then filtering or trapping the target particles in a miniaturized space and finally delivering them to the detecting area. The Fido hand-held unit (Fig. 4.6b) from Nomadics, which shows a much better detecting capability than today's most sensitive laboratory-based IMS instruments, is the perfect example demonstrating how a preconcentrator helps in achieving high performance.

The proper design of such a preconcentration unit involves the detailed consideration and engineering of several critical components. A particularly important point to be taken into account is that explosive molecules have adhesive properties.

TABLE 4.4 Effect of Different Concentrations on Quenching Percentage

Concentration, mass percentage	Quenching Percentage, %			Integration Time, s
	10 s	20 s	30 s	
0.13%	16	32	47	2.5
0.07%	22	45	53	3
0.03%	30	56	68	5
0.02%	30	63	87	8

They stick to every bit of metal, plastic, and ceramic they encounter before arriving at the target area. Although this property may appear to be a negative factor, in fact, it can be used to advantage in a well-designed system.

Perhaps the most complete discussion regarding the preconcentration assembly is given in the patent of Nomadics Inc. filed in May 2003 (35), where the following components or factors are included or considered in their Fido model (Fig. 4.13):

1. a vacuum pump 14 to assist in drawing a large volume of surrounding air with an optimum airflow rate through flow meter 78;

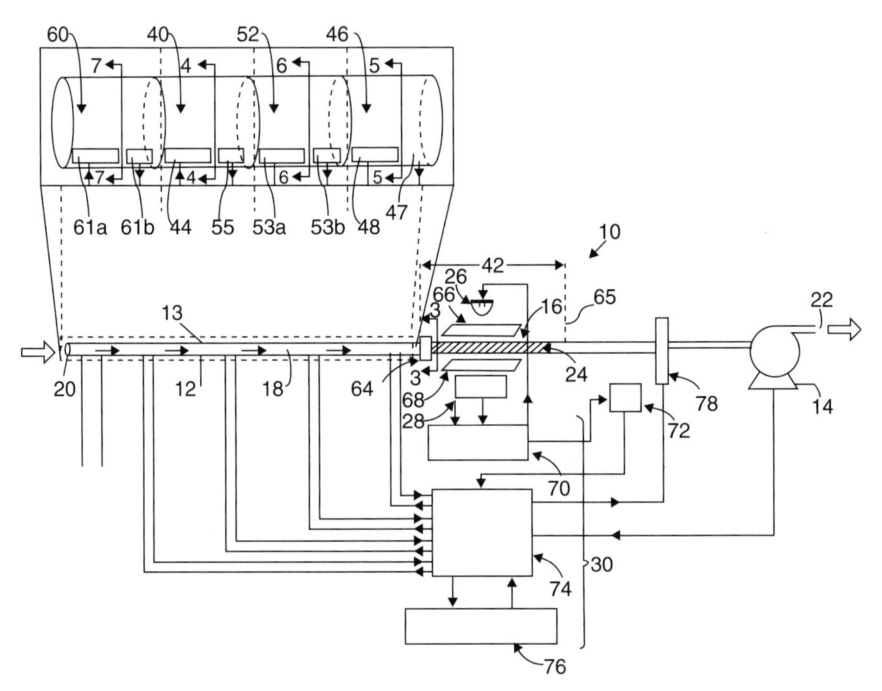

FIGURE 4.13 A preconcentration assembly summarized in US Patent No. 6,558,626 B1. Key elements in this figure: 10, detector; 12, glass capillary housing; 13, capillary; 14, air pump; 16, sensing assembly; 18, enclosed sensing volume; 20, inlet; 22, outlet; 24, sensing unit; 26, excitation light source; 28, light detector; 30, signal processing or converter assembly; 40, adsorption/desorption zone; 42, sensing zone; 52, chromatography zone; 44, first temperature control assembly; 46, equilibrium zone; 47, second temperature sensor; 48, second temperature control assembly; 53a, fourth temperature control assembly; 53b, fourth temperature sensor; 60, preconditioning zone; 61a, fifth temperature control assembly; 61b, fifth temperature sensor; 64, third temperature controller; 65, AFP binding site; 66, first (excitation) light filter; 68, second (emission) filter; 70, synchronous detector; 72, A/D converter; 74, microcontroller unit; 76, graph and/or log of data points; 78, flow meter. The inset is the sensing zone intercepted at 3–3, where 5–5 is the equilibrium zone; 6–6 is the chromatography zone; 7–7 is the preconditioning zone.

2. an adsorption/desorption zone 40 to trap/screen/release the explosive vapor and the coating on the internal wall for maximum performance;
3. further separation of multiple analytes by establishing a GC zone 52;
4. several temperature controllers in different zones for managing heating/cooling cycles (44, 48, 53a, 61a, and 64) and the speed from cooling to heating;
5. removal of the possible unwanted vapor form quenchers other than the target analyte, which may trigger a false alarm;
6. the proper formation of a "pulse" form highly concentrated vapor package in the adsorption/desorption process;
7. ambient light rejection at the location of inlet 20;
8. an equilibrium zone 46 to match the temperatures of both vapor and AFP to eliminate fluctuation of the signal caused by the temperature variation.

The proper combination and consideration of all elements and factors in this long list is fundamental if one expects to have a system that is able to detect a ppt concentration of explosive vapor with a minimum probability of false alarms.

4.7.1 Adsorption/Desorption Zone 40

This zone is the first screening point intended to partly separate the target vapor from unwanted components in the intake. To maximize the adsorption of target vapor, the inner glass wall can be coated with a specific film such as dimethylpolysiloxane/*diphenylpolysiloxande* mixtures or trifluoropropylmethylpolysiloxanes. This film is selected for its strong interaction with target molecules. As a result, it slows down the migration of these molecules. Some of the other unwanted molecules that are less affected by this film will migrate at the same speed as the carrier gas. The selective absorption performance of this film strongly depends on the following factors: the chemical architecture of the film, its thickness, length of the adsorption/desorption zone, the temperature of the zone, and the flow rate of the moving carrier gas. A detailed adjustment among these factors can maximize the analyte selection at this stage. A temperature-control assembly 44 and sensor 45 is needed in this zone to provide for regulation of the process. As the different components contained in the carrier gas usually have different boiling points, at a low temperature, compounds with low boiling point tend to pass through the zone unaffected while compounds with high boiling point will be held back. A thermal trap is thus created. The detailed adjustment may result in an ideal situation such that once the temperature rises to a certain point, the trapped targeted molecules can form a "pulse" of vapor in a kind of package containing a high concentration of target molecules. Selection of a proper time-shift between low- and high temperatures is critical to the formation of an ideal length of "pulse" to match the active area of AFP, and thus produce the highest possible quenching efficiency.

4.7.2 Equilibrium Zone 46

Positioned between the adsorption/desorption zone and the sensing zone containing AFP, the equilibrium zone is used to synchronize the temperature of the AFP and the target vapor package. All fluorescence-based technology is highly temperature sensitive, and thus requires a constant temperature throughout the entire assay process. Ensuring this constant temperature in the equilibrium zone is the role of the third temperature controller 64. The equilibrium zone also has specific coating to interact with the incoming analytes. The coating material can be polydimethylsiloxane, diphenyl, cyanopropylphenyl, trifluoropropyl, or polyethylene glycol. In contrast to the procedure in the adsorption/desorption zone, temperature in the equilibrium zone is adjusted so that the interaction with the coating allows the analytes to pass through this zone stably and quickly.

4.7.3 Chromatography Zone 52

As mentioned before, the adsorption / desorption zone can only partly separate the target vapor from unwanted components. The chromatography zone uses another principle—GC—to separate explosive molecules from other molecules. A special coating in this zone interacts with the "pulse" of vapor. Variation in the strength of such interactions among the contained components leads to different migration speeds and causes the components to emerge at the end of this zone at different times. Again, temperature is a critical factor, which is adjusted through the fourth control assembly 53a and the fourth sensor 53b.

4.7.4 Preconditioning Zone 60

This is the zone that interacts with the external world and with the adsorption/desorption zone. Temperature control, performed by the fifth temperature control assembly 61a and the fifth sensor 61b, is critical in this zone, whose function is to raise the vapor temperature from the point initialized by the ambient environment up to the level that minimizes the adsorption of analytes in this zone. A nonretentive coating is used in this zone to further reduce the adsorption of vapor. This coating can be chosen from a number of alkyl silanes such as octadecyltrichlorosilane (OTS), 1,1,1,3,3,3-hexamethyldisilazane (HMDS), and decyltrichlorosilane. Ambient light rejection should also be considered in this stage. This can be achieved by the use of opaque coating material.

4.7.5 Sensing Zone 42

In this zone, temperature remains at the same constant value as in equilibrium zone 46 through the shared third temperature control assembly 64, ensuring the minimum temperature fluctuation between the incoming analyte vapor and the AFP coating. The constant temperature of the AFP also stabilizes the initial fluorescent signal when the system is in standby mode.

If we consider all these stages, we see that formation of the sharp and short-duration vapor "pulse" in the adsorption/desorption zone is highly critical for the overall performance of the system. The combination of low temperature, proper coating, and an adequate accumulation time causes a short-term buildup of vapor analytes. After this gathering time, a sudden rise of temperature within 3 s from a typical value of the ambient temperature to 200°C releases the vapor and forms a quick moving vapor "pulse." Research from Nomadics Inc. scientists indicates that this "pulse" has an annulus shape, which means that most of the analytes will naturally build up at the outer ring of the pulse. This ring will create a maximum interaction with the capillary inner wall. AFP coated on the inner wall will receive the maximum possible preconcentrated vapor and thus ensure peak system performance.

A secondary means, offered by the short "pulse" and elapsed time of different analytes, may be provided to prevent a false alarm. After passing the adsorption/desorption zone and the chromatography zone, different analytes are separated and will reach the AFP at different times. The elapsed time between the moment when the "pulse" is formed in the adsorption/desorption zone and the time when the AFP response is detected can be compared with the elapsed time for the target analyte, which is known and stored in the system. A false response caused by unwanted analytes can be identified when the detected elapsed time is different from this prestored value.

The length of the AFP, which is called the *AFP band*, should be shorter than the vapor "pulse" before AFP zone rather than being the same. This is because AFP strongly interacts with or "traps" analytes when the vapor "pulse" passes by, so that fewer target vapor molecules reach the far end of the AFP zone. As a result, when there is a long AFP band, a significant number of illuminated AFP molecules at the far end are only partly quenched, leading to a strong unwanted fluorescent background. A shorter AFP with a sufficient detectable fluorescent signal level is thus preferable.

4.8 FUTURE DIRECTIONS AND CONCLUSIONS

Not long ago in the detection field, IMS and canines were the unrivaled "gold standards" of explosive detection because of their combined advantages of a detection limit down to the ppt level and a quick response time of mere seconds or fractions of a second. However, an IMS instrument, even in its miniaturized version, is still bulky, has a high cost, and requires well-trained personnel to operate it. The major challenge in miniaturization is that the core element of an IMS instrument, the drift tube, is a high precision component with a complicated design and is thus difficult to reduce in size. It has been found that peak broadening caused by field inhomogeneity, space-charge effects, and diffusion becomes more significant at small dimensions, limiting the resolution of a miniaturized drift tube (36–38). Research on a hand-held IMS device for detecting explosives and other species is still underway in a number of organizations, including Sandia National Laboratories and the University of Arizona (39), with the objective of achieving performance comparable with that of the bench-top counterpart. The motivation behind these continuing efforts is that in principle, IMS and dogs can sense an unlimited or at least a wider scope of explosive types and other nonexplosive species than other sensors. Promising results in miniaturizing IMS have been reported but still have a long way to go to realize genuinely high performance, hand-held, and cost-effective versions of IMS instruments. As for dogs, as mentioned before, they need to be well trained by professionals through a costly and time-consuming process. They can also get tired during a lengthy field operation or confused when various scents appear on scene simultaneously. However, to ensure the safety of operation personnel, dogs have obviously unique advantages and are irreplaceable when dangerous chemical species or an awkward and unsafe scene is involved. There is no doubt that the emergence of AFP has challenged both IMS devices and dogs by offering a detection limit another order of magnitude lower. This fact, combined with its miniaturized size, low cost, and ease of operation, gives the AFP sensor a powerful advantage over the two competing techniques for TNT explosive identification.

A particular feature of AFP-based fiber-optic sensing that distinguishes it from other technologies is its optical signal variation based on amplified fluorescent quenching. The fact that these variations can naturally be transferred via optical fiber offers an ideal solution for *in situ* remote detection of explosives. Existing fiber-optic network technologies can be adapted to create an explosive-detection network by multiplexing signals from a large number of fiber-optic explosive-detection sensors. This capability is particularly important for monitoring large expanses of critical infrastructure such as airports and government buildings. In the near future, it is expected that AFP-based explosive detectors, miniaturized IMS sensors, and canines will figure as the three most sensitive explosive-detection technologies and will coexist for different application scenarios. It is also expected that a properly designed explosive-vapor preconcentration unit will continue to be one of the most important elements in an AFP-based system in order to minimize the risk of false alarms.

REFERENCES

1. The 9/11 Commission, Final Report of the National Commission on Terrorist Attacks upon the United States. Executive Summary, The 9/11 Commission Report, The National Commission on Terrorist Attacks Upon the United States, Washington, DC, 2004 July. http://govinfo.library.unt.edu/911/archive/index.htm.
2. GAO. *Aviation Security: Federal Action Needed to Strengthen Domestic Air Cargo Security. Report to Congressional Requesters*. Washington, DC: United States Government Accountability Office; 2005.
3. Shea DA, Morgan D. Detection of explosives on airline passengers. CRS Report for Congress, Library of Congress RS21920; 2006 Aug 9.
4. Department of the Treasury, Bureau of Alcohol, Tobacco and Firearms. Commerce in explosives: list of explosive materials. Fed Reg 2002;67(81):20864–20866.

5. Shaw J, Seldomridge N, Dunkle D, Nugent P, Spangler L, Bromenshenk J, Henderson C, Churnside J, Wilson J. Polarization lidar measurements of honey bees in flight for locating land mines. Opt Express 2005;13:5853–5863. Available at http://www.opticsinfobase.org/oe/abstract.cfm?URI=oe-13-15-5853 (accessed year 2005).
6. Kathy B. Better Than Dogs—INEEL Technologies "Sniff Out" Smuggled Explosives. Feb 2003. http://www.inl.gov/featurestories/2003-02-28.shtml.
7. Yinon J. Field detection and monitoring of explosives. Trends Anal Chem 2002;21:292–301.
8. Gaft M, Nagli L. Standoff laser-based spectroscopy for explosive detection. Proc SPIE Conf Ser 2007;6739:673903.
9. Jimenez AM, Navas MJ. Chemiluminescence detection systems for the analysis of explosives. J Hazard Mater 2004;106A:1–8.
10. Williams VE, Lemieux RP, Thatcher GRJ. Substituent effects on the stability of Arene-Arene complexes: an AM1 study of the conformational equilibria of cis-1,3-diphenylcyclohexanes. J Org Chem 1996;61:1927–1933.
11. Yang JS, Swager TM. Fluorescent porous polymer films as TNT chemosensors: electronic and structural effects. J Am Chem Soc 1998;120:11864–11873.
12. Fido Explosives Detectors Technical Overview, FLRS Systems, Inc. http://gs.flir.com/uploads/file/products/brochures/fido%20technical%20overview.pdf.
13. Shriver-Lake LC, Donner BL, Ligler FS. On-site detection of TNT with a portable fiber-optic biosensor. Environ Sci Technol 1997;31:837–841.
14. Lim DV. Detection of microorganisms and toxins with evanescent-wave fiber-optic biosensors. Proc IEEE 2003;91:902–907.
15. Jung CC, Saaski EW, McCrae DA, Lingerfelt BM, Anderson GP. RAPTOR: A fluoroimmunoassay-based fiber-optic sensor for detection of biological threats. IEEE J Sensors 2003;3:352–360.
16. Shriver-Lake LC, Anderson GP, Golden JP, Ligler FS. The effect of tapering the optical fiber on evanescent-wave measurements. Anal Lett 1992;25:1183–1199.
17. Shriver-Lake LC, Breslin KA, Charles PT, Conrad DW, Golden JP, Ligler FS. Detection of TNT in water using an evanescent-wave fiber-optic biosensor. Anal Chem 1995;67:2431–2435.
18. Albert KJ, Walt DR. High-speed fluorescence detection of explosives-like vapors. Anal Chem 2000;72:1947–1955.
19. Albert KJ, Myrick ML, Brown S, James D, Milanovich F, Walt DR. Field-deployable sniffer for 2,4-dinitrotoluene detection. Environ Sci Technol 2001;35:3193–3200.
20. Alan EG, Judy KP, Mark LS, Ray MVW, William KR, Jani CI, Gregory DL. Detection device for high explosives. US Patent 5,157,261. 1992 Oct 20.
21. Orghici R, Willer U, Gierszewska M, Waldvogel S, Schade W. Fiber-optic evanescent field sensor for detection of explosives and CO2 dissolved in water. Appl Phys B Lasers Opt 2008;90:355–360.
22. Schade W, Orghici R, Willer U, Waldvogel S. Fiber-optic evanescent-field sensor device for CO2 and explosive detection. In: Sampson D, Collins S, Oh K, Yamauchi R, Editor. 19th International Conference on Optical Fibre Sensors, Proceedings of SPIE; 2008. 7004: 700431–1. http://dx.doi.org/10.1117/12.786017.

23. Dakin J, Culshaw B, Editors. *Optical fiber sensors*. Boston, London: Artech House Inc.; 1997;4:Ch. 7.
24. Utzinger U, Richards-Kortum R. Fiber-optic probes for biomedical optical spectroscopy. J Biomed Opt 2003;8:121–147.
25. Plaza P, Quy Dao N, Jouan M, Fevrier H, Saisse H. Simulation et optimisation des capteurs à fibres optiques adjacentes. Appl Opt 1986;25:3448–3454.
26. Probes that make sense, Ocean Optics, Inc. www.oceanoptics.com.
27. Nguyen H, Li X, Ma J, Bock WJ, Wang ZY. Fiber-optic detection of explosives using readily available fluorescent polymers. Macromolecules 2009;42:921–926.
28. Gang Q, Li X, Wang ZY. Visible and near-infrared chemosensor for colorimetric and ratiometric detection of cyanide. J Mater Chem 2009;19:522–530.
29. Ma J, Li X, Nguyen H, Bock WJ, Wang ZY. Dramatic enhancing of fluorescent light collection efficiency from a very thin layer of polymer on a planar glass substrate. In: Vallée R, Piché M, Mascher P, Cheben P, Côté D, LaRochelle S, Schriemer HP, Albert J, Ozaki T, Editors. Photonics North 2008. Proceedings of SPIE; 2008 Jun 2–4; Montreal, Quebec, Canada; 7099: 70991Z-1-6. http://dx.doi.org/10.1117/12.804776.
30. Snyder AW, Love JD. *Optical Waveguide Theory*. London: Chapman and Hall Ltd.; 1983.
31. Ma J, Bock WJ, Cusano A. Insights into tunnelling rays: outperforming guided rays in fiber-optic sensing device. Opt Express 2009;17:7630–7639. Available at http://www.opticsinfobase.org/abstract.cfm?URI = oe-17-9-7630.
32. Ma J, Bock WJ. Reshaping a sample fluid droplet: toward combined performance enhancement of an evanescent-wave fiber-optic fluorometer. Opt Lett 2007;32:8–10.
33. Organofunctional silanse for powerful connections. Wacker Chemie AG. http://www.wacker.com/cms/media/publications/downloads/6085_EN.pdf.
34. Frank JC. Hand held explosive detection system. US Patent No. 5,138,889. 1990 Oct 11.
35. Craig AA, Colin JC, Mark EF, Michael JF, Marcus JL, Dennis KR, Mark GR, Eric ST. Vapor sensing instrument for ultra trace chemical detection. US Patent No. 6,558,626 B1. 2003 May 6.
36. Spangler GE, Vora KN, Carrico JP. Miniature ion mobility spectrometer cell. J Phys E: Sci Instrum 1986;19(3):191–198.
37. Baumbach I, Berger D, Leonhardt JW, Klockow D. Ion mobility sensor in environmental analytical chemistry—concept and first results. Int J Environ Anal Chem 1993;52(1–4):189–193.
38. Xu J, Whitten W, Ramsey JM. Space charge effects on resolution in a miniature ion mobility spectrometer. Anal Chem 2000;72(23):5787–5791.
39. Babis JS, Sperline RP, Knight AK, Jones DA, Gresham CA, Denton MB. Performance evaluation of a miniature ion mobility spectrometer drift cell for application in hand-held explosives detection ion mobility spectrometers. Anal Bioanal Chem 2009;395:411–419.

CHAPTER FIVE

Photonic Liquid Crystal Fiber Sensors for Safety and Security Monitoring

TOMASZ WOLINSKI
Faculty of Physics, Warsaw University of Technology, Warszawa, Poland

5.1 INTRODUCTION

In the past decade, photonic crystal fibers (PCFs) (1–3) have attracted an increasing scientific interest owing to a great number of potential applications, including that for fabrication of new class of in-fiber devices. PCFs are manufactured as an array of silica tubes and rods, which are then heated to around 2000°C and drawn down to the fiber. Their core is usually made by a defect in the periodical structure of the PCF cross section (missed or additional rod/tube). In this way, either a hollow-core PCF or a solid core PCF can be manufactured. Optical wave guiding in a PCF is governed by one of the two principal mechanisms responsible for light trapping within the core. While the first one is a classical propagation effect based on the modified total internal reflection (mTIR or index guiding) phenomenon, which is well known and similar to the wave guiding effect within a conventional fiber, the second, referred to as the *photonic band gap (PBG) effect*, occurs if the refractive index of the core is lower than the mean reflective index of the cladding region.

Infiltrating the air holes with different materials allows for creation of a special class of infused PCFs with enhanced optical properties. In this context, the application of liquid crystals (LCs) has gained particular attention, resulting in new highly

Photonic Sensing: Principles and Applications for Safety and Security Monitoring, First Edition.
Edited by Gaozhi Xiao and Wojtek J. Bock.

tunable photonic structures, called *photonic liquid crystal fibers* (PLCFs) (4), in which light propagation conditions are determined by both: an LC "guest" material and a PCF "host" structure. While the properties of effective PLCF structures strongly depend on LC molecular orientation, the techniques for efficient (e.g., UV-light-induced) orientation of an LC within PCF holes have to be developed (5). In this way, PLCFs benefiting from a merger of passive PCF host structures with active LC guest materials are responsible for the diversity of new propagation, spectral, and polarization properties.

Light guiding dynamics, including switching between both mechanisms of propagation, can be simply achieved in PLCFs, thanks to the unique properties of an infiltrating LC. Owing to high electro-, magneto-, and thermo-optic responses of LCs, their refractive indices may be relatively easily changed either by temperature or external physical fields. In this context, thermal and electrical tuning along with unusual spectral and polarization properties of PLCFs have been studied over the last few years (6–15). After the demonstration of tunable PBGs using thermo-optic tuning of the LC (4), one of the most spectacular phenomena was successful realization of temperature-induced switching between two mechanisms of light propagation (7). Recently, tunability in highly birefringent (HB) solid core PLCF has been also demonstrated (10, 11). The latter achievement is related to polarization properties of the analyzed photonic structures that can be dynamically tuned via external-field-induced reorientation of LC molecules. While analyses of polarization-dependent phenomena are extremely important, polarization and depolarization phenomena were intensively investigated (14–16), also in view of potential applications for a compensator of polarization mode dispersion. A tunable polarizer (17), tunable single-polarization operation (9), and also tunable birefringence (10, 18, 19) have been experimentally demonstrated. Owing to high sensitivity to external conditions, PLCFs were also used for switching and sensing applications, (20–23), when subjected to the influence of temperature, electric field, or hydrostatic pressure. Resultant devices can find potential applications as threshold sensors, in birefringence measurements, temperature and pressure sensors, all-optical multiparameter sensors, tunable filters, and all-fiber attenuators. For this purpose, PCFs infiltrated with doped nematic liquid crystals (NLCs), including cholesteric dopant mixtures, azo-compound, and dyes, can also be successfully used (14, 24). Moreover, an intensive work on long-period fiber-grating sensors based on LC microstructures has been initiated (25, 26). In addition, research on high-index-glass-based PLCFs has also been performed (27). It allows for index-guiding propagation in a wide range of LC guest materials and enables continuous and repeatable broadband birefringence tuning, as well as continuous adjustment of polarization-dependent losses (which can be potentially used for the development of a new type of all-in-fiber polarization controllers).

The aim of this chapter is to review and demonstrate the most spectacular features of PLCFs that are used for dynamically controlled and tunable photonic devices that might be used for safety and security monitoring. The chapter is organized as follows: Section 5.2 summarizes all the guest (LC) and host (PCF) structures along with experimental setups used to investigate PLCF properties.

Section 5.3 (Principle of operation) discusses the propagation and LC molecules orientation in PLCFs. Tuning possibilities of PLCFs are presented in Section 5.4, and prospective photonic devices on PLCFs including environmental sensors, tunable attenuators, polarizers, filters, and phase shifters are demonstrated in Section 5.5. Final conclusions on future perspectives (Section 5.6) summarizes the chapter.

5.2 MATERIALS AND EXPERIMENTAL SETUPS

In order to study the optical properties of PLCFs, different types of both silica glass and high index glass PCF structures have been used, which are summarized in Table 5.1. The first type of the PCFs is an isotropic prototype silica glass fibers manufactured at the Maria Curie-Skłodowska University (UMCS) in Lublin, Poland (positions 1–4 in Table 5.1). PCF hosts of the second type are commercially available anisotropic HB fibers: Blazephotonics PM-1550-01 PCF and LMA-PM5 PCF (produced by NKT Photonics, Denmark). The third type of host fibers is a specially designed high index PCF that was manufactured at the Institute of Electronics Materials Technology (ITME), Warsaw (Poland). According to the ITME internal name system used to distinguish fibers, the fiber is denoted as PCF14(6) and is made by the stack-and-draw technique from lead-bismuth-gallate (Pb-Bi-Ga)-based glass designated as PBG08 (28). The transparent window of this glass

TABLE 5.1 Specification of Host PCFs with Pictures of Their Cross Sections

No.	Name	Physical Parameters
1	070124	$\Lambda = 6.5$ μm; $d_h = 3.9$ μm; $d_h/\Lambda = 0.6$ μm; $m = 3$
2	070123	$\Lambda = 6.5$ μm; $d_h = 4.1$ μm; $d_h/\Lambda = 0.63$ μm; $m = 6$ outside diameter: 119–127 μm
3	061221P2	$\Lambda = 6.2$ μm; $d_h = 3.8$ μm; $d_h/\Lambda = 0.61$ μm; $m = 8$ outside diameter: 126–138 μm
4	1023	$\Lambda = 6.5$ μm; $d_h = 4.8$ μm; $d_h/\Lambda = 0.74$; $m = 9$ outside diameter: 125 μm
5	PM-1550-01	d_h of big holes = 4.5 μm; d_h of small holes = 2.2 μm; noncircular core combined with the large air–glass refractive index step creates strong birefringence
6	LMA-PM5	$d_c = 5.0$ μm combines stress rod applied birefringence $\sim 1.5 \cdot 10^{-4}$
7	PCF14(16)	$\Lambda = 7.6$ μm; $d_h = 5.2$ μm; $d_h/\Lambda = 0.68$; $m = 6$ outside diameter: 125.4 μm

1) 2) 3) 4) 5) 6) 7)

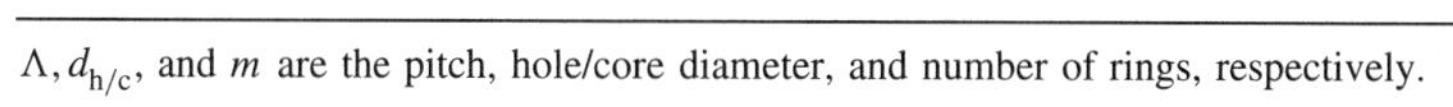

$\Lambda, d_{h/c}$, and m are the pitch, hole/core diameter, and number of rings, respectively.

spreads from 450 (UV absorption cutoff) to 4800 nm (multiphonon IR absorption edge) corresponding to its attenuation of an order of 10 dB/m (28).

As an "active" element for the fabrication of the PLCFs presented in this chapter, we used three different classes of LCs that exhibited nematic, chiral nematic, and ferroelectric smectic C phases. Each of the LC mixtures used was synthesized at the Military University of Technology (MUT) in Warsaw (Poland).

NLCs have only a long-range orientation order of the molecules long axes, but their centers of gravity are randomly distributed. To fabricate PLCFs, we used NLCs with low- (LB), medium- (MB), and high- (HB) birefringence Δn. Table 5.2 provides a summary of the electro-optical properties for all the NLC materials used. NLCs of low birefringence have been prepared in the form of prototype mixtures designated as 1800, 1110, 1550, and 1550A (obtained by doping the 1550 material with propyl propenylbicyclohexane). These LB NLC mixtures are especially interesting for silica glass fibers, as their ordinary refractive index (n_0) in the specific temperature ranges is lower than the refractive index of the silica glass (n_{SiO_2}). As the NLCs with medium birefringence, two commonly known nematics, namely, 5CB and 6CHBT were chosen. The thermal characteristics of the refractive indices for these LB and MB NLCs are presented in Figure 5.1. In

TABLE 5.2 Electro-optical Properties of the Investigated Nematic LCs (Measured at 589 nm at ~25°C)

No.	Symbol	T_c(°C)	n_0	n_e	$\Delta n = n_e - n_0$	$\Delta\varepsilon$
1	1110	42	1.459	1.506	0.047	<0
2	1550	77.3	1.461	1.523	0.062	3.2
3	1550A	79	1.464	1.529	0.065	—
4	1800	63	1.460	1.518	0.058	—
5	6CHBT	43	1.52	1.67	0.15	8
6	5CB	33	1.54	1.71	0.16	16.1
7	1294-1b	150	1.50	1.81	0.31	>0

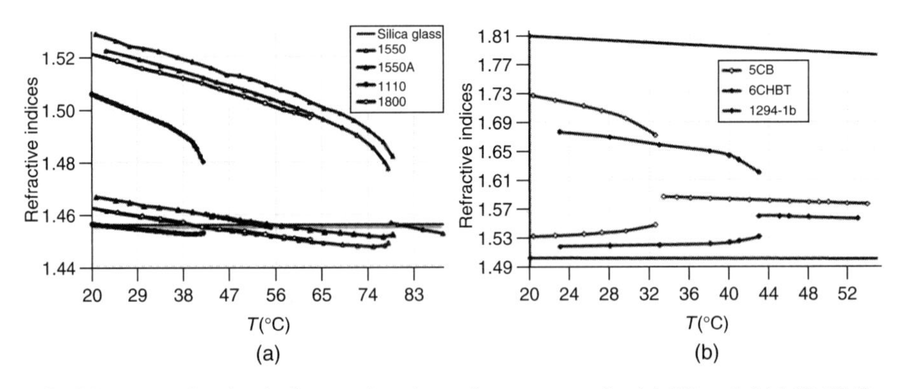

FIGURE 5.1 Refractive indices as functions of temperature for (a) LB and (b) MB NLCs.

addition, three experimental NLC mixtures with high birefringence, abbreviated as 1294-1b, 1658A, and 1679, were chosen, and so far there have been limited data on their refractive indices.

Host PCFs were also filled with chiral nematic liquid crystals (ChNLCs), that is, nematics doped with chiral compounds—optically active dopants (OADs), for example, cholesteric dopants, azo-compounds, and dyes that can modify the switching and sensing possibilities of PLCFs. ChNLCs), owing to their spatial periodicity, are characterized by unique optical properties such as selective Bragg reflection, circular dichroism, and optical activity. The selective Bragg reflection is a total reflection of circularly polarized light (having the same handedness as the helical pitch of the chiral structure), occurring at the wavelength $\lambda = \overline{n}P$ where $\overline{n}$ is the mean refractive index and P the pitch (defined as a distance it takes for the director to rotate one full turn in the helix). For ChNLCs, we used LC mixtures marked as: PW500, PW600, PW700, PW1000, and PW1500, which are characterized by significantly reduced temperature sensitivity of the selective Bragg reflection (Fig. 5.2a). They were composed of alkylcyclohexyl bicyclohexylbenzene nitryles and isothiocyanates base mixture W-1816, doped with optically active dopants (Fig. 5.2a and b), and formulated in concentrations presented in Table 5.3.

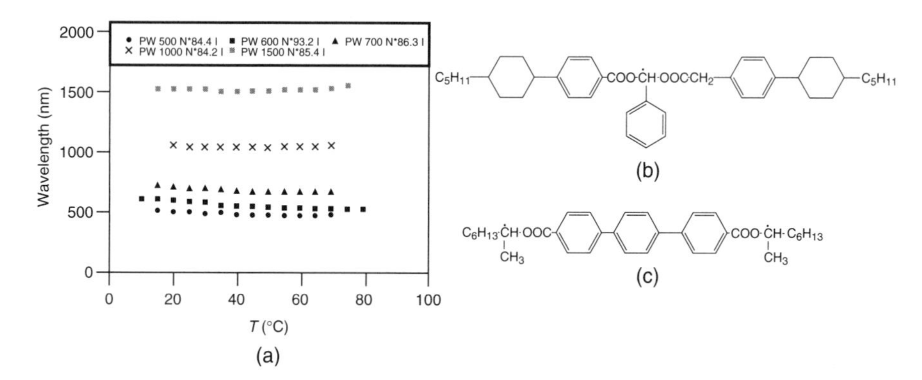

FIGURE 5.2 (a) The selective Bragg reflection for ChNLCs optically active dopants, (b) OAD1 is right handed and with helicoidal twisting power (HTP) = 67.2 μm^{-1}, (c) OAD2 is left handed and with HTP = 30.7 μm^{-1}.

TABLE 5.3 Parameters of ChNLCs

ChNLCs	λ_{max} (nm)	N*- Iso	OAD1 (%weight)	OAD2 (%weight)	Pitch (μm)
PW500	500	84.4	10	0	0.31
PW600	600	83.2	9.4	0	0.40
PW700	700	86.3	5	0	0.43
PW1000	1000	84.2	1.3	7	0.60
PW1500	1500	85.4	0	6	0.95

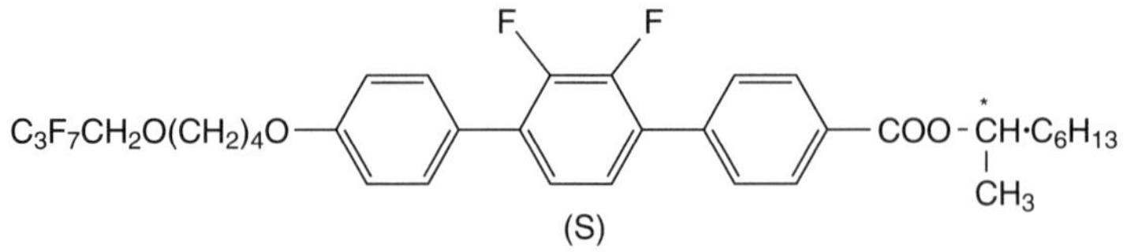

FIGURE 5.3 Chemical structure of the MT2_B15 ferroelectric liquid crystal molecule.

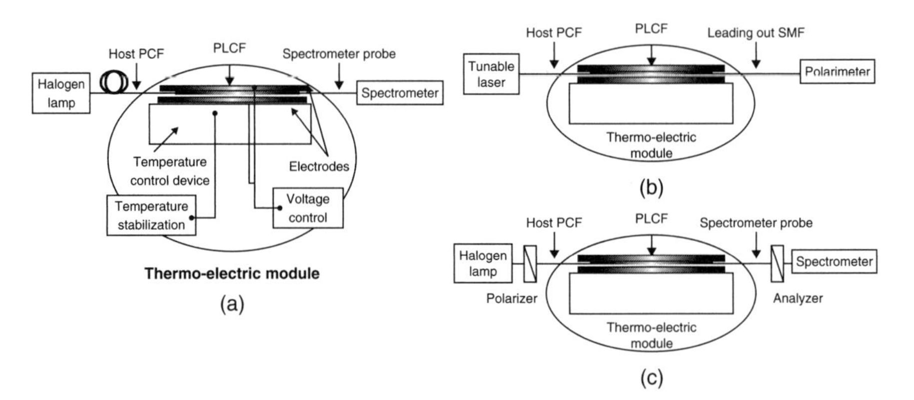

FIGURE 5.4 (a–c) The experimental setup for temperature and external electric-field-induced measurements.

In addition, chiral smectic C* ferroelectric liquid crystals (FLCs) were investigated as a guest material. The FLC material was an experimental mixture MT2_B15 (chemical structure shown in Figure 5.3), and it has a smectic C* phase at room temperature up to 63°C.

The next part of the section describes relevant experimental procedures to combine PCFs with LCs. Figure 5.4 presents experimental setups to conduct temperature/electric effects measurement properties. The terminal section of the PCF filled with an LC was placed into a thermo-electric module.

Temperature regulation was in the range of 10–120°C with the Testo 735 precise thermometer (0.05°C resolution and 0.1°C accuracy). Each measurement was performed after the stabilization of temperature in the Peltier modules. In order to apply an external electric field to the samples, a function generator (1 μHz to 10 MHz) with an amplifier and two electrodes was used. Then electrical control was achieved within a typical range of 0–1000 V with a frequency of 2 kHz. On the basis of the requirements, two different sources of light were used during the experiments: a halogen lamp for the VIS part of the spectrum and a tunable laser source (Tunics Plus CL) operating at the third optical window (spectral range 1500–1640 nm) for the IR range. An output optical signal coming out of the PLCF was analyzed by the Ocean Optics HR4000 fiber-optic spectrometer. The measurement apparatus also included a tunable laser. In order to analyze the polarization properties, the polarization analysis of the outgoing light from the PLCFs was

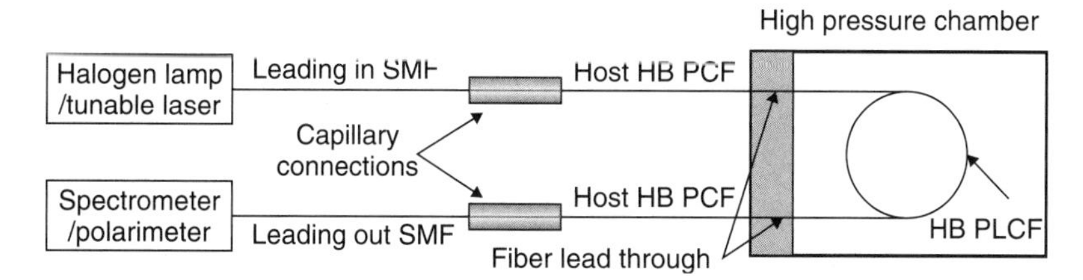

FIGURE 5.5 The experimental setup for hydrostatic-pressure-induced measurements.

conducted in two configurations as shown in Figure 5.4b and c. In addition, the state of polarization (SOP) could be visualized on the Poincare sphere by using the PAT 9000B polarimeter (Tektronix).

A special sensing head was used to investigate the influence of hydrostatic pressure on light propagation in the PLCFs (Fig. 5.5). The first section of ~20 mm of a 35-cm-long PCF was filled with an LC, and then the LC was moved to the middle part of the PCF by using pressurized air. The PLCF was next introduced into a high pressure chamber through a specially designed lead through system. The PLCF sample was connected to the white light source and to the spectrometer by using capillary connections and single-mode (SM) leading fibers. Finally, hydrostatic pressure (ranging from 20 to 70 MPa) was applied to the chamber, modified, and controlled with a dead-weight piston manometer.

5.3 PRINCIPLE OF OPERATION

5.3.1 Mechanism of Propagation in a PLCF

An LC introduced into a PCF significantly changes its guiding properties. According to the optical fibers theory, a PCF can be approximated by a step-index single-mode fiber (SMF) and described by its core refractive index (equal, in this case, to the refractive index of the silica glass from which the PCF is made, n_{PCF}) and an average index of refraction of the holey cladding that takes into account the holes region in the cladding $n_{\mathrm{PCF,\,LC}}$.

Let us denote a difference between the core and the cladding refractive indices for the PLCF as Δn_i, where $i = 1, 2, 3$. Then the following three cases can be distinguished (also illustrated in Figure 5.1):

Case 1: The value of the cladding region's refractive index decreases compared to the PCF with air holes, but light is still being propagated by index guiding also described as the mTIR.

Case 2: The value of the cladding region refractive index is equal to Δn_{PCF}, and the propagation disappears.

Case 3: The difference between the core and cladding region indices is negative and light propagation is only possible owing to the PBG mechanism. Hence, in this case only selected wavelengths are being guided.

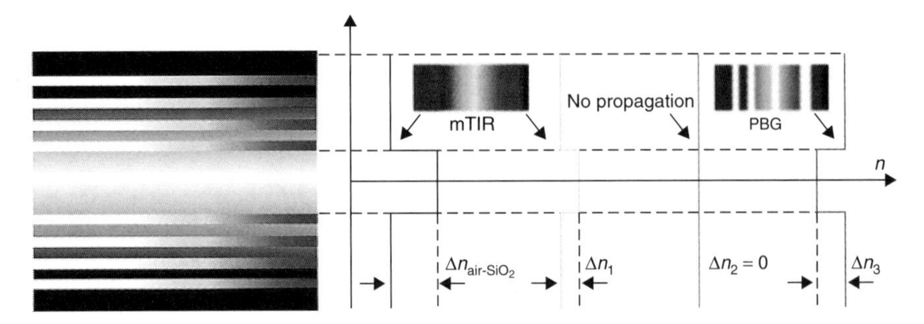

FIGURE 5.6 Shaping the contrast between core and cladding refractive indices of the PCF (from $\Delta n_i > 0$ to $\Delta n_i < 0$) by filling it with the LC characterized by different magnitude of refractive indices, n_{LC}.

According to Figure 5.6, depending on the type of the LC material tested, different propagation effects could be observed in the PLCF: from the mTIR mechanism (e.g., occurring for the 1800 LC mixtures) to the PBG effect.

In addition, taking into consideration the isotropic properties of the host LC, one more case should be discussed: when the PCF is made of glass whose refractive index is below the extraordinary and above the ordinary refractive index of the LC. Numerical simulations have proved that in such fibers, guiding properties strongly depend on the alignment of LC molecules and polarization of light. In particular, if all molecules are perpendicular to the fiber axis, one polarization is guided by the mTIR mechanism, because for orthogonal polarization, only selected wavelengths can be guided by the PBG effect (Fig. 5.7). In this situation, the mechanism of light propagation is referred to as the *hybrid-guiding mechanism*, for which the mTIR guiding and the PBG effect act together to confine light in a single guided mode.

5.3.2 LC Arrangement in PCF

The light propagation mechanisms inside PLCFs and their tuning range essentially depend on the orientational configuration of an LC within the cladding air holes and the photonic structure itself. In order to analyze the LC molecules orientation in the PCF, each of the microholes filled with an LC can be treated as a single capillary. For cylindrical confinement, there are several well-known possible configurations: planar, transverse, and a combination of two previous structures, the radial-escaped or axial geometry (Fig. 5.8). Typically, the PCFs are infiltrated with LCs by capillary action, just by immersing one end of the fiber in the container with an LC for several minutes. For this technique, the LC orientation near the microhole walls is determined by the anchoring conditions, while in the central part of the microholes, the LC molecules are oriented parallel to the axis of the capillary (flow-induced orientation). Thus, in practice, planar or axial orientations dominate the radial alignment.

However, the LC molecule orientation within the fiber holes might differ with time and temperature of infiltration. It may also vary in the adjacent holes if their

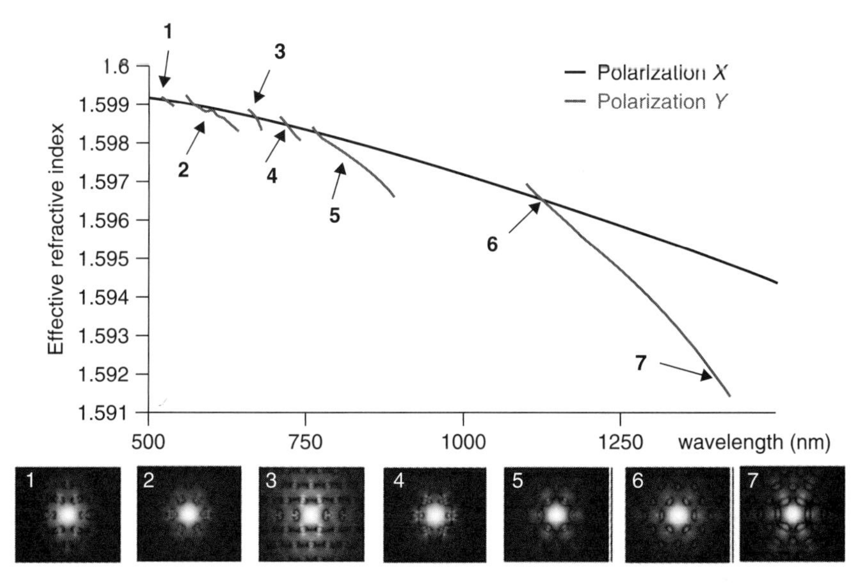

FIGURE 5.7 Effective indices and PBG mode profiles calculated for IEMT-PCF18 fiber filled with 5CB with the assumption of transverse alignment of LC molecules—simultaneous polarization-dependent propagation of the PBG and mTIR modes.

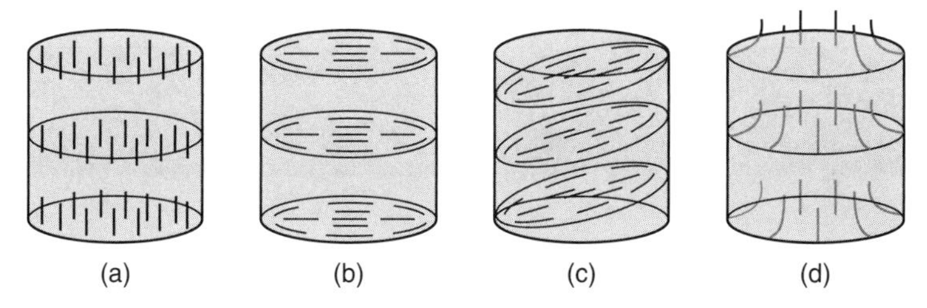

FIGURE 5.8 LC alignment configurations realized in cylindrical microcapillaries with (a–c) planar and (d) homeotropic anchoring conditions: (a) axial- (b) transversal- (c) tilted- (d) escaped radial- configuration.

diameters were slightly different owing to fluctuations during the drawing process. To ensure a high level of repeatability and stability in PCFs infiltration with LCs, novel technologies for LC molecules orientation control should be applied. The LC alignment in PCF tubes has some peculiarities owing to the fact that cylindrical geometry is relatively "unusual" for LCs, as compared to standard planar geometry normally used for display applications. First, the LC trapped in narrow tubes is strongly and elastically deformed. Thus together with anchoring, elastic forces, depending on the shape and size of tubes as well as LC orientational elasticity, strongly influence the resultant alignment configuration. Secondly, the inner surface of these tubes is not accessible for mechanical (rubbing, molding, etc.) and particle beam alignment treatments (29). The photoalignment (5, 30, 31) seems to be the unique method for this case, because light is capable to reach the tubes passing through the glassy body of the PCF.

In order to examine the LC arrangement in PCF, we used cylindrical capillaries of 8–25 μm in diameter, made of silica glass (manufactured at Maria Curie-Skłodowska University, Lublin, Poland). By coating the internal capillaries surfaces with an appropriate aligning material, we were able to set homeotropic and planar boundary conditions. Homeotropic anchoring has been realized by using *n*-octadecyltriethoxysilane (Aldrich). In the tubes with planar anchoring, LC alignment was generated and manipulated with a photoalignment technique, and as a basic photoaligning material, we used polyvinyl cinnamate (Aldrich) and cellulose 4-pentryloxycinnamate (CPC) (32) (synthesized in Institute of Bioorganic and Petrol Chemistry of NASU, Kyiv, Ukraine). In addition, we used selected diazodyes (33) and bis-methacrylic polymers (34) that also showed a good photoaligning effect. As a result, a variety of LC orientational configurations could

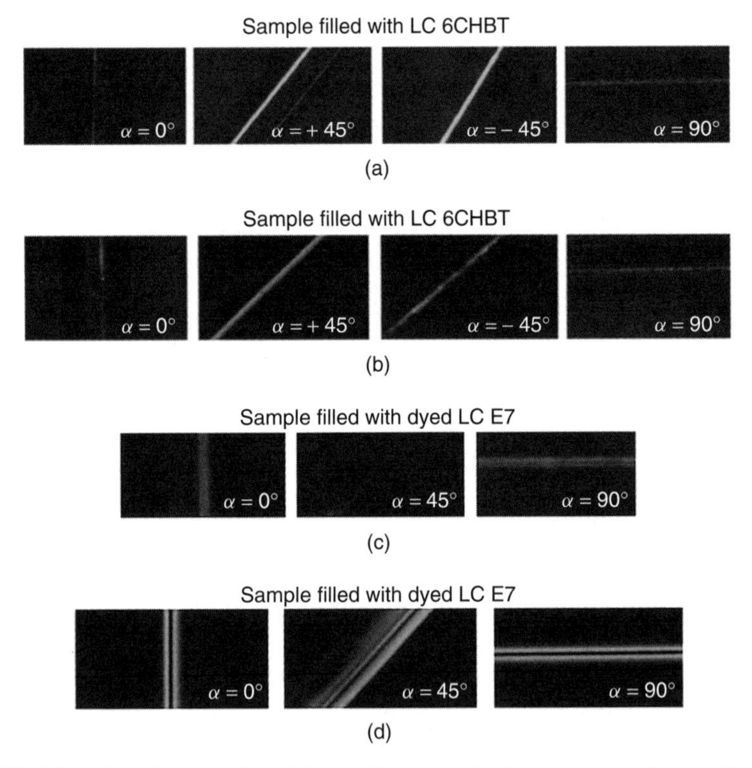

FIGURE 5.9 Microphotographs of the capillary samples between two of crossed polarizers having, accordingly, vertical and horizontal polarization axis, α. By setting the planar boundary conditions insides capillaries, (a) the planar, (b) transversal, and (c) tilted alignment configuration could be realized. In capillaries with homeotropic boundary conditions, (d) the radial alignment configuration was observed.

be realized. The realized alignment structures were studied by means of polarizing microscopy involving the host–guest effect in LC systems. In the case of homeotropic anchoring, an escaped radial configuration is realized with radial (homeotropic) and parallel alignment in the periphery and central part, respectively (Fig. 5.8d). In the capillaries with planar anchoring, the alignment configuration was controlled by the photoalignment technique. On the basis of the polarization direction of light, planar, transversal, and tilted alignment structures were realized (Fig. 5.8a–c). In Figure 5.9, the microphotographs of the capillary samples with a different LC alignment configuration are presented. We predict that the field of LC structures obtained by the photoalignment technique is much broader, and our further attempts will be aimed at their realization. The approach developed in this experiment can be further applied to control LC alignment in PLCFs, showing a big promise for various modern photonic applications.

5.4 TUNING POSSIBILITY

5.4.1 Thermal Tuning

Refractive indices of LCs are temperature dependent, and therefore by using LCs, continuously tunable functionalities of the PLCF can be obtained. Thus temperature-dependent refractive indices of LCs are fundamentally interesting and practically important for optimizing the photonic sensors employing LCs.

A great majority of LCs are characterized by both refractive indices higher than the refractive index of the silica glass used for the PCF fabrication process. As a result, in PLCFs, the effective index of cladding is usually higher than the refractive index of the fiber core, so only propagation based on the PBG effect is possible. Then the positions of PBGs in the transmission spectrum of the PLCF are determined by the refractive indices of the LC and the background material (silica). As optical properties of LCs can be relatively easily modified with temperature, the thermal tuning of the position of PBGs in the transmission spectrum of PLCF is possible. The examples of such thermal tuning of PCF filled with MB LC can be found in References 4, and 35. After the demonstration of tunable PBGs using thermo-optic tuning of the MB LC, one of the most spectacular phenomena was the successful realization of temperature-induced switching between two mechanisms of light propagation. By tuning the special design PLCF with temperature, it was possible to obtain reversible changes from PBG guiding to mTIR guiding. Such a behavior results from a specially designed and reoriented LB LC whose ordinary refractive index decreases with temperature from the value above the refractive index of silica glass (PBG guiding), through the value that matches the index of silica glass, and finally to the value below the refractive index of silica glass (index guiding) (7).

Recently, there has been new interest in infiltrating PCFs with chiral nematics (35). ChNLCs are very interesting materials to infiltrate PCFs, as they are characterized by unique optical properties such as the selective Bragg reflection, circular

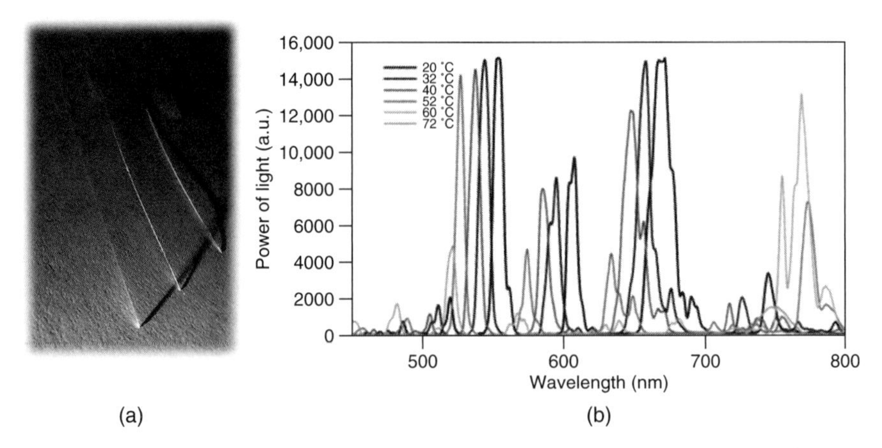

FIGURE 5.10 (a) Side views of the investigated PCFs (from left to right) 5—ring PCF (empty), PCF + PW700 ChNLC, PCF + PW500 ChNLC. (b) Transmission spectra of PCF with three rings of holes filled with PW500 (2.5 cm).

dichroism, and optical activity. Figure 5.10a presents the selective Bragg reflection observed along the 070124 PCF (Table 5.1) with five rings of holes filled either with PW500 or PW700, which selectively reflect light at the wavelengths 500 and 700 nm, respectively. As the mean refractive index of the ChNLC is higher than the effective refractive index of the fiber core, light can be guided owing to the PBG propagation mechanism. We have also observed PBGs tuning under the influence of temperature in the PCFs filled with a ChNLC, which is typical for PLCFs. The PCF with three rings of holes (Table 5.1) was filled by capillary forces with the PW500 LC in a 2.5 cm PCF section. We noticed that PBGs were moving into shorter wavelengths when the temperature was increasing. The propagation for this sample was very good, and the PBG tuning phenomenon was repeatable (Fig. 5.10b). It can be pointed out that during a change in temperature, the width of PBGs (for 500–700 nm range) was stable. Moreover, the PCF filled with the PW500 mixture demonstrates a possibility of optical filtering by temperature-induced tuning of PBGs and attenuation of the fiber in a wide range of temperatures from at least 20 to 60°C is very low. Comparing the spectrum of light for the same PCF filled with 5CB LC nematic with different ChNLCs: PW500, PW600, PW700, PW1000, or PW1500 (Fig. 5.11), the power of light for PBGs is lower for 5CB than that for ChNLCs. Consequently, this property may be used to propose a new idea of a fiber-optic filter that can be prospectively used in optical processing systems.

By filling the PCF structure with chiral smectic C* ferroelectric liquid crystals (FLCs), the PBG position in the transmission spectrum may be tuned via temperature as well. The molecules of an FLC were infiltrated into an experimental isotropic PCF 070123 (Table 5.1). The temperature of the photonic ferroelectric liquid crystal fiber (PFLCF) sample was varied between 30 and 55°C to investigate the tuning properties of PBGs in the smectic C* phase as a function of temperature (Fig. 5.12). By increasing the temperature of the sample, we noticed the blueshift

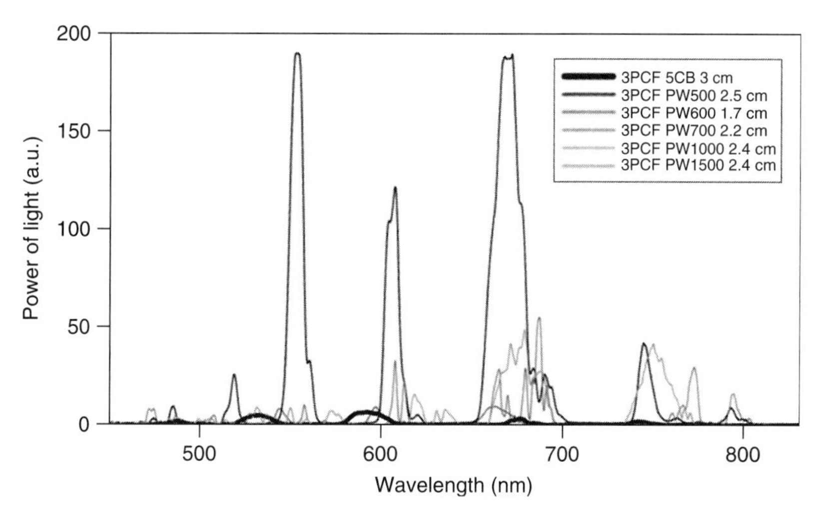

FIGURE 5.11 Comparison of selective wavelengths in transmission spectra of PCFs with three rings of holes filled with PW500, PW600, PW700, PW1000, PW1500, and 5CB.

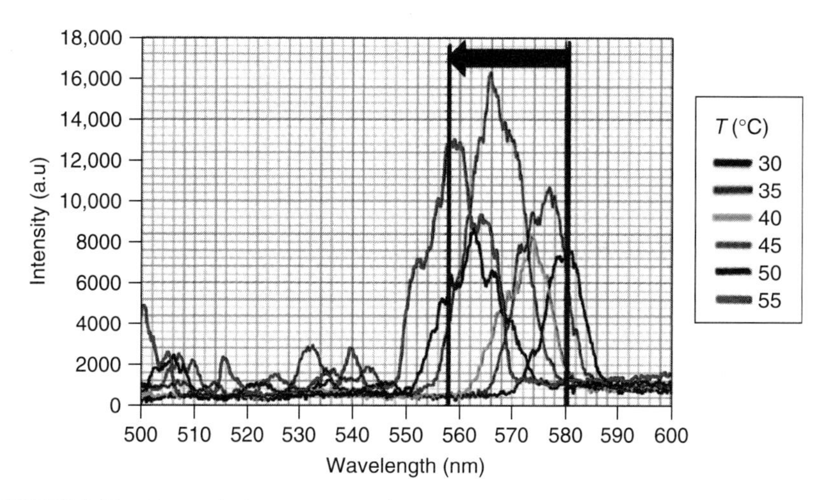

FIGURE 5.12 Transmission spectrum of a PFLCF under the influence of temperature.

effect in the transmission spectrum. The observed shift was ~20 nm from 580 (in 30°C) to 560 nm (in 55°C).

5.4.2 Electrical Tuning

PBGs positions in a PLCF depend on the guest LC refractive index. This value can be adjusted not only by temperature changes but also by molecules orientation modification. The simplest way of dynamical modification of LC orientation is applying an external electric field. The first electrical tuning in a hollow-core PLCF was demonstrated in 2004 (11). By applying external voltage to the PLCF, an electrically tunable fiber-optical switch with over 30 dB attenuation at $60V_{rms}$ for a He–Ne laser beam was achieved. Later, the electric tuning of PBGs in solid core PLCF was presented, for example, in Reference 12.

How the position of the PBGs in the transmission spectrum of the PLCF can be controlled by an external electric field is demonstrated in Figure 5.13. Particularly, the PLCF is able to operate between two states of the PBGs, namely, the off- and on-voltage state. This behavior significantly differs from temperature tuning of the PLCF, where the position of PBGs can be smoothly changed.

The first effect that was observed after increasing the voltage was the decay of optical power transmitted in the PLCF. For the PCF filled with 5CB, increasing the voltage to 41 V (2 kHz) caused almost complete decay of light at the output of the fiber. In the PLCF manufactured by using the 1294-1b mixture, the maximum decay of the output optical power was observed at 160 V. A further voltage increase resulted in the rise of optical power transmitted in the PLCF but the positions of PBGs were different from that in the off-voltage state (Fig. 5.13).

The influence of electric field on PLCFs is different for below and above the specified threshold voltage (V_T) values. For a PLCF filled with 5CB, $V_T \approx 41$ V, and for one filled with 1294-1b, $V_T \approx 160$ V (in practice, value of the threshold voltage V_T can be adjusted by using different PCFs and LCs). Initially, an increase in voltage to V_T causes gradual decay of optical power. However, a further voltage increase above the threshold value induces electrically tuned PBGs in PLCFs (as seen in Figure 5.13). As a result, the influence of the E-field strongly depends on the operating wavelength. The examples of responses to increasing voltage for selected wavelengths are shown in Figure 5.14.

The choice of the operating wavelength may depend on perspective applications. From the practical point of view, there are four useful choices of operating wavelengths.

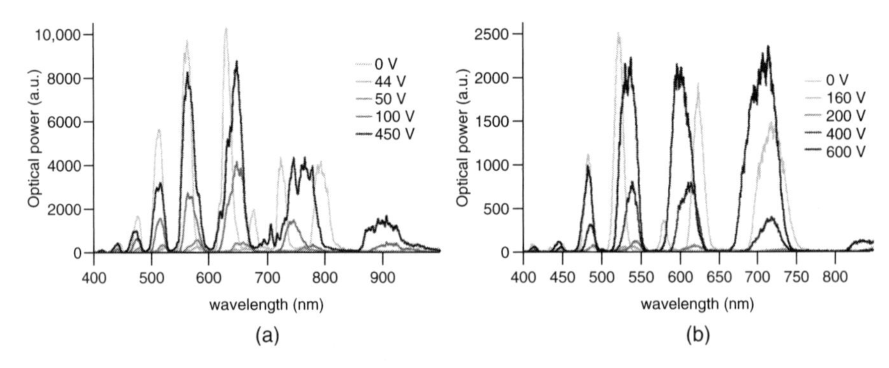

FIGURE 5.13 (a and b) Selective light propagation in PLCF filled with 5CB (electrical tuning of PBGs position in the PLCF) and with 1294-1b mixture (applied voltage results in decay of the optical power transmitted in PLCF).

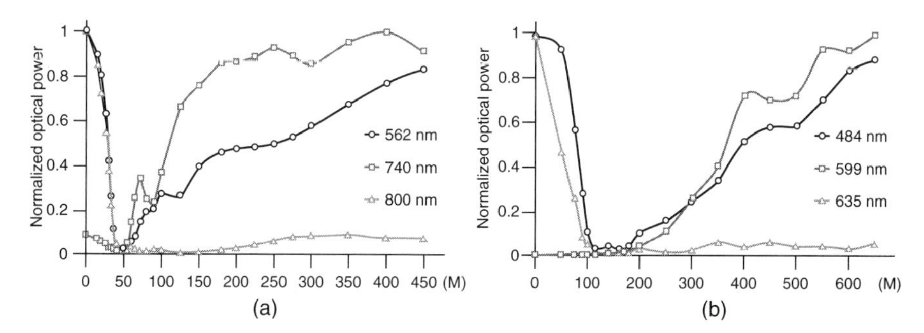

FIGURE 5.14 Normalized optical power for selected wavelengths in function of applied voltage for PCF filled (a) with 5CB and (b) with 1294-1b mixture.

1. With "linear" dependence in the range 0 V–V_T—to measure low intensities of the E-field (562 and 800 nm in Figure 5.14a or 635 nm in Figure 5.14b).
2. With "linear" dependence for voltages > V_T—to measure high intensities of the E-field (562 nm in Figure 5.14a or 484 nm in Figure 5.14b).
3. With a low level of the signal in the ranges 0 V–V_T—in automatics, to signalize that the E-field value is higher than V_T (740 nm in Figure 5.14a or 599 nm in Figure 5.14b).
4. With a low level of the signal for voltages >V_T—in automatics, similar to (3).

It must be mentioned that there are some potential issues that must be discussed in the context of practical applications. The first one is that the refractive indices of LCs are usually highly temperature dependent, which may result in unpredictable behavior with changing environmental conditions. This issue can be solved by using special LC mixtures with reduced thermal sensitivity (as 1294-1b LC mixture presented in Figure 5.1). The second potential issue concerns the so-called "memory effects" that may occur in PLCFs subjected to a high electric field—the field-reoriented LC molecules may be "frozen," even if the E-field is no longer present. This problem can be eliminated by using special aligning layers within the fiber holes. Such layers with high anchoring energy can lead to one of the specified molecular orientation (Fig. 5.8) (31).

Furthermore, the electrical tuning of the PCFs filled with dual-frequency (DF) LCs was obtained and presented in Reference 13. By applying an external electric field to such PLCFs, the DF LC molecule will start to reorient toward being parallel or perpendicular to the field, depending on which frequency is applied to the electrodes. In addition, frequency tunability of the solid core PCFs filled with nanoparticle-doped LCs has been reported recently in Reference 36. The efficiency of electrical tuning of PLCFs could also be significantly improved by using FLCs, mainly owing to their fast response to an external electric field (37). In our experiment, the infiltration process was performed by dipping one end of the PCF into a container with the FLC material heated up to 150°C. Owing to capillary forces,

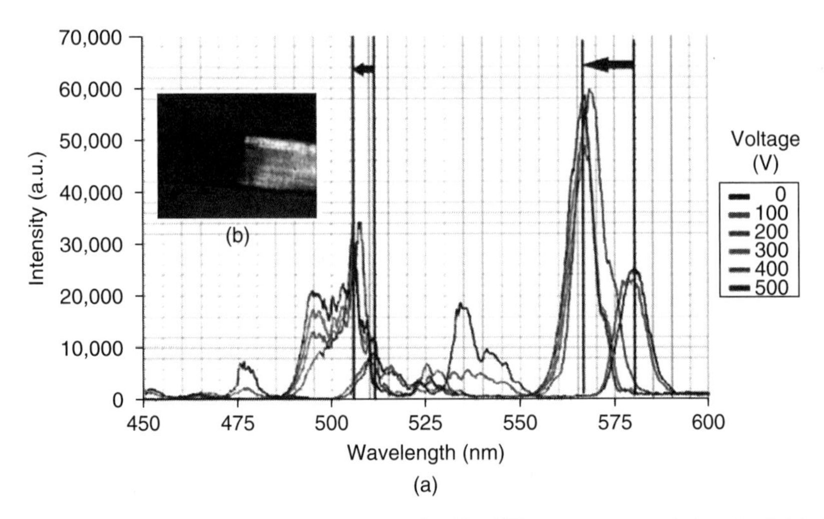

FIGURE 5.15 (a) Transmission spectrum of a PFLCF versus external electric field. (b) PCF with filled and nonfilled part under a microscope with crossed polarizers.

the molecules of the FLC got drawn into the PCF structure. In Figure 5.15a, we present a PCF fiber partially filled with molecules of the MT2_B15 FLC observed under a microscope with crossed polarizers. According to our observations of the PFLCF structure under a polarization microscope, we noticed that the orientation of the molecules in the PCF structure appears to be random and that an external electric field can reorient FLC molecules. As shown in Figure 5.15b, the PBGs in the transmission spectrum are observed. By increasing the voltage, a blueshift effect in two regions of the spectrum was noticed. The shift of one PBG was 5 nm, from 510 nm (0 V) to 505 nm (500 V), and 14 nm for the second PBG, from 579 nm (0 V) to 566 nm (500 V). It should be pointed out that in the case of electric tuning of PFLCF, the observed blueshift of PBG positions seems to be a switchlike shift. Up to 100 V between electrodes, no significant shift in the PBG spectrum has been observed. After exceeding 200 V, a rapid blueshift has been noticed. By increasing the voltage from 200 to 500 V between electrodes, once more no significant shift has been noticed.

5.4.3 Pressure Tuning

Hydrostatic pressure can also change the optical properties of the PLCF. The positions of PBGs maxima in transition spectra depend not only on the refractive index of the holes' infiltrating material but also on the geometry (diameter) of the capillary. Subsequently, any change in the holes' diameter induces a change in the PBG wavelength.

In our experiment, we used a LMA-PM-5 PCF (35 cm in length) filled with 6CHBT LC (section of 2 cm). The 6CHBT LC is characterized by both refractive indices higher than the refractive index of the silica glass used for the HB PCF

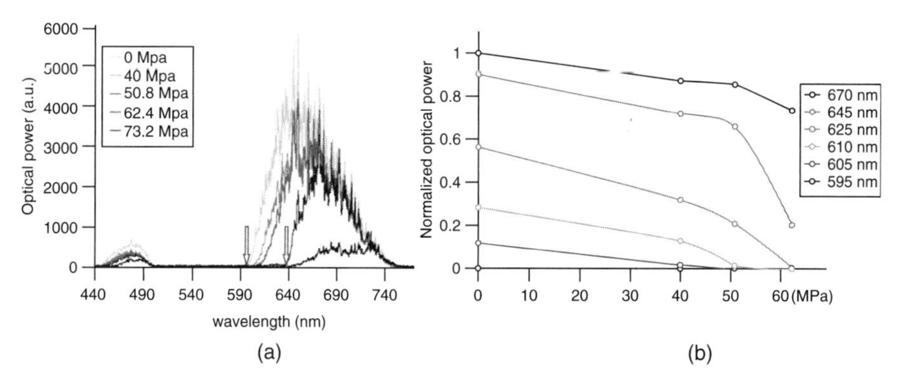

FIGURE 5.16 (a) Transmission spectra for LMA-PM-5 PCF infiltrated with 6CHBT LC mixture versus hydrostatic pressure (b) and normalized optical power for selected wavelengths as a function of hydrostatic pressure.

fabrication process. It means that the propagation in the fiber core is possible only for the wavelengths corresponding to PBGs formed in the cladding. The HB PLCF sample was connected to the white light source and to the spectrometer. In transmission spectra without any external perturbation, two possible ranges for the light propagation were observed. It was observed that the hydrostatic pressure resulted in narrowing the changes in the direction of longer wavelengths. This effect was especially evident for the range of 600–750 nm (Fig. 5.16a), in which the observed narrowing effect was about 40 nm. The preliminary results obtained are especially interesting in view of our previous research on the influence of the external perturbations on PLCF. Under hydrostatic pressure, we did not observe the PBGs tuning effect. Instead, we observed one-side narrowing effect of the PBGs in transmission spectra (Fig. 5.16a). The sensitivity of the PLCF to hydrostatic pressure can provide the basis for prospective construction of a fiber-optic pressure sensor. It is evident that the sensor response to hydrostatic pressure must depend on the operating wavelength (Fig. 5.16b). Hence by injecting only a selected wavelength into the PLCF and observing the output light, we are able to scale the sensor head to measure the hydrostatic pressure.

The polarization properties of the HB PCF infiltrated with an LC are also strongly sensitive to hydrostatic pressure, which was also investigated. The polarimeter PAT 9000B with the POL 9320FIR module served as an analyzer. In the Poincare sphere representation, we observed repeatable hydrostatic-pressure-induced evaluations of the polarization state (Fig. 5.17). This effect shows that the hydrostatic pressure can tune polarization properties of the sensor head based on HB PLCF. The results obtained suggest that the phase birefringence changes under the influence of hydrostatic pressure. Relying on the experimental data, we calculated the hydrostatic pressure sensitivity parameter K of the phase birefringence by using the following formula:

$$K = \frac{2\pi}{T_{\mathrm{p}}L} \cong 0.18(\mathrm{rad}/(\mathrm{m}*\mathrm{MPa})), \tag{5.1}$$

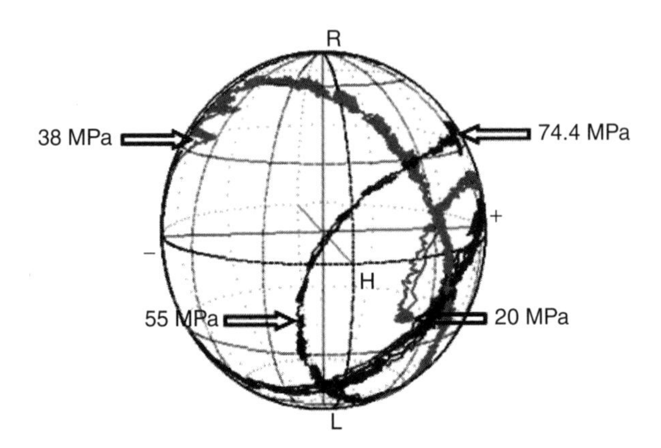

FIGURE 5.17 Change in polarization state in LMA-PM-5 PCF with 6CHBT LC under hydrostatic pressure—visualization of the Poincare sphere.

where T_p is the value of hydrostatic pressure responsible for 2π phase shift between two polarization eigenmodes and L the fiber length. Further experiments are in progress.

5.4.4 Optical Tuning

All-optical modulation in PCF filled with dye-doped NLC (14) was reported in 2004. A similar tuning technique of the PLCF is presented in Reference 38, where the PCF is filled with an LC doped with azobenzene. This all-optical modulator based on PLCF exhibited the modulation range for the 632 nm wavelength at 10 dB and the response time for switching was less than 1 s by manually obstructing the pumping light path. The light guiding mechanism can also be tuned in PCFs filled with NLCs by increasing optical power (39, 40). In particular, when the refractive index of an infiltrating LC material is higher than that of the silica surround, the analyzed PLCFs with a spatial periodicity of refractive index distribution may be considered as a matrix of waveguide channels. This architecture, analogical to a multicore optical fiber, allows for discrete light propagation. When the power of a propagating beam increases, the refractive index of the excited LC waveguides is modified by optical nonlinearity. Positive Kerr-like nonlinearity ($n_0 > 0$) analyzed here corresponds to reorientational nonlinearity observed in LCs. In such a case, an increase in optical power results in stronger guiding of light within LC rods, which may lead to light self-localization. If the input beam power is high enough, nonlinearity balances the diffractive broadening and light propagates in a limited region of the fiber (here it is the first ring of the holes).

Here, a numerical analysis of light propagation versus optical power in PLCFs is presented. The LC13/7 PCF structure under investigation is described in Table 5.1. As a guest material for the analyzed PLCF, we chose a prototype NLC mixture, named 1550 (Table 5.2). Numerical simulations were performed by implementing the finite difference beam propagation method (FD-BPM) (41) for solving an electromagnetic wave equation (obtained with an assumption of paraxial and slowly varying envelope approximation) in the following form:

$$2ik_0 n_0 \frac{\partial E}{\partial z} = \nabla_{\perp}^2 E + k_0^2 (n^2 - n_0^2) E, \tag{5.2}$$

where n_0 is the effective refractive index, k_0 the free-space wave vector, z the propagation direction, E the slowly varying complex amplitude of the electric field, and $n = n(x, y, z)$ the refractive index distribution in the fiber cross section. The latter may depend on the optical power of injected light, and in our case, the standard Kerr-like nonlinearity has been considered with a nonlinear refractive index: $n_{\mathrm{NL}} = n_L + n_2 I$, where n_{L} is the linear refractive index (i.e., obtained for the low power excitations), n_2 the nonlinear (Kerr) coefficient, and I the intensity of the propagating beam. It has to be also noted that the analyzed scalar problem described in Equation 5.2 was simplified by assuming that the initial NLC refractive index distribution within PCF holes is uniform. In addition, by supposing that the infiltration of PCFs by using the capillary effect results in the planar orientation of NLC molecules within fiber capillaries (with possible enhancement by special orientation treatments, e.g., by the method described in Section 5.3.2), an ordinary refractive index, n_0, of NLC was applied in calculations.

Our simulations, based on the FD-BPM, revealed the importance of space discretization in numerical algorithms applied. In particular, Figure 5.18 presents low

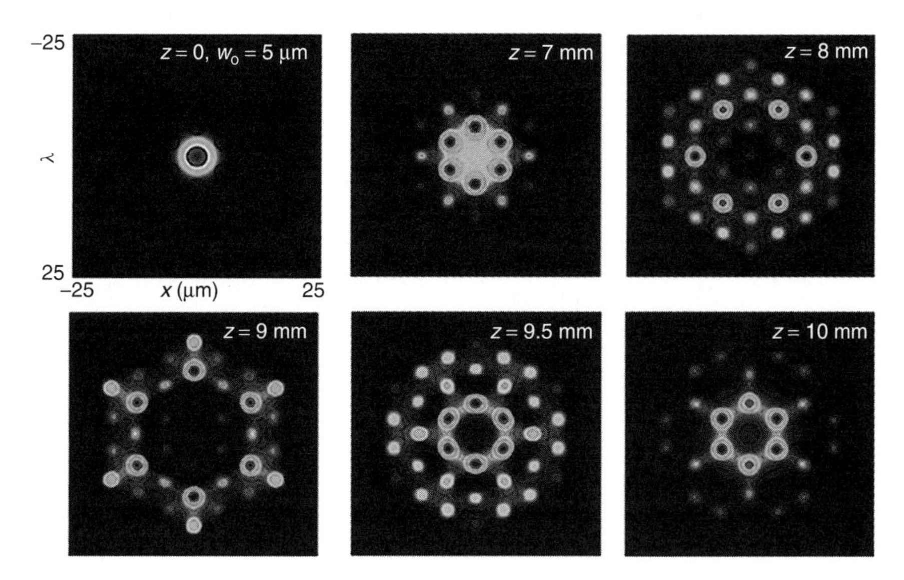

FIGURE 5.18 Beam transverse intensity profiles (i.e., $\sim |E|^2$) along propagation direction z. Numerical results presented obtained for the square-shaped mesh with the nods separated by $h = 0.4$ μm.

power light propagation for the Gaussian beam injected into the glass core of a PLCF structure with the results obtained for the square discrete mesh with a grid period of $h = 0.4$ μm. In addition, the limited size of the holey region of the PCF causes the light to reach the border of the periodic structure and come back to the central part of the fiber. As a result, the light beam is alternately broadening and narrowing on the way of propagation. A detailed look at the intensity distributions shown in Figure 5.18 reveals their imperfections. Unsymmetrical beam broadening (narrowing) was still observed for numerical simulations performed with twice the denser mesh grid, that is, $h = 0.2$ μm. It has provoked additional analyses of uniform discrete mesh application in the FD-BPM algorithm. One of the possible approaches applied in the FD-BPM is to use the square grid with a resolution suited the considered numerical problem. However, it should be noted that for PLCFs with hexagonal symmetry, even the implementation of a highly dense square mesh does not help to avoid differences in distances between adjacent holes, along with their size and shape obtained in the digitalization process. The lack of symmetry in the numerical distribution of fiber holes and differences in hole sizes and shapes result in different coupling coefficients for the neighboring NLC rods. It means that numerically a periodical matrix of the NLC channels does not consist of identical elements. It causes asymmetrical beam intensity distribution in the fiber cross section on the way of propagation. It becomes more and more of a problem with the propagation distance and/or with optical power. When the power of a propagating beam increases, the refractive index of the excited NLC waveguides is modified by optical nonlinearity. The problem of strong asymmetry of beam profiles in the case of a square mesh may be solved by applying a triangular grid. The numerical results, showing the influence of focusing nonlinearity on light beam propagation, obtained for both: square and triangular mesh, are shown in Figure 5.18. It seems clear that the triangular discrete mesh shown is a better numerical representation when compared to the standard square grid used for modeling PCFs with hexagonal symmetry (Fig. 5.19). By applying a triangular grid, it is possible to have the center of each NLC rod in a grid node and to have equal distances between neighboring holes, as well as their size and shape after numerical representation (the lower column in Figure 5.18).

5.4.5 Birefringence Tuning

One of the most interesting features of PLCFs is that their polarization properties can be dynamically tuned owing to external-field-induced changes in the value of the LC guest refractive index. A tunable polarizer (27), tunable single-polarization operation (28), and tunable birefringence (6, 8) have been already obtained in the silica-based PCFs filled with LCs. In this section, we briefly present the experimental results of the effects induced by temperature and electric field in the birefringence of the PLCFs.

To investigate birefringence properties of an "empty" silica-based PCF, we used the PM-1550-01 HB PCF (42, 43). In order to characterize the birefringence of

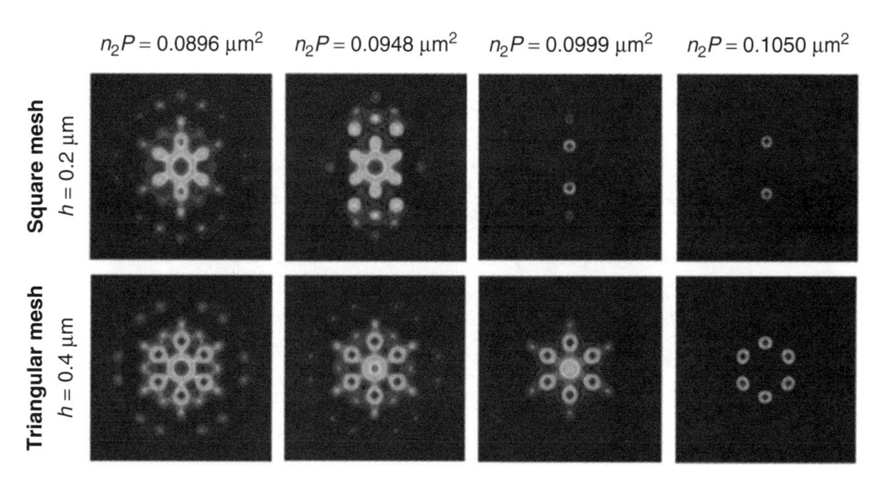

FIGURE 5.19 Light intensity distribution for the propagation distance of 2.5 mm as a function of the input optical power P obtained for the square (top) and triangular (bottom) mesh. A 5-μm-waist Gaussian beam is initially launched into the glass core of the fiber, and the initial refractive index of NLC (i.e., for $P \sim 0$) was taken as n_0. It is worth noting that for the Kerr coefficient $n_2 = 10^{-10}$ m²/ W, the product $n_2P = 0.1$ μm² corresponds to an optical power of 1 mW. The size of the numerical window presented is the same as that in Figure 5.18.

this fiber, both wavelength-dependent: phase birefringence $B(\lambda)$ and group birefringence $G(\lambda)$, are defined as

$$B = \left(\frac{\lambda}{2\pi}\right)(\beta_y - \beta_x) \tag{5.3}$$

$$G = B - \lambda\left(\frac{dB}{d\lambda}\right) \tag{5.4}$$

should be measured, as they generally have different values. The PM-1550-01 fiber is highly birefringent; thus the output state of the polarization is wavelength dependent, and as a result, modulation in the transmission spectrum can be observed. The transmission spectra measured for PM-1550-01 HB PCF lengths L smaller or equal to 71 mm are presented in Figure 5.20a. The wavelength spacing between neighboring peaks (maxima), $\Delta\lambda$, depends on the group birefringence G and L. Any decrease in the L leads to an appropriate increase in the distance between maxima. This suggests that a wavelength scanning technique can be used for measuring the group beat length resulting from polarization modes coupling of the fundamental mode. To calculate G (Eq. 5.4) for the HB PCF from the typical interferograms, the wavelength dependence of the modal phase birefringence B (Eq. 5.3) must be determined. To determine B, an assumption concerning the wavelength dependence of the modal phase birefringence is necessary. For conventional HB fibers with stress-induced birefringence, this is simplified, because B is mostly independent of

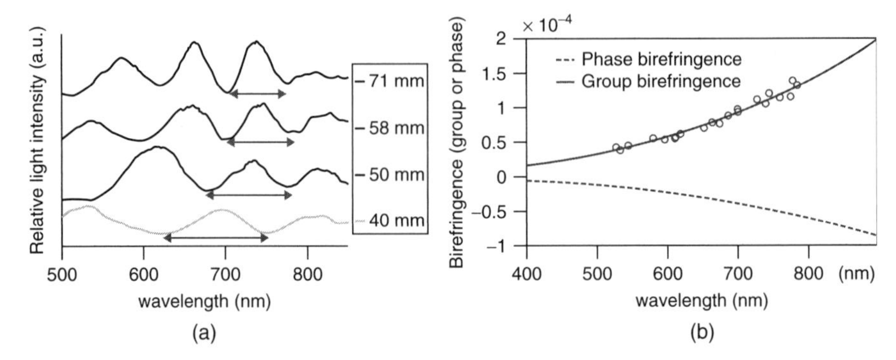

FIGURE 5.20 (a) Optical spectra observed for different lengths of the Blazephotonics PM-1550-01 HB PCF—a decrease in the fiber length increases its spectral period. (b) Group and phase birefringence of empty PM-1550-01 fiber.

wavelength, and therefore, the derivative term in Equation 5.3 vanishes and hence G and B are equal. However, in the case of HB PCFs, the wavelength dependence cannot be ignored while evaluating B. It has been observed that the wavelength dependence of the modal birefringence in HB PCFs often follows the empirical power law dependence of the form $B = \alpha\lambda^k$ (where α and k are the constants to be determined). Substituting this assumption in Equation 5.4 gives

$$G = -\alpha(k - 1)\lambda^k \tag{5.5}$$

By fitting the power law function $x\lambda^k$ to the experimental results of G, the constants α and k can be determined and the value of B could be calculated. The calculated group and phase birefringence of empty PM-1550-01 fiber are shown in Figure 5.20b.

In the next step, we selectively infiltrated a 2-cm section of a 48-cm-long PM-1550-01 P with the 1110 NLC mixture CF. The ordinary refractive index of the 1110 LC mixture is lower than the refractive index of silica glass. As a result, after infiltration all wavelengths can still be guided by the mTIR mechanism in the PLCF. The procedure to achieve selective infiltration of two large holes in the HB PCF structure was described in Reference 35. It is worth noting that filling only selected microholes of the HB PCF with an LC changes the shape of the propagation mode profile. This means that in the HB PLCF section, birefringence axes can rotate compared to the HB PCF section, leading to the situation in which the "fast" axis of the PCF corresponds to the "slow" axis of the PLCF and vice versa.

Generally, HB PCFs are characterized by low temperature dependence of their birefringence. However, the HB PLCF can be very sensitive to temperature owing to the thermal sensitivity of the LC guest. At room temperature, irregular modulation of the spectra was observed for the HB PLCF studied here. After heating the filled PCF section, significant changes in the modulated spectra were observed. It

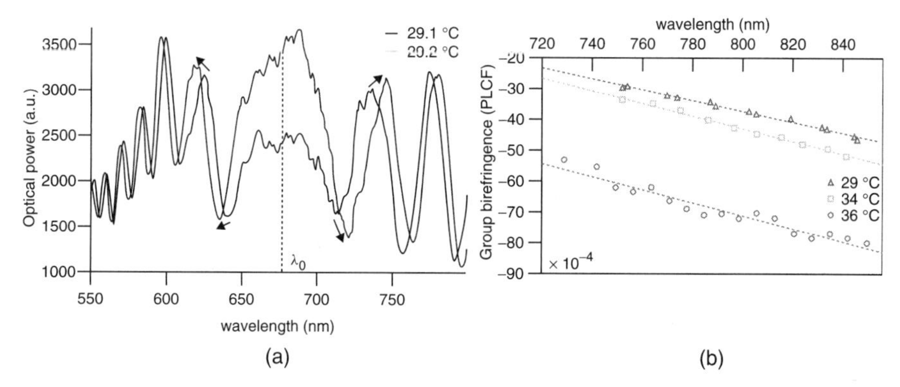

FIGURE 5.21 (a) Typical spectra of the PLCF based on the 1110 LC mixture collected at similar temperatures. (b) Group and phase birefringence of empty PM-1550-01 fiber.

is worth noting that the total birefringence is in fact the birefringence of a tandem configuration comprising two different HB fiber sections: a 46-cm-long section of an empty PM-1550-01 PCF and a 2-cm-long section of the PLCF. In Figure 5.21, typical spectra observed at narrow temperature ranges are shown. As it can be seen, the density of the peaks increases with the distance from the characteristic wavelength λ_0. The results obtained by us suggest that for the wavelength $\lambda_0 \approx 675$ nm, the optical phase delay (OPD) of a 46-cm-long section of the empty PM-1550-01 PCF was equal to the that of a 2-cm-long section of the same fiber, only when filled with the 1110 NLC mixture (at temperature about 29°C, for different temperatures λ_0 may change, at 42°C $\lambda_0 \approx 675$ nm). It means that the birefringence of the PLCF is 20–30 times higher than the birefringence of an empty PCF (as measured at a wavelength of 675 nm), and the birefringence axes are reversed. To calculate the exact value of the group birefringence of this sample, Equation 5.4 cannot be used, as the measured sample was composed of two different HB fibers, a 40-cm-long section of the empty PM-1550-01 PCF and a 20-mm-long section of the PLCF. The phase difference between two orthogonal polarization modes induced by the cascaded configuration of two different HB fiber sections can be expressed as follows:

$$\Delta\varphi = \Delta\varphi(B_1L_1) + \Delta\varphi(B_2L_2) = \frac{2\pi}{\lambda}(B_1L_1) + \frac{2\pi}{\lambda}(B_2L_2), \tag{5.6}$$

where B_i and L_i signify phase birefringences and lengths of both fiber sections ($i = 1$ for PCF and $i = 2$ for PLCF). By differentiating Equation 5.6 with respect to the wavelength, we obtain

$$\frac{\partial\Delta\varphi(\lambda)}{\partial\lambda} = \left(\frac{2\pi}{\lambda^2}\right)(-L_1G_1 - L_2G_2), \tag{5.7}$$

where G_i is the group birefringence. Taking into account the fact that the distance between two peaks $\Delta\lambda$ corresponds to a 2π phase change, the group birefringence of the PLCF section G_2 can be expressed as follows:

$$G_2 = sgn(\Delta\varphi)\left(\frac{\lambda^2}{\Delta\lambda L_2}\right) - \left(\frac{L_1 G_1}{L_2}\right) \tag{5.8}$$

The function $sgn(\Delta\varphi)$ introduced in Equation 5.8 is necessary, as $\Delta\varphi$ can be either positive or negative if B_1 and B_2 are of the opposite sign. In a special case, $\Delta\varphi$ can be equal to zero, when the optical path difference (OPD) introduced by the first fiber (PCF) is compensated by the second fiber (PLCF) ($\lambda_0 \approx 675$ nm in Figure 5.21a). Hence it is possible to calculate the group birefringence of the PLCF section if the group birefringence of an empty PCF section is known. By using Equation 5.5 and the results shown in Figure 5.20b, we calculated the group birefringence at three different temperatures (Fig. 5.21b). It was confirmed that the temperature sensitive group birefringence of the PLCF is ~20–30 times higher than the group birefringence of the PCF and is of the opposite sign.

One of the biggest issues concerning light propagation within PLCFs is the high level of its attenuation. So far, experimental works have been focused on a silica-glass-based PCF filled with LCs. However, refractive indices of a great majority of LCs are higher than the silica glass refractive index (~1.46), and thus in silica glass PCFs filled with LCs, light can be generally propagated owing to the PBG phenomenon. When the PBG mechanism occurs in the PLCF, its losses are on the order of a few, or even a few tens, of decibels per centimeter (6). Low loss propagation and continuously tunable birefringence have been obtained very recently in high refractive index PCFs manufactured from multicomponent glasses and filled with NLCs (27).

Our simulations indicate that filling the PCF14(6) with an LC 5CB does not significantly introduce changes in confinement losses (theoretical losses connected with an imaginary part of the effective index), which for both the empty PCF and the filled fiber are almost the same (44). The low level of confinement losses is connected with the fact that the mode is well located in the core area, and moreover, the mode profile remains almost unchanged in the whole tuning range. To measure the attenuation of an empty PCF14(6) filled with a 5CB LC, we decided to use the cutback technique. In this technique, the output power was measured for the fiber whose length was systematically shortened. Attenuation is defined as

$$A = \left(\frac{1}{x}\right) 10 \log\left(\frac{P(x)}{P_{\text{in}}}\right), \tag{5.9}$$

where P_{in} is the power at the input of the fiber, $P(x)$ the power measured at the output of the fiber, and x the length of the fiber. Equation 5.9 can be also expressed as

$$10 \log P(x) = Ax + 10 \log P_{\text{in}} \tag{5.10}$$

Hence attenuation can be calculated by finding the slope of the 10 log P_{in} function. Figure 5.22 shows the functions plotted for both: an empty PCF14(6)

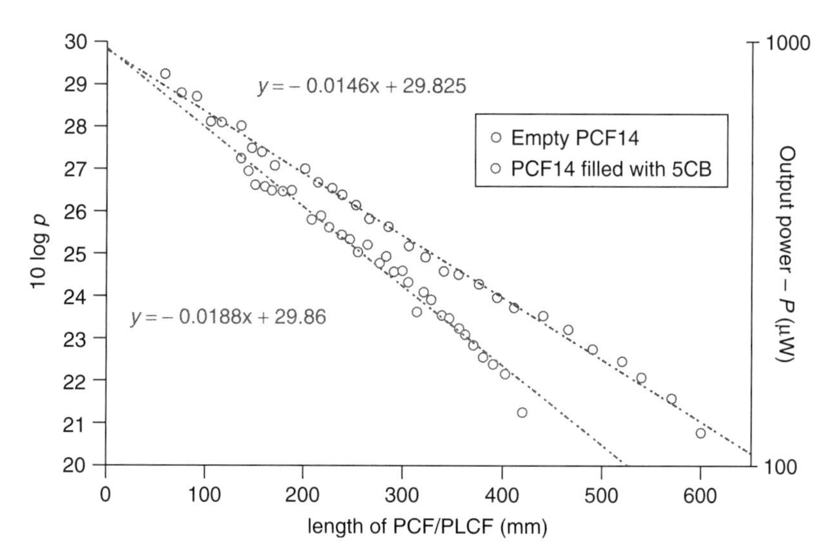

FIGURE 5.22 Attenuation measurement of the empty PCF14(6) and the PCF filled with 5CB.

and for a PCF14(6) filled with a 5CB, as well as their linear fit expressions. The measured values of attenuation of an empty PCF14(6) is ~0.15 dB/cm, and it increases to ~0.19 dB/cm after infiltration with a 5CB (the measurement was performed by using a He–Ne laser operating at the wavelength of 632.8 nm).

To investigate birefringence tuning we have used three different PLCF samples based on the high index PCF14(6) filled with 5CB, 1679, and 1658A LCs. Each PLCF sample, which is ~15 cm long, was placed between two 12-cm-long flat electrodes according to the experimental setup presented in Figure 5.4. The distance between both electrodes was limited by the PCF14(6) diameter and equal to 125 μm. Both ends of the sample were connected face-to-face with leading-in and leading-out SMFs. A tunable laser operating in the range of 1500–1640 nm (Tunics Plus) was used as a light source and changes in the output signal were analyzed by the PAT9000B polarimeter. It has been noticed that the changes in the electric field led to continuous and repeatable changes in the SOP in the output of the PLCF sample (Fig. 5.23a–c). For each sample, an increase in electric field resulted in circular traces on the Poincare sphere. It means that continuous phase changes between two polarization components of the guided mode have been electrically induced. The phase difference between two orthogonally polarized components of the fundamental mode after propagation through the HB fiber can be expressed as follows:

$$\delta\varphi = \left(\frac{2\pi}{\lambda}\right) BL \qquad (5.11)$$

It has to be noted that when the electric field is increasing, only the phase birefringence is changing in the PLCF placed between electrodes. Assuming that

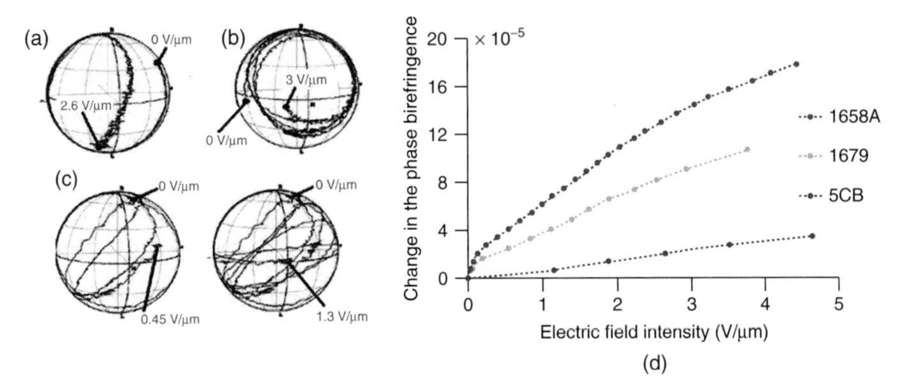

FIGURE 5.23 Electrically induced changes in the state of polarization (SOP) of the PCF14(6) filled with (a) 5CB; (b) 1679; (c) 1658A. In (d), the comparison of the phase birefringence tuning at 1640 nm in the high index PCF14(6) filled with medium (5CB) and high birefringence (1679 and 1658A) LCs is presented.

the length of electrodes (LEs) is constant, the phase difference induced by the electric field E can be expressed as follows:

$$\Delta\delta\varphi(E) = \left(\frac{2\pi}{\lambda}\right) L_{\mathrm{E}} \Delta B(E) \tag{5.12}$$

As $\Delta\delta\varphi(E)$ can be easily measured from the Poincare sphere traces (one full circle corresponds to $\Delta\delta\varphi = 2\pi$), a change in the phase birefringence ΔB can be easily calculated by using the following formula:

$$\Delta B(E) = \left(\frac{\Delta\delta\varphi(E)}{2\pi}\right)\left(\frac{\lambda}{L_{\mathrm{E}}}\right) \tag{5.13}$$

The comparison of birefringence tuning ranges of all three PLCF samples is presented in Figure 5.23d. The tuning range of the PLCF sample filled with 1679 was $\sim 10^{-4}$, and for PLCF sample filled with 1658A, it was almost 2.10^{-4}. Additional simulations based on limited information about 1679 and 1658A refractive indices confirmed that the tuning range depends on the value of the LC birefringence. If the difference between both LC ordinary and extraordinary refractive indices is higher, the same reorientation of molecules induces higher birefringence in the fiber.

5.5 PHOTONIC DEVICES

As presented in Section 5.4, the infiltration of air holes of the PCFs with LC materials allowed a diversity of novel and uncommon spectral and polarization properties. Owing to the highest level of sensitivity to external physical fields, effective PLCFs technology can be used for highly tunable advanced photonic devices. Some prototypes and demonstrators of such devices, developed at the Warsaw University of Technology, are presented in this paper. It is worth noting that related projects have been realized in close collaboration with the Marie Curie-Sklodowska University (UMCS, Lublin, Poland) and the Military University of Technology (MUT, Warsaw, Poland). Scientific groups from these institutions provide customized (not commercially available) PCFs and LC materials (e.g., LB and HB LC mixtures), respectively. The examples of resultant devices are presented in the following sections.

5.5.1 Electrically Tuned Phase Shifter

The experimental tests performed in the setup shown in Figure 5.24a have suggested a possible application of PLCFs as dynamically tuned fiber-optic-based phase shifters. The phase shifters demonstrated here were achieved for two different configurations with the 070124 PCF (Table 5.1) used as a host structure of the PLCF. In the first configuration, the 1800B LC was applied and a 5-cm-long section of the PCF was successfully infiltrated. As a light source, a laser diode, operating at a wavelength of 633 nm, was used and the operating temperature was set to 61°C. Total losses for this fiber sample were estimated as −35 dB. Eventually, the SOP was changing under the influence of electric field, and a full 2π-phase-shift was observed for the electric field of 2.3 V/μm generated at a frequency of 1 kHz (Fig. 5.24d). In the second configuration, the PCF was filled with a 1550A LC, where the LC section was 3 cm in length. A tunable fiber laser set to the wavelength of 1540 nm was used as a light source with operating temperature kept constant at around 30°C. The total losses of the measured sample were −30 dB. While the influence of the external electric field on the SOP was tested, the 2π-phase-shift was observed at the frequency of 1 kHz and 2 V/μm of electric

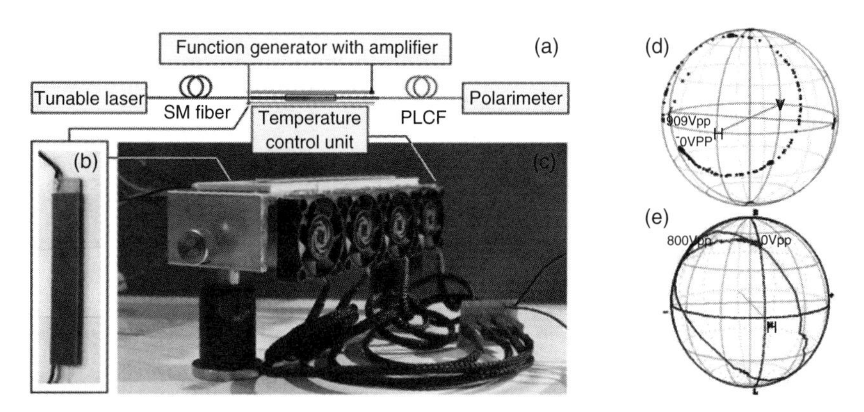

FIGURE 5.24 The experimental setup (a) with a system of electrodes (b) and temperature control unit including the Peltier modules, radiators, and fans. (c) Experimentally observed changes in the state of polarization under the influence of electric field for 070124 PCF infiltrated with: (d) 1800B LC and (e) 1550A LC.

field applied (Fig. 5.24e). To conclude, the maximum range of phase birefringence changes available in the described geometry has been determined to be 1.3×10^{-5} for the 1800B LC and 5.1×10^{-5} for the 1550A LC used as an infiltration. The results obtained have proven that the proposed devices enable a smooth change of the phase delay between two mutually orthogonal components of mode polarization (in a full range from 0 to 2π).

5.5.2 Thermally/electrically Tuned Optical Filters

The principle of this device operation is similar to that of the thermally tuned PLCFs presented in Reference 35 and has been adapted here in order to obtain an optical filter based on a PLCF, that is, tunable over a wide portion of the spectrum, such as the visible portion.

A corresponding demonstrator has been achieved by providing a PLCF composed from 061221P2 PCF filled with 6CHBT LC (presented in Figure 5.25). An important aspect of the device functionality was the presence of a resistive heater along the fiber. The latter was provided by a metal coil with electrical connectors. In this way, by varying the electrical current, a change in and control of temperature were obtained. Such a construction of the heater is very simple and results in compactness and robustness of the device as a whole. In addition, the heating system in such a form is characterized by low power consumption, for example, to heat the sample to 34°C (starting from the room temperature) only 0.6 W is required. In Figure 5.25a the transmission spectra of the analyzed PLCF are shown, and they consist of series of PBG-related peaks located at a specific spectral range. An advantage of this PBGs-shifting-based operation is the possibility of achieving electrically and thermally induced tuning of the shape of the PLCF spectrum. Even if the temperature change was limited to the range from 24 to 43°C (owing to the phase transition of the LC used), the obtained results indicate that such a small,

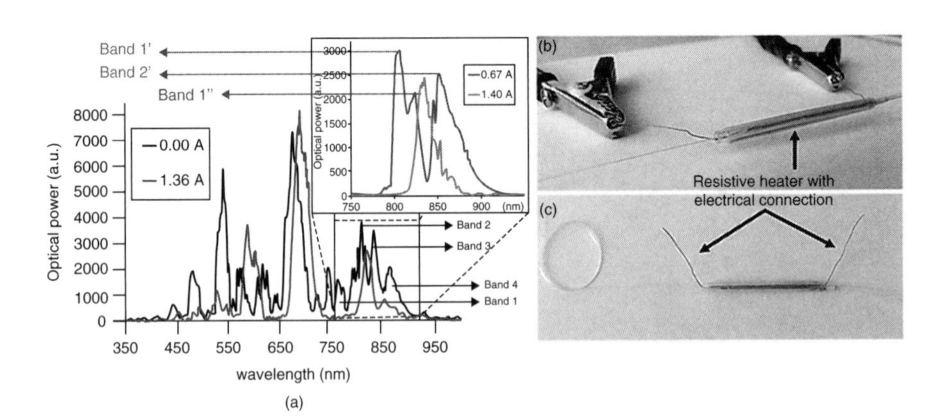

FIGURE 5.25 (a) Transmission spectra of the PLCF-based device versus the current flowing through the resistive heater. (b and c) The optical filter based on the PLCF placed inside a package heater with electrical connectors.

high performance PLCF-based device may be used as a dynamic filter exhibiting a high level of tunability for the required spectra wavelengths range.

5.5.3 Electrically Controlled PLCF-based Polarizer

The main principle of the PLCF-based polarizer operation presented here is a change of optical properties (i.e., rotation of the optical axis) of the device due to an external electric field applied (Fig. 5.26a and b). An appropriately chosen combination of the host structure (PCF) and the guest material (LC) parameters can result in a fiber whose properties can be changed in such a way that only one particular polarization can be guided and transmitted through the fiber. For this purpose, an external electric field is used to control the fiber parameters and to induce an optical birefringence affecting the propagating beam. For a specific value of the electric field applied, the birefringence induced in the fiber enables only (for selected wavelengths) the single polarization of the propagating beam for the exiting fiber (Fig. 5.26c).

5.5.4 Thermally Tunable Attenuator

In this section, we present our attempt to make an all-in-fiber thermally tunable attenuator. The effect of optical attenuation controlled by temperature has been recently reported (45), so we have decided to build a simple device by using similar materials. The device was based on the 070124 PCF filled with a low birefringent 1800 LC mixture. The ordinary refractive index of such a mixture was lower than the refractive index of silica glass, and its value was decreasing with temperature (increasing thus the contrast between refractive indices of the core and infiltrated holes). As a consequence, the attenuation of the PLCF was decreasing with temperature. We decided to optimize the designed attenuator for

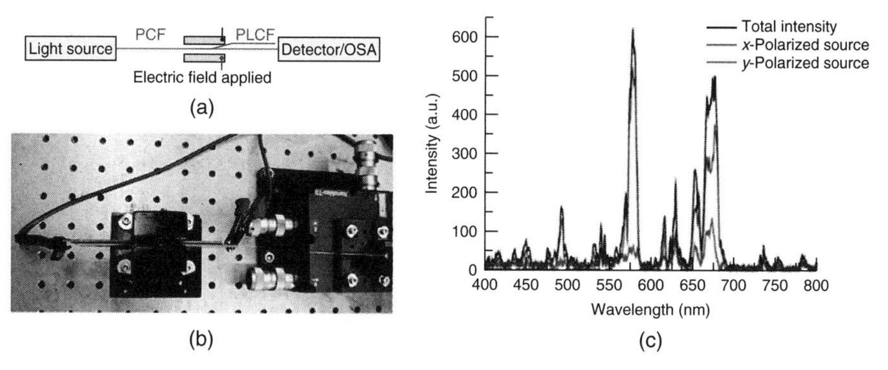

FIGURE 5.26 (a) Scheme of the electrically controlled PLCF polarizer. (b) A picture of the PLCF polarizer unit. (c) Change between two orthogonal states of polarization induced by electric voltage of 297 V for selected wavelengths (429 nm, 576 nm, and 680 nm). Level of changes exceeding 85% is achieved.

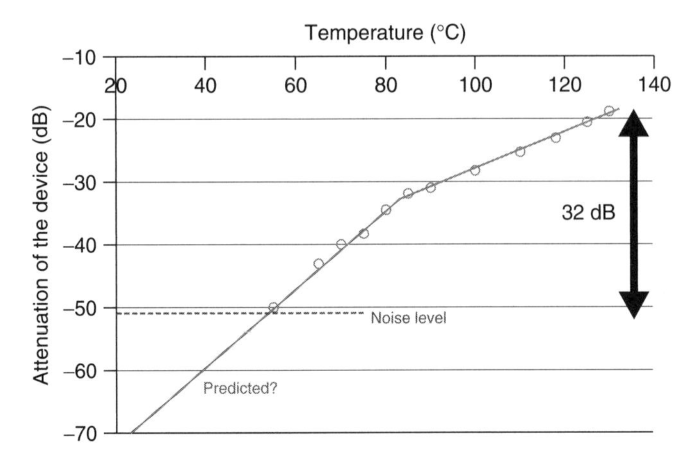

FIGURE 5.27 Experimental results for thermally tunable attenuator achieved in PLCF.

operation at the wavelength emitted by a He–Ne laser (i.e., 632.8 nm), and for this purpose, we connected a 50-mm-long section of the PLCF to two pieces of the SM600 fiber (of the length of about 1 m). In Figure 5.27, we present the thermal characteristic of the device. It can be noticed that the lowest value of attenuation is about 18 dB (mainly caused by high losses at the connections). From our measurements, we can predict that the tuning range of the device is at least 32 dB; however, we were not able to measure the attenuation of the device at temperatures lower than 50°C because of the output signal being at the noise level of the optical power meter used. Nevertheless, from measurements made with "unconnected" PLCF, we know that attenuation at lower temperatures should be much higher (see, for example, Reference 45), so we can expect that, in fact, the tuning range is broader and reaching ~50 dB. In Figure 5.27, we can also notice that the slope of the characteristic (i.e., attenuation vs temperature) is significantly changing for a temperature of about 80°C. For lower temperatures, this slope is of about 0.6 dB/°C, whereas for higher ones, it is about two times lower (i.e., ~0.3 dB/°C). This change of slope can be related to the phase transition of the LC.

5.6 PHOTONIC LIQUID CRYSTAL FIBER SENSORS FOR SENSING AND SECURITY

PLCFs benefiting from a merge of "passive" PCFs host microstructures with a variety of "active" LC guest materials create a new challenge for both fiber optics and LC photonics. PLCFs introduce new levels of tunability to PCFs and boost the performance of these fibers owing to a diversity of new propagation, spectral, thermo-optic, electro-optic, and polarization properties, simultaneously demonstrating that such optical fibers can be more special and unique than previously thought. Apart from high sensitivity of the PLCFs to influences from external fields as temperature, electrical/magnetic/optical fields, and hydrostatic pressure, the use of different LC molecular orientation "scenarios" within the microholes can determine either index guiding (mTIR) or PBG propagation mechanisms and reversible switching between them.

PLCFs—owing to infiltration of LCs into PCFs—form a novel and common platform that addresses the need for an all-in-fiber configuration suitable to be used in optical sensing and security. The possibility of sensing applications arises, in particular, from the presence of LCs characterized by high optical tunability with external and internal factors, such as temperature, electric, magnetic, and optical fields; deformations; and introduction of dopants, dyes, nanoparticles, or other chemicals.

Consequently, the resulting PLCFs are highly sensitive to external physical fields such as temperature, pressure, electric/magnetic field, so they can be used in various types of prospective optical fiber sensors. A novel technology for all fibers based on temperature, electric field, and hydrostatic pressure has been already proposed and developed (20) by using different PLCF structures that are based on particular dynamic and tunable properties of LCs as anisotropic media. As a host material, both isotropic and/or highly birefringent PCFs can be used, whereas an "active" element of the PLCF is generally a NLC with either low or with moderate birefringence. Changes in the output optical spectrum induced by temperature, electric field, and hydrostatic pressure can be used for measurement of these quantities. Observed spectral shifts of PBG wavelengths are repeatable and are generally typical for PLCFs operating in the PBG regime. It is evident that by using different combinations of PCFs and LCs, a new method of optical fiber sensing with a wide sensitivity ranges and operating regions can be implemented. Moreover, this particular new sensing method allows measuring external factors distribution in a specified area by using an array of PLCFs sensing elements. The main advantage of proposed solution is its compactness, simplicity, and relatively low cost.

Another area of PLCFs for sensing and security applications is envisaged in tunable all-fiber components that will be used in optical fiber sensing setups. These include, for example, PLCF-based tunable waveplates and phase shifters that are extremely important all-fiber components in polarimetric optical fibers and sensors (45) to be used for dynamic adjusting of polarization of light coupled to the sensing fiber. Tunable PLCF-based polarizers as well as PLCF polarization controllers could find applications to fiber Bragg gratings (FBGs) written in highly birefringent fibers enabling simultaneous dynamic switching between orthogonal polarizations (46).

Prospective PLCF-based sensing devices may also find their applications in so-called lab-on-a-chip structures. Moreover, particular use of LCs as an infiltration for a new class of optofluidic devices allows for many additional distinguished applications (while highly sensitive for external physical fields). LCs thus play a fundamental role in upgrading and innovation of already existing optofluidic systems (where typically isotropic fluids are used) by allowing them to exploit fully the unique and remarkable properties of anisotropic fluidic infiltration. The latter directly results in the exceptional efficiency, precision, and unique functionality, which, in particular, may cover a wide spectral range. PLCF-based optofluidic devices can be used in various physical, chemical, and biological sensing schemes because of their low cost, capability of multiplexed measurements, fast response and also compatibility with a wide range of surface chemistries that can enable high specificity and sensitivity. These structures allow for measuring the influence of the external factors, such as temperature, electric field, and pressure, as well as the "lab-in-a-fiber" biochemical analysis.

Unquestionably, the impact of the PLCF-based devices in terms of their applications for optical sensing will be significant. Importantly, PLCF-based devices may find application not just in photonics; similar technology can be potentially used to contribute to the solutions of some of the most important problems in biology and chemistry, including possibility of detecting small traces of harmful chemicals.

5.7 CONCLUSION

In this chapter, we have discussed and reviewed the basic physics and innovative technology of PLCFs including some of impressive sensing and highly tunable photonic devices applications demonstrated over the past few years at the Warsaw University of Technology. Importantly, the proper choice of PLCF components, together with a precise adjustment of their parameters may result in various advanced photonic elements for different uses. All these developments open up great possibilities of PLCF sensors for safety and security monitoring applications. Current research trends open up new and unique perspectives toward the development of practical alternatives to existing conventional solutions not only for a set of specific goals in terms of sensing but also for solving some of the most challenging problems in photonics, both from a fundamental point of view (e.g., propagation properties tuning of the photonic crystal structures) and also a more applicative one (e.g., LC molecules orientation control within host optical fiber microstructures). In a broader context, this will contribute to high technology PLCF-based devices with a novel and versatile design, opening the way for fascinating and still unexplored research areas.

ACKNOWLEDGMENTS

The author is much indebted to many collaborators from the Warsaw University of Technology (WUT, Poland), Military University of Technology (MUT, Poland) in Warsaw, Maria Curie-Skłodowska University in Lublin (UMCS, Poland), ITME-Warsaw, Université du Québec en Outaouais (UQO, Canada), Universidad Politécnica de Madrid (UPM, Spain), Nanyang Technical University (NTU, Singapore), Ghent University (GU, Belgium), Vrije Universiteit Brussel (VUB, Belgium), and National Academy of Sciences of Ukraine (NASU, Ukraine) represented by Professors: Wojtek J. Bock (UQO), Roman Dąbrowski (MUT), Andrzej Domański (WUT), Kristiaan Neyts (GU), Jose Oton (UPM), Ping Shum (NTU), Hugo Thienpont (VUB); Doctors: *Jan Wójcik* (UMCS), Ryszard

Buczyński (ITME), Daniel Budaszewski (WUT), Aleksandra Czapla (WUT), Sławomir Ertman (WUT), Piotr Lesiak (WUT), Paweł Mergo (UMCS), Tomasz Nasiłowski (MUT), Edward Nowinowski-Kruszelnicki (MUT), Oleg Yaroshuk (NASU), Dariusz Pysz (ITME), Katarzyna Rutkowska (WUT), Marek Sierakowski (WUT), and Katarzyna Szaniawska (WUT), and also Ph. D. students from WUT: Miłosz Chychłowski, Katarzyna Nowecka, and Marzena Tefelska for fruitful collaboration. The valuable assistance of Dr. Aleksandra Czapla during the preparation of the manuscript is gratefully acknowledged. This work was partially supported by the MISTRZ Programme of the Foundation for Polish Science.

REFERENCES

1. Russell PStJ. Photonic crystal fibers. Science 2003;299:358.
2. Knight C. Photonic crystal fibers. Nature 2003;424:847.
3. Larsen TT, Bjarklev A, Hermann DS, Broeng J. Optical devices based on liquid crystal photonic bandgap fibers. Opt Express 2003;11:2589.
4. Ertman S, Woliński TR, Czapla A, Nowecka K, Nowinowski-Kruszelnicki E, Wójcik J. Liquid crystal molecular orientation in photonic liquid crystal fibers with photopolymer layers. Proc SPIE 2007;6587:658706.
5. Wolinski TR, Szaniawska K, Bondarczuk K, Lesiak P, Domanski AW, Dabrowski R, Nowinowski-Kruszelnicki E, Wojcik J. Propagation properties of photonic crystals fibers filled with nematic liquid crystals. Opto-Electron Rev 2005;13(2):59–64.
6. Wolinski TR, Szaniawska K, Ertman S, Lesiak P, Domanski AW, Dabrowski R, Nowinowski-Kruszelnicki E, Wojcik J. Influence of temperature and electrical fields on propagation properties of photonic liquid crystal fibers. Meas Sci Technol 2006;17:985–991.
7. Wolinski TR, Ertman S, Lesiak P, Domański AW, Czapla A, Dąbrowski R, Nowinowski-Kruszelnicki E, Wójcik J. Photonic liquid crystal fibers—A new challenge for fiber optics and liquid crystals photonics. Opto-Electron Rev 2006;14(4):329–334.
8. Woliński TR, Ertman S, Czapla A, Lesiak P, Nowecka K, Domanski AW, Nowinowski-Kruszelnicki E, Dabrowski R, Wójcik J. Polarization effects in photonic liquid crystal fibers. Meas Sci Technol 2007;18(10):3061–3069.
9. Woliński TR, Czapla A, Ertman S, Tefelska M, Domański AW, Nowinowski-Kruszelnicki E, Dąbrowski R. Tunable highly birefringent solid-core photonic liquid crystal fibers. Opt Quantum Electron 2007;39(12–13):1021–1032.
10. Lu Du YQ, Wu S-T. Electrically tunable liquid-crystal photonic crystal fiber. Appl Phys Lett 2004;85:2181–2183.
11. Haakestad MW, Alkeskjold TT, Nielsen MD, Scolari L, Riishede J, Engan HE, Bjarklev A. Electrically tunable photonic bandgap guidance in a liquid-crystal-filled photonic crystal fiber. IEEE Photonics Technol Lett 2005;17(4):819–821.
12. Scolari L, Alkeskjold TT, Riishede J, Bjarklev A, Hermann D, Anawati A, Nielsen M, Bassi P. Continuously tunable devices based on electrical control of dual-frequency liquid crystal filled photonic bandgap fibers. Opt Express 2005;13:7483–7496.
13. Alkeskjold T, Lægsgaard J, Bjarklev J, Hermann D, Anawati A, Broeng J, Li J, Wu S-T. All-optical modulation in dye-doped nematic liquid crystal photonic bandgap fibers. Opt Express 2004;12:5857–5871.

14. Budaszewski D, Domański AW, Czapla A, Ertman S, Woliński TR, Nasilowski T, Thienpont H. Depolarization of light in microstructured fibers filled with liquid crystals. Opto-Electron Rev 2009;17:156–160.
15. Budaszewski D, Cieślak R, Domański AW. Birefringence measurements of photonic liquid crystal fiber by use of the depolarization method. Mol Cryst Liq Cryst 2009;502:47–55.
16. Woliński TR, Ertman S, Tefelska M, Czapla A, Budaszewski D, Domański AW, Dąbrowski R, Nowinowski-Kruszelnicki E, Wójcik J. Polarizing and depolarizing optical effects in photonic liquid crystal fibers. Mol Cryst Liq Cryst 2008;489:69–182.
17. Ertman S, Wolinski TR, Pysz D, Buczynski R, Nowinowski-Kruszelnicki E, Dabrowski R. Tunable broadband in-fiber polarizer based on photonic liquid crystal fiber. Mol Cryst Liq Cryst 2009;502:87–98.
18. Ertman S, Czapla A, Woliński TR, Nasiłowski T, Thienpont H, Nowinowski-Kruszelnicki E, Dąbrowski R. Light propagation in highly birefringent photonic liquid crystal fibers. Opto-Electron Rev 2009;17:150–155.
19. Lesiak P, Wolinski TR, Slusarz K, Ertman S, Domanski AW, Dabrowski R. Temperature tuning of polarization mode dispersion in single-core and two-core photonic liquid crystal fibers. Opto-Electron Rev 2007;15:27–31.
20. Woliński TR, Czapla A, Ertman S, Tefelska M, Domański A, Wójcik J, Nowinowski-Kruszelnicki E, Dąbrowski R. Photonic liquid crystal fibers for sensing applications. IEEE Trans Instrum Meas 2008;57(8):1796–1802.
21. Hu JJ, Shum P, Ren G, Yu X, Wang G, Lu C, Ertman S, Wolinski TR. Investigation of thermal influence on the bandgap properties of liquid-crystal photonic crystal fibers. Opt Commun 2008;281(17):4339–4342.
22. Woliński TR, Tefelska MM, Chychłowski MS, Godyń K, Dąbrowski R, Wójcik J, Nasiłowski T, Thienpont H. Multi-parameter sensing based on photonic liquid crystal fibers. Mol Cryst Liq Cryst 2009;502:220–234.
23. Scolari L, Gauza S, Xianyu H, Zhai L, Eskildsen L, Alkeskjold TT, Wu S-T, Bjarklev A. Frequency tunability of solid-core photonic crystal fibers filled with nanoparticle-doped liquid crystals. Opt Express 2009;17:3754–3764.
24. Czapla A, Bock WJ, Miculic P, Woliński TR. Towards tuning of thermal sensitivity of the long period fiber gratings using a liquid crystal layer. Bull Pol Acad Sci Tech Sci 2010;58(4):503–508.
25. Czapla A, Woliński TR, Dąbrowski R, Nowinowski-Kruszelnicki E, Bock WJ. An electrically tunable filter based on a long-period fiber grating with a thin liquid crystal layer. Photon Lett Poland 2010;2(3):116–118.
26. Ertman S, Wolinski TR, Pysz D, Buczynski R, Nowinowski-Kuruszelnicki E, Dabrowski R. Low-loss propagation and continuously tunable birefringence in high-index photonic crystal fibers filled with nematics liquid crystals. Opt Express 2009;17(21):19298–19310.
27. Rutkowska KA, Woliński TR. Modeling of light propagation in photonic liquid crystal fibers. Photon Lett Poland 2010;2(3):107–109.
28. Takatoh K, Hasegawa M, Koden M, Itoh N, Hasegawa R, Sakamoto M. *Alignment Technologies and Applications of Liquid Crystal Devices*. Taylor & Francis, London and New York 2005.
29. Shannon PJ, Gibbons WM, Sun ST. Patterned optical properties in photopolymerized surface-aligned liquid-crystal films. Nature 1994;368:532–533.

30. Chychłowski MS, Ertman S, Woliński TR. Analysis of liquid crystals orientation in microcapillaries. Photon Lett Poland 2010;2(1):28–33.
31. Kwon SB, Kim KJ, Choi YS, Iranovich GI, Dyadyusha A, Reznikov YA. Liquid crystal display device. US patent No. 6,399,165. Issued on June 4, 2002.
32. Chigrinov V, Prudnikova E, Kozenkov V, Kwok H, Akiyama H, Kawara T, Takada H, Takatsu H. New developments in liquid crystal photo-aligning by Azo-Dyes. Liq Cryst 2002;29:1321–1327.
33. Vretik L, Syromiatnikov V, Zagniy V, Paskal L, Yaroshchuk O, Dolgov L, Kyrychenko V, Lee C-D. Polymethacryloylarylmethacrylates: New concept of photoalignment materials for liquid crystals. Mol Cryst Liq Cryst 2007;479:121–134.
34. Woliński TR. Photonic liquid-crystal fibers: new sensing opportunities. *Optical Waveguide Sensing and Imaging, NATO Science for Peace and Security Series B: Physics and Biophysics*. Bock WJ, Gannot I, Tanev S, editors. Dordrecht, The Netherlands, 2008. p 51–72.
35. Tefelska MM, Chychłowski MS, Woliński TR, Dąbrowski R, Rehmer WJ Wójcik. Propagation effects in photonic liquid crystal fibers with a complex structure. Acta Phys Pol A 2010;118(6):1259–1261.
36. Budaszewski D, Woliński TR, Geday MA, Otón JM. Photonic crystal fibers infiltrated with ferroelectric liquid crystals. Photon Lett Poland 2010;2(3):110–112.
37. Hsiao VKS, Ko C-Y. Light-controllable photoresponsive liquid-crystal photonic crystal fiber. Opt Express 2008;16:12670–12676.
38. Brzdąkiewicz KA, Laudyn UA, Karpierz MA, Woliński TR, Wójcik J. Linear and nonlinear properties of photonic crystal fibers filled with nematic liquid crystals. Opto-Electron Rev 2006;14(4):287–292.
39. Rutkowska KA, Laudyn UA, Rutkowski RT, Karpierz MA, Woliński TR, Wójcik J. Nonlinear light propagation in photonic crystal fibers filled with nematic liquid crystals. Proc SPIE 2007;6582:658215.
40. Mitchell AR, Griffiths DF. *The Finite Difference Method in Partial Differential Equations*. London: John Wiley and Sons; 1980.
41. Woliński TW, Czapla A, Ertman S, Tefelska M, Domański AW, Nowinowski-Kruszelnicki E, Dąbrowski R. Tunable highly birefringent solid-core photonic liquid crystal fibers. Opt Quantum Electron 2007;39:1021–1032.
42. Ertman S, Czapla A, Woliński TR, Nasiłowski T, Thienpont H, Nowinowski-Kruszelnicki E, Dąbrowski. Light propagation in highly birefringent photonic liquid crystal fibers through the fiber. Opto-Electron Rev 2009;17(2):150–155.
43. Ertman S, Woliński TR, Buczyński R, Pysz D, Nowinowski-Kruszelnicki E, Dąbrowski R, Wójcik J. Photonic liquid crystal fibers with low-, medium-, and high refractive indices of the core. Photon Lett Poland 2010;2(3):113–115.
44. Tefelska MM, Chychlowski MS, Wolinski TR, Dabrowski R, Wojcik J. Tunable attenuation in photonic liquid crystal fibers. Photon Lett Poland 2009;1(2):97–99.
45. Woliński TR. Polarimetric optical fibers and sensors. In: Wolf E, editor. Volume XL, *Progress in Optics*. Amsterdam: North Holland; 2000. p 1–75.
46. Ertman S, Tefelska MM, Chychłowski MS, Rodríguez AH, Pysz D, Buczyński R, Nowinowski-Kruszelnicki E, Dąbrowski R, Woliński TR. Index guiding photonic liquid crystal fibers for practical applications. IEEE/OSA J Lightw Technol 2012;30:1208–1214.

CHAPTER SIX

Miniaturized Fiber Bragg Grating Sensor Systems for Potential Air Vehicle Structural Health Monitoring Applications

HONGLEI GUO
Microwave Photonics Research Laboratory, University of Ottawa, Ottawa, ON, Canada

GAOZHI XIAO
Institute for Microstructural Sciences, National Research Council Canada, Ottawa, ON, Canada

NEZIH MRAD
Department of National Defense, National Defense Headquarters, Ottawa, ON, Canada

JIANPING YAO
Microwave Photonics Research Laboratory, University of Ottawa, Ottawa, ON, Canada

6.1 INTRODUCTION

Aircraft operators are facing requirements of reducing the operating and maintenance cost, while at the same time improving the safety and reliability of air

Photonic Sensing: Principles and Applications for Safety and Security Monitoring, First Edition.
Edited by Gaozhi Xiao and Wojtek J. Bock.

vehicles (1). For military aircrafts, the need to serve beyond their designed life cycle poses further pressure on the already heavy maintenance and inspection burdens (2, 3). Although conventional periodic on-ground inspection has successfully ensured the safety and reliability of air vehicles so far, it introduces high expenses (4). In addition, the conventional methods might require disassembling and reassembling components, which could possibly introduce unexpected damage and degradation to structures and auxiliary systems (5). The concept of structural health monitoring (SHM) is proposed and developed to reduce the complexity and the costs associated with these conventional methods by the implementation of on-board and real-time monitoring sensors (1). Air vehicle SHM consists of two critical aspects, that is, operational load monitoring and impact damage inspection (6). Operational load monitoring is used to estimate the structural life and provide information about the possible structural damage. Two methods are used to achieve this. One is the direct measurement of local stresses/strains; the other is the inferring from global flight parameters and structural usage data, including flight speed, altitude, mass, and acceleration (7, 8). The emergence of effective and reliable strain sensors in the past 20 years have made it possible to monitor the operational load by the combination of the two methods. In such case, a limited number of strain sensors are mounted at critical locations for the direct strain measurement, while flight parameters are used to estimate the operational load at other locations. In fact, the operational load monitoring using this method is the only proven technique in air vehicle SHM. However, it has limitations to address the SHM issues related to the impact damages resulting from cracks and corrosion in metals, and debonding/delamination in composite materials, which have led to the loss of the Shuttle Columbia (1) and numerous commercial aircraft tragedies, for example, the 2002 China Airlines Crash (9). Therefore, impact damage inspection techniques have to be introduced and studied to directly identify the occurrence, size, and location of damages. The technique studied most for this is the actuation-sensing method using the ultrasonic/acoustic waves (10).

Varieties of sensing technologies have been studied for the two aspects of air vehicle SHM. For operational load monitoring, electrical strain gauges, accelerometers, and fiber optic sensors are the main choices. Both strain gauges and accelerometers are relatively mature, but each sensor needs its own dedicated wires. This makes it difficult to multiplex them, and hence limits the number of measurement locations. Because most end users would like to see more locations being covered by sensors without increasing the system complexity and cost, an alternative method of using fiber Bragg grating (FBG) sensors is proposed and studied for operational load monitoring (11). An FBG load sensor operates on the principle that its Bragg wavelength changes monotonically with the strain applied on it. By monitoring the Bragg wavelength drift, any measurand inducing the change can be accurately detected. One of the main advantages of FBG sensors is that tens or hundreds of FBG sensors operating at different wavelengths can be fabricated and multiplexed in a single fiber strand and monitored remotely (12). This strong multiplexing capability can significantly alleviate the wiring issue and has made FBG sensors particularly attractive for large-scale structural SHM applications

(13, 14). For impact damage inspection, the actuators and sensors used to perform the actuation-sensing process are usually made of piezoelectric (PZT) materials. These devices have the same wiring challenge as strain gauges. Furthermore, a preamplifier is usually required for the signal amplification, aggravating the case by adding two more wires and extra cost. The use of FBG sensors to pick up ultrasonic/acoustic waves has been demonstrated for impact damage inspection, where multiple FBG sensors are employed and no additional amplifier is required (15). In addition, FBG sensors offer other great features required by air vehicle SHM, such as immunity to electromagnetic interference (EMI), lightweight, and small size. Besides these, as optical component prices have kept falling for the past few years, the cost of FBG sensors has also significantly decreased (16). A growing number of global enterprises, including National Instrument, Luna Innovations, New Focus, Micron Optics, SmartTec, and SmartFibre, are now developing and commercializing FBG sensor systems (17).

To fully exploit the benefits of FBG sensors in air vehicle SHM, several challenges need to be addressed. One of them is to find a suitable sensor interrogator to monitor the Bragg wavelength shift caused by the measurands. Currently, optical spectrum analyzer (OSA) is the standard instrument for FBG sensor interrogation. It provides large measurement range and high accuracy, but it is bulky, heavy, and not suitable for field applications. Several types of compact size interrogation units have been reported in recent years to address the field application issues (18). Among them, tunable Fabry–Perot (FP) filter is the most mature one and is widely used. It employs the principle of a tunable optic filter of very narrow wavelength bandwidth and is fabricated with either bulk optics (19) or fiber optics (20). Other techniques include arrayed waveguide grating (AWG) (21), diffractive grating with a charge coupled device (CCD) array (22), Mach–Zehnder interferometer (23), all fiber-based optical filters (24), chirped FBG (25), long period grating (LPG) (26), and Fourier transform spectroscopy (27). For air vehicle SHM, requirements addressed to sensor interrogators for operational load monitoring and impact damage inspection are different. For operational load monitoring, it is required to measure large changes (thousands of $\mu\varepsilon$) at a low speed (static). While for impact damage detection, the capability of measuring acoustic waves, that is, small changes (tens or hundreds of $\mu\varepsilon$) at an ultrafast speed (50–500 kHz), is usually required for a standard FBG sensor with a 10-mm grating length (28). On the basis of the review of interrogation techniques classified by the interrogation speed in Reference 29, tunable FP filter is used when the required speed is under 1 kHz, and the combination of diffractive grating and CCD array is suitable for the speed from 1 to 20 kHz (30). For the speed from 20 up to 500 kHz, three methods have been reported, that is, the use of a laser diode with its output wavelength fixed at the middle of one FBG sensor spectrum slope (15); an FP filter with a broadened and fixed spectrum (31); and an AWG with a fixed spectrum (32). Recent development of FBG sensors has made it feasible for FBG sensors to perform both operational load monitoring and impact damage inspection (33). This is ideal for air vehicle SHM. Therefore, an efficient and effective FBG sensor system used in air vehicle SHM would require the interrogator possessing the capabilities to address both load monitoring and impact damage detection. In fact, such an interrogator is not commercially available currently. Nevertheless, planar lightwave circuits (PLC)-based demultiplexer, including AWGs and Echelle diffractive gratings (EDG), have shown great potential (21, 32). PLC-based devices are fabricated by the standard silicon foundry tools, materials, and processes and offer the features of small size, lightweight, low cost, and excellent reliability required by the stringent telecommunication standards. The use of PLC in an interrogation unit inherits these features (34) and make it ideal for air vehicle SHM applications (33, 35).

In recent years, we have initiated the development of a miniaturized and field deployable FBG sensor system (mainly the interrogators) targeting air vehicle SHM applications. Different from the studies of AWG-based systems in References 21 and 32, where the spectrum of an AWG is fixed, we have introduced spectrum tuning techniques, that is, thermally tuning and mechanically tuning, to our AWG/EDG-based interrogation units to increase the measurement range and resolution. In this chapter, the development of AWG-based FBG sensor system with fixed spectrum is briefly introduced first. Then, spectrum tuning PLC-based FBG sensor interrogation units are presented, including the theoretical analyses on the two types of spectrum tuning techniques and their applications in wavelength interrogation. Finally, a dual function PLC-based FBG sensor system targeting both load monitoring and impact damage detection is discussed. In this system, two operation modes are introduced, that is, the sweeping mode for operational load monitoring and the parked mode for impact damage inspection.

6.2 SPECTRUM FIXED AWG-BASED FBG SENSOR SYSTEM

AWG is a type of PLC-based device initially developed to increase the transmission capacity of the dense wavelength division multiplexing (DWDM) optical communication system (36, 37). It has an array of narrow bandwidth optical filters and can be used as either multiplexers or demultiplexers for telecom application (36, 37). Moreover, these narrow bandwidth optical filters, defined as AWG channels, also have found applications in wavelength interrogations (21, 32).

6.2.1 Operation Principle

A typical AWG consists of input/output waveguides, two focusing slab regions (input/output coupler), and an array of multiple channel waveguides connecting the two slab regions with a constant path difference of ΔL between the adjacent waveguides. Different wavelengths experience different phase retardations within each channel waveguide and are spatially separated at the second focusing slab region. The phase retardations of two light beams passing through the ith and $(i + 1)$th channel waveguide are analyzed within the region of the second focusing slab (output coupler) as illustrated in Figure 6.1.

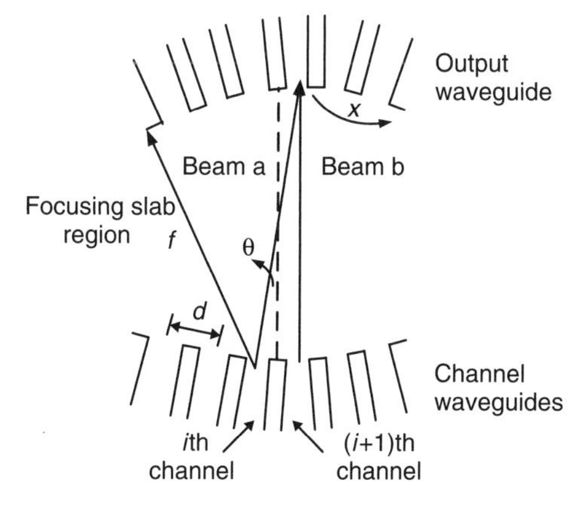

FIGURE 6.1 Enlarged view of the second focusing slab region.

Figure 6.1 shows the schematic configuration of the second focusing slab region. The channel waveguide separation is d and the radius of curvature is f. After the light beams, beam a and b, travel through the ith and $(i + 1)$th channel waveguide, respectively, and they constructively interfere at the focal point x. The geometrical distance of the two beams could be expressed by

$$f + \frac{d}{2}\sin\theta \approx f + \frac{dx}{2f}, \tag{6.1}$$

$$f - \frac{d}{2}\sin\theta \approx f - \frac{dx}{2f}, \tag{6.2}$$

where Equations 6.1 and 6.2 are for beam a and b, respectively. In order to satisfy the condition of constructive interference, the difference between the total phase retardations of the two beams passing through the ith and $(i + 1)$th channels should be an integer multiple of 2π. Considering that the first focusing slab region (not shown) has the same configuration as that of Figure 6.1, the interference condition can be written as (37)

$$\begin{aligned}&\left\{\beta_s(\lambda_c)\left(f_1 - \frac{d_1 x_1}{2f_1}\right) + \beta_a(\lambda_c)[L_0 + (i+1)\Delta L] + \beta_s(\lambda_c)\left(f + \frac{dx}{2f}\right)\right\}\\ &- \left\{\beta_s(\lambda_c)\left(f_1 + \frac{d_1 x_1}{2f_1}\right) + \beta_a(\lambda_c)[L_0 + i\,\Delta L] + \beta_s(\lambda_c)\left(f - \frac{dx}{2f}\right)\right\} = 2m\pi.\end{aligned} \tag{6.3}$$

The terms in the first and second { } represent the phase retardations in beam a and b, respectively. d_1 and f_1 are the channel waveguide separation and radius of curvature in the first focusing slab region, respectively, x_1 is the input light position, β_s and β_a denote the propagation constants in the slab region and channel waveguide,

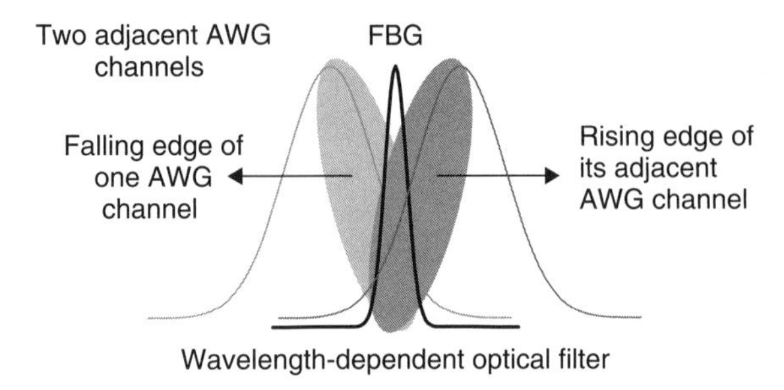

FIGURE 6.2 Illustration of wavelength interrogation using a spectrum fixed AWG.

L_0 is the minimum channel waveguide length, λ_c is the center wavelength of the AWG, and m is an integer.

Using this design, wavelengths coming out of each output waveguides are different and the difference (named wavelength spacing) between two neighboring AWG channels are the same. This corresponds to the generation of a series of narrow bandwidth optical filters. Wavelength interrogation can be achieved by measuring the light power ratio from two adjacent AWG channels as shown in Figure 6.2.

In this method, two adjacent AWG channels are used to build a wavelength-dependent optical filter, which consists of the falling-edge of one AWG channel and rising-edge of its adjacent AWG channel. The power ratio of the two AWG channels has a unique relationship with respect to the Bragg wavelength position of FBG sensor (21). Thus, by measuring the light power ratio from these two AWG channels, the wavelength of FBG sensor can be obtained.

6.2.2 Applications

Reference 38 reports the first use of a spectrum fixed AWG in FBG sensor system, which is for a distributed FBG sensor configuration. Then, a detailed theoretical and experimental study is described in Reference 21, where eight AWG channels are used to perform the wavelength interrogation of seven FBG sensors. It is concluded that as long as the bandwidth of an FBG sensor is sufficiently smaller than that of the AWG, this method could be used for the interrogation of FBG sensors. The resolution and accuracy are dependent on the bandwidth and channel spacing of the AWG. Since this method does not have any moving parts, the interrogation speed is only determined by the speed of photodetectors and data processing. Thus, it can be applied to detect acoustic waves with a frequency up to 300 kHz (32). In this application, FBG sensors and PZT actuators are used to monitor the integrity of composite structures in aircrafts as an impact damage inspection system. The specimen is a practical aircraft wing box structure composed of CFRP as shown in Figure 6.3. The FBG sensors and PZT actuators are bonded to the flanges of the stringers as shown in Figure 6.4.

In a standard actuation-sensing process, the acoustic waves generated from the PZT actuators are picked up by the sensors. In this chapter, the FBG sensors are

FIGURE 6.3 Image of the test specimen. *Source:* Reprinted from Reference 32 with permission from SPIE.

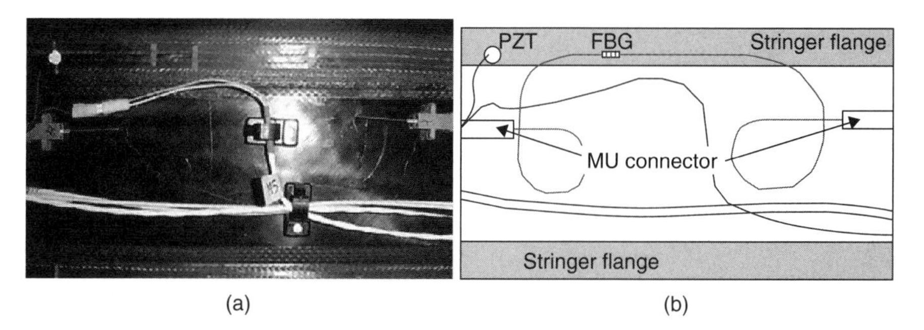

FIGURE 6.4 (a) Layout of the FBG sensors and (b) PZT actuators in the specimen. *Source:* Reprinted from Reference 32 with permission from SPIE.

implemented instead of PZT sensors due to the unique features of FBG sensors discussed above. An AWG-based interrogation unit is used to convert the received signal into wavelength changes of FBG sensors, which are further analyzed to reconstruct the propagation of the acoustic waves in the structure. Initial results show that the system integrity is obtained, and the use of the hybrid FBG/PZT with the AWG-based interrogation unit can be used to identify the impact damage.

Limited by the channel spacing of the AWG (0.8 nm is the typical value), the measurement range reported in References 21 and 32 for each FBG sensor is ~0.75 nm, which corresponds to ~750 $\mu\varepsilon$ for strain measurement. This range is small for load monitoring. Furthermore, before each measurement, the wavelength of an FBG sensor has to be placed at the middle of the two adjacent AWG channels to achieve an optimum performance, which introduces extra calibration work before each measurement. The measurement range can be improved by using the interferometric wavelength shift detection technique (39). A better way is to use multiple AWG channels for the wavelength interrogation (40). Figure 6.5 shows the user interface developed for this purpose.

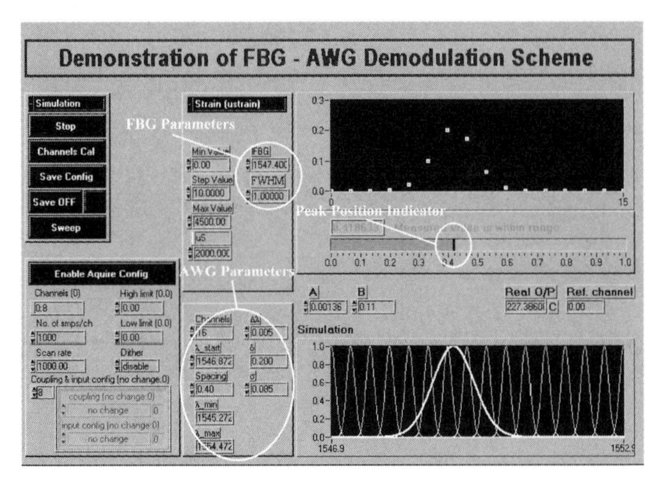

FIGURE 6.5 User interface of the AWG-based interrogation unit. *Source:* Reprinted from Reference 40 with permission from IEEE. Copyright © [2003] IEEE.

In Figure 6.5, 16 AWG channels with a channel spacing of 0.4 nm are used to interrogate an FBG sensor applied with a strain of 2000 $\mu\varepsilon$ at room temperature. During the data processing, spectrum profiles of AWG channels and FBG sensor are simulated with Gaussian shapes. The peak position of the FBG sensor is calculated by a quadratic function through the three points around the maximum channel output. Initial results show that the strain values measured by this system accords well with the result obtained from the reference strain gauges.

The measurement resolution of a spectrum fixed AWG-based FBG sensor system is counterproportional to the AWG channel spacing and channel bandwidth (21). Although standard commercial off-the-shelf AWGs offer proved reliability and lower cost, their measurement resolution is limited by the available channel spacings (usually fixed as either 0.4 or 0.8 nm) and the channel bandwidth. To address this, Cheben et al. (41) design a high resolution AWG-based interrogation unit. In this unit, the AWG is on a silicon-on-insulator (SOI) platform, having 50 channels with the channel spacing of 0.18 nm and a footprint of 8 mm × 8 mm. The authors demonstrated the use of this high resolution SOI AWG-based interrogator for the monitoring of a tilted fiber Bragg grating (TFBG)-based refractive index sensor.

6.3 SPECTRUM TUNING AWG-/EDG-BASED FBG SENSOR SYSTEMS

Although decreasing the AWG channel spacing can improve the measurement resolution of a spectrum fixed AWG-based FBG sensor system, the method poses great fabrication difficulties on the AWG and significantly increases the device cost. It also tends to reduce the measurement range of the interrogator. To address this, we introduce spectrum tuning techniques in the PLC-based interrogation units. In our designs, two types of spectrum tuning techniques are used, that is, thermal tuning and mechanical tuning. In recent years, we have developed and demonstrated AWG- and EDG-based interrogation units based on the two spectrum tuning techniques. In this session, theoretical analyses and applications in FBG sensor interrogations are described.

6.3.1 Principle of Spectrum Tuning AWG

Spectrum tuning techniques have been applied on both AWG- and EDG-based interrogation units. Since an EDG has a similar operation principle and the same spectrum profile as those of an AWG (42), only the theoretical analyses of spectrum tuning AWG are discussed in this session.

6.3.1.1 Thermal Spectrum Tuning In optical communication systems, temporal wavelength shift of an AWG channel is usually minimized by the accurate control of the AWG chip temperature (43). It is reported that the transmission wavelength of an AWG channel has a linear relationship with respect to the AWG temperature (44). The temperature-induced wavelength shift coefficient is approximately 0.01 nm/°C for a silica-based AWG as measured in our previous study (45). This linear relationship between the transmission wavelength of an AWG channel and the AWG temperature can be used for the wavelength interrogation.

From Equation 6.3, the center wavelength of an AWG can be determined as

$$\lambda_c = \frac{n_a \Delta L}{m}, \tag{6.4}$$

where n_a is the effective refractive index of the arrayed waveguide. Owing to the temperature dependency of both n_a and ΔL, the center wavelength shift is obtained by differentiating temperature on both sides of Equation 6.4 (17, 37)

$$\frac{d\lambda_c}{dT} = \frac{1}{m}\frac{d(n_a \Delta L)}{dT}. \tag{6.5}$$

For an AWG fabricated with the material of silica, we have (44)

$$\frac{d(n_a \Delta L)}{dT} = 10^{-5} \times \Delta L. \tag{6.6}$$

Substituting Equation 6.6 into Equation 6.5, we obtain

$$\lambda_c(T) = \lambda_c(T_0) + \frac{10^{-5}\Delta L}{m}\Delta T, \tag{6.7}$$

where $\lambda_c(T_0)$ and $\lambda_c(T)$ are the transmission wavelength of an AWG channel at the temperature T_0 and T, respectively, ΔT is the temperature difference between T_0 and T. From Equation 6.7, it is seen that the transmission wavelength shifts

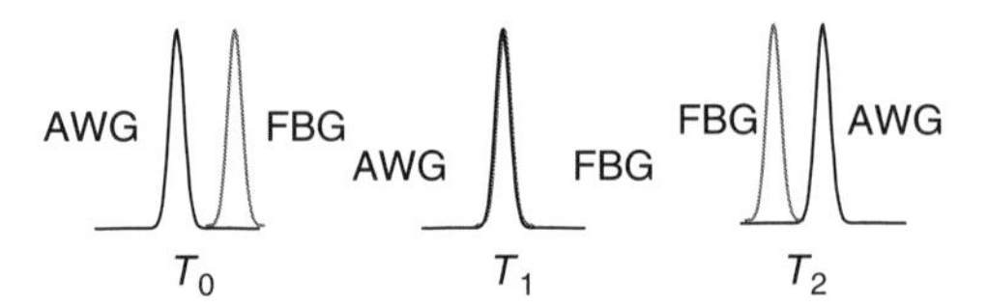

FIGURE 6.6 Illustration of the peak matching method using a tunable AWG spectrum.

linearly with the AWG temperature. According to the AWG operation principle, all AWG channels possess the same wavelength tuning capability.

Using the peak matching technique, as illustrated in Figure 6.6, the spectrum tuning AWG can be used to measure the wavelength of an FBG sensor. Supposing that the transmission wavelength of AWG channel A is shorter than the wavelength of the FBG sensor at temperature T_0, the light power from channel A is received as P_0 by an optical power meter. When increasing the temperature of the AWG, the transmission wavelength of channel A shifts. At temperature T_1, the transmission wavelength of channel A overlaps with the wavelength of the FBG sensor with the light power from channel A measured as P_1. Obviously, from temperature T_0 to T_1, the received light power starts from P_0 and reaches the maximum value of P_1 at temperature T_1. If continuously heating the AWG, the transmission wavelength of channel A becomes larger than the wavelength of the FBG sensor, causing the received light power from channel A to decrease. Thus, by selecting the suitable transmission channel of an AWG and knowing the temperature corresponding to the maximum value of the received light power, the wavelength of the FBG sensor can be precisely interrogated using the relationship between the chip temperature and the transmission wavelength discussed in Equation 6.7.

According to the relationship between the silica chip temperature and the AWG transmission wavelength (46), it needs to tune the chip temperature for 80°C in order to cover the 0.8 nm channel spacing of a standard commercial AWG. Since all the AWG channels have the same wavelength tuning capability, the 80°C temperature tuning range will cover the free spectral range (FSR) of an AWG. The wavelength measurement resolution is now dependent on the temperature tuning resolution rather than the channel spacing. Hence, the spectrum tuning AWG can increase the measurement resolution without scarifying the measurement range.

6.3.1.2 Mechanical Spectrum Tuning and Space-to-Wavelength Mapping

The mechanical tuning technique is based on the fact that the transmission wavelength of an AWG channel changes linearly with the beam position along the AWG input coupler (first focusing slab region). The spectrum tuning is achieved by either manually tuning with a position stage or automatically tuning with a PZT motor. This technique provides a mapping from the spatial position of the input light beam to the transmission wavelength of an AWG channel, which is referred as the *space-to-wavelength mapping*.

Recall that Equation 6.3 describes the fundamental principle of an AWG, it could also be rearranged as (37)

$$\beta_s(\lambda_c)\frac{d_1 x_1}{f_1} - \beta_s(\lambda_c)\frac{dx}{f} + \beta_a(\lambda_c)\Delta L = 2m\pi. \tag{6.8}$$

By differentiating Equation 6.8, we obtain the relationship between the output focal point x and the wavelength λ for fixed input position x_1, shown as

$$\frac{\Delta x}{\Delta\lambda} = -\frac{n_a f_1 \Delta L}{n_s d\lambda_c}, \tag{6.9}$$

where n_s is the refractive index of the slab region. Because an AWG is a reciprocal device, the dependence of the wavelength λ on the input position x_1 for a fixed output focal point x_1 can be expressed by

$$\left|\frac{\Delta\lambda}{\Delta x_1}\right| = \frac{n_s d_1 \lambda_c}{n_a f_1 \Delta L}. \tag{6.10}$$

Equation 6.10 shows that the transmission wavelength of an AWG channel can be tuned by the scanning of the input light beam along the AWG input coupler. A typical simulation result of a selected AWG channel tuning spectrum obtained using the beam propagation method is shown in Figure 6.7.

In the simulation, the AWG parameters are set by the default values in the BeamPro software, and the input light beam positions are set with six values (0, 20, 40, 110, 140, and 160 μm) as shown in Figure 6.7. The spectrum of one designated AWG channel is calculated under these values. It is seen that a tuning spectrum is achieved by changing the input light beam position and a much larger

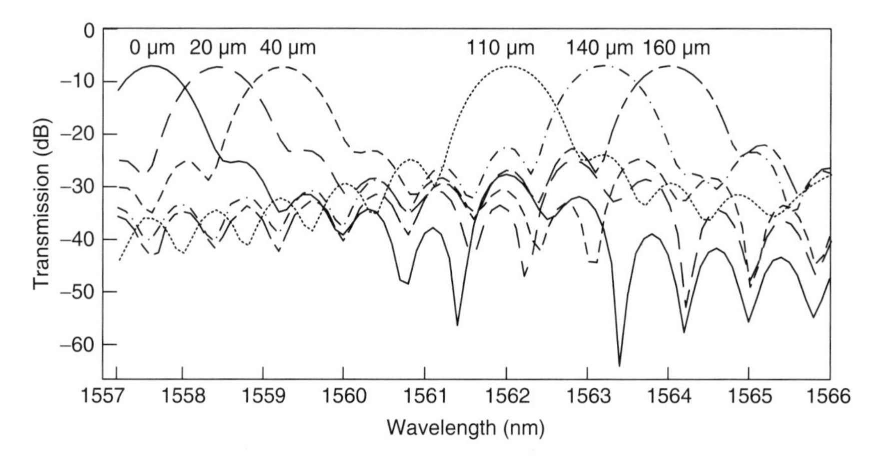

FIGURE 6.7 Theoretical simulation of the spectrum shifts of a designated AWG channel with the input light beam position changes.

tuning range can be achieved than the thermal tuning method. On the basis of the simulation, it is feasible to build a multichannel interrogation unit with each AWG channel covering the entire C and L band.

Although butt-to-butt coupling between an optical fiber and planar waveguide is a relatively mature technique and variety of methods can be used to minimized the transmission loss (47), it is challenging to couple light from an optical fiber to an AWG input coupler as the geometrical profile of the AWG input coupler is a Rowland Circle. Direct butt-to-butt coupling between an optical fiber and the input coupler with the Rowland Circle profile results in significant insertion loss. To circumvent this, we cut the Rowland Circle into a slab profile. The shape change of the input coupler and an illustration of the proposed space-to-wavelength mapping is presented in Figure 6.8.

Figure 6.8 illustrates the mechanical spectrum tuning and the space-to-wavelength mapping technique. The wavelength interrogation of an FBG sensor is achieved by measuring the input light beam position along the modified AWG input coupler, which corresponds to the maximum light power received from the designated AWG output channel. The input light beam position is then converted into the transmission wavelength of the designated AWG channel using the space-to-wavelength mapping. According to the peak matching principle, this obtained transmission wavelength has the same value as the wavelength of the FBG sensor.

6.3.2 Applications of Spectrum Tuning PLC

The PLC type FBG sensor interrogation units based on the spectrum tuning techniques have been verified in a variety of applications. The initial experimental results are summarized in this session.

6.3.2.1 AWG Interrogator Based on the Thermal Spectrum Tuning We first studied FBG sensor interrogation using the thermal tuning AWG. Details can be found in References 48 and 49. The AWG used is a standard commercial product from NKT Integration. It has 32 output channels with a channel spacing of 0.8 nm. An illustration of the AWG is shown in Figure 6.9.

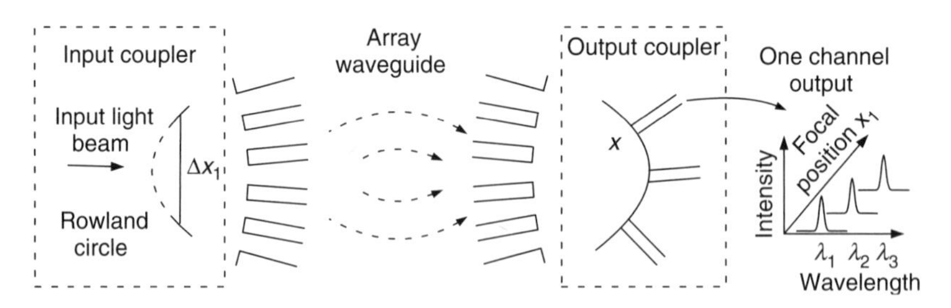

FIGURE 6.8 Illustration of the mechanical spectrum tuning and the space-to-wavelength mapping technique.

FIGURE 6.9 An illustration of the AWG chip used for FBG sensor interrogation.

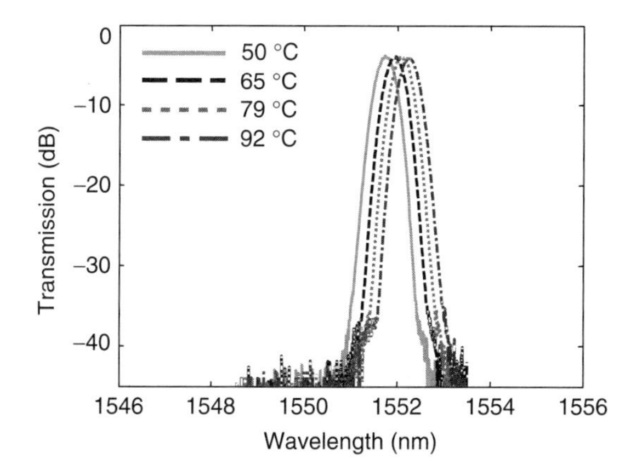

FIGURE 6.10 Spectrum of a designated AWG channel with different AWG temperature.

The thermal tuning performance is first tested by monitoring the transmission spectrum of one designated AWG channel. The result is shown in Figure 6.10. It can be seen that when changing the AWG temperature, the transmission wavelength of the designated AWG channel shifts without distorting the spectrum profile.

In order to verify the linear relationship between the AWG temperature and the transmission wavelength of the AWG, six AWG channels are selected and tested. The results of their temperature-induced wavelength shift coefficients are shown in Figure 6.11. It is clear that the transmission wavelength of the AWG channel changes linearly with the AWG temperature at a rate of 0.011 nm/°C. Furthermore, all the six AWG channels have the same spectrum tuning rate, validating the theoretical assumption that all the AWG channels have the same spectrum tuning performance.

This thermal tuning AWG is used to interrogate six FBG sensors with the experimental setup shown in Figure 6.12 (49). The setup includes a broadband light source, an optical circulator, six FBG sensors, and the AWG-based interrogation

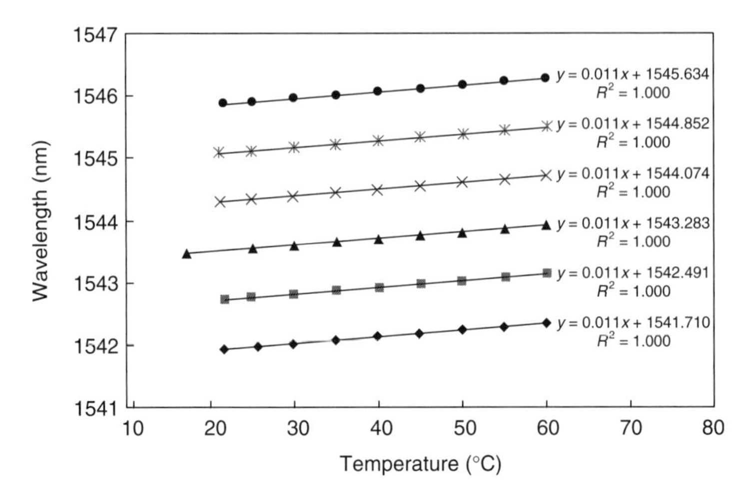

FIGURE 6.11 Relationship between the AWG temperature and the transmission wavelengths of the AWG channels. *Source:* Reprinted, by permission, from Reference 49.

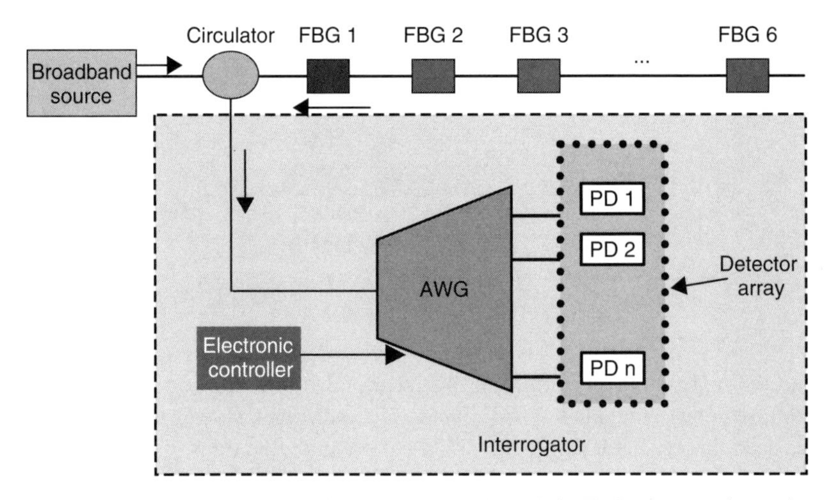

FIGURE 6.12 Experimental setup for the interrogation of six FBG sensors using a thermal tuning AWG.

unit, which consists of a 32-channel AWG with 0.8 nm channel spacing, an 8-channel photodetector array, and an electronic controller. The temperature tuning is performed by a film heater that is glued to the back of the AWG chip. A resistance temperature detector (RTD) is used to monitor the AWG temperature, and the AWG temperature is precisely adjusted with the assistance of the electronic controller. The six FBG sensors are supplied by Avensys Inc., with the wavelength spacing between neighboring FBG sensors as 0.8 nm. All the six FBG sensors

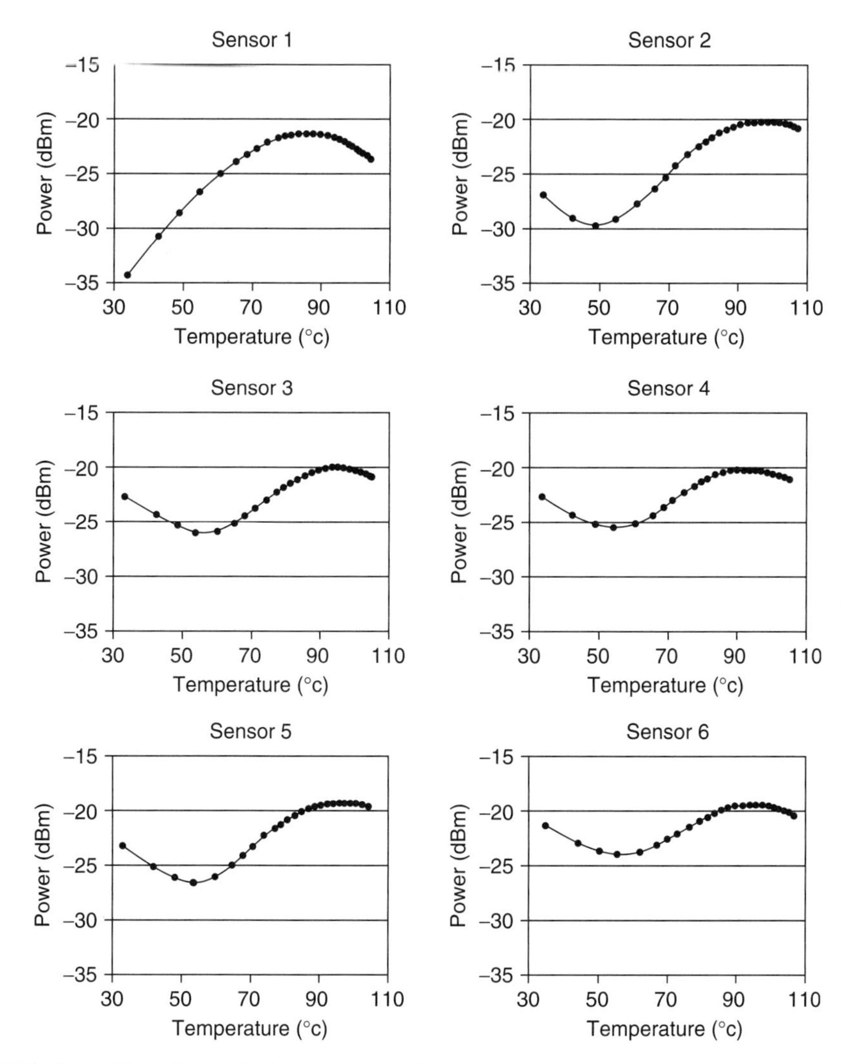

FIGURE 6.13 Measured light power with respect to the AWG temperature. *Source:* Reprinted by permission from Reference 49.

are separated with a distance larger than the coherent length of the light reflected by any FBG sensor. A Labview program is used to perform the data acquisition (DAQ), temperature control, and data processing. The experimental result is shown in Figure 6.13.

The obtained results are the measured light power from six photodetectors with respect to the AWG temperature. Each photodetector is connected to one of the six AWG channels and is used to monitor one FBG sensor. According to the wavelength interrogation principle, as discussed above, the temperatures corresponding to the peak light power are first obtained, and then their values are converted to

TABLE 6.1 Comparison between the Bragg Wavelengths of the Six FBG Sensors Measured by the Developed Interrogation Unit and the Data Provided by the Manufacturer[a]

Sensors	Peak Temperature, °C	Sensor Wavelength, nm	Sensor Wavelength Provided by the Manufacturer, nm
1	86.42	1542.661	1542.650
2	94.81	1542.661	1543.520
3	93.79	1543.534	1544.30
4	92.94	1544.315	1545.060
5	96.49	1545.096	1545.900
6	93.18	1546.659	1546.660

[a] Reference 49.

the wavelengths using the linear relationship as shown in Figure 6.11. All those temperature values and the measured wavelengths are listed in Table 6.1. For comparing reason, the sensor wavelengths provided by the manufacturer are also shown in the same table. It can be seen that the measured results agree well with the data provided by the manufacturer. The small variation between the two sets of data is believed to be attributed to the difference in measurement environment, such as temperature and strain.

The thermal tuning AWG is also used for the interrogation of an LPG sensor with the experimental setup shown in Figure 6.14 (50).

In this test, the AWG temperature is changed from 25.4°C to 98.8°C in order to achieve a tuning spectrum range of ~0.8 nm. On the basis of the uniformity of the AWG (37) and experimental results discussed above, all the AWG channels have the same wavelength tunability. Therefore, the 32-channel AWG is able to cover a 25-nm measurement range, which is broad enough for the interrogation of an LPG sensor.

When using the thermal tuning AWG in the LPG sensor interrogation, we take the measurement of the temperature corresponding to the minimum light power, which is different from that of the FBG sensor interrogation. Although 32 channels are used to cover a 25-nm measurement range, only the AWG channel, which detects the minimum light power, is of the interest as shown in Figure 6.15.

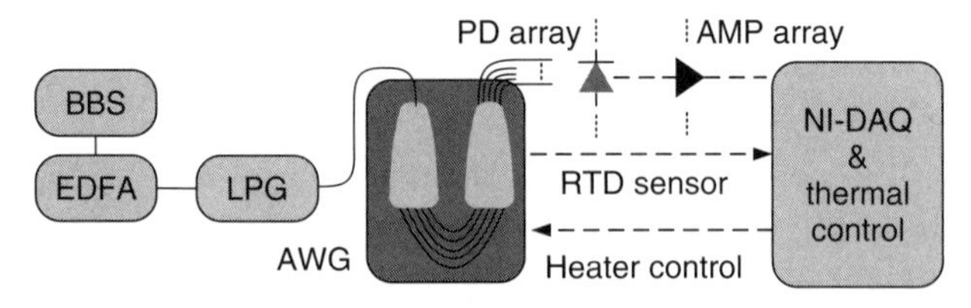

FIGURE 6.14 Experimental setup for the interrogation of an LPG sensor using a thermal tuning AWG. *Source:* Reprinted, by permission, from Reference 50.

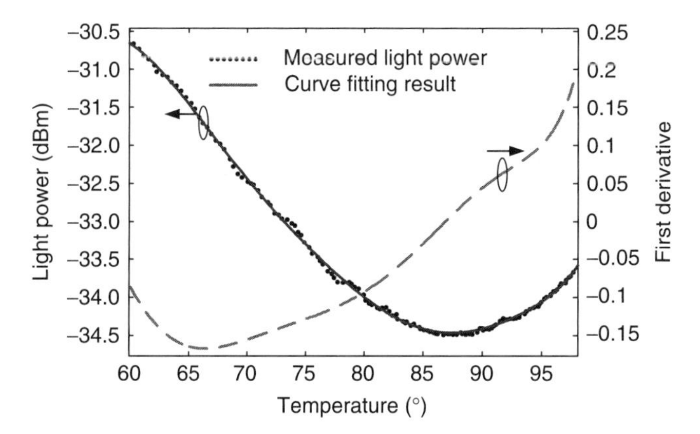

FIGURE 6.15 Measured light power, curve fitting, and the first-order derivative with respect to the AWG temperature. *Source:* Reprinted, by permission, from Reference 50.

In the data processing, curve fitting is first applied to the measured light power with respect to the temperature. Then, the desirable temperature is obtained by analyzing the first-order derivative as shown in Figure 6.15 (the dashed line). The interrogated center wavelength of the LPG sensor is found agreeing well with the value provided by the manufacturer.

We have found that better than 0.01°C temperature resolution can be achieved in the laboratory conditions (49). This corresponds to a wavelength interrogation resolution of approximately 0.1 pm as indicated by the results shown in Figure 6.11. The use of a film heater to control the AWG temperature is a proved approach, as its reliability has already been tested and verified by the stringent telecommunication application requirements, where a commercial AWG chip uses a film heater to maintain its temperature at 70°C (or other preferred temperatures) to avoid the wavelength drift introduced by the ambient temperature changes (43, 46). In addition, the interrogation setup shown in Figure 6.12 can be designed and packaged into a hand-held all-solid device, as the AWG chip is very small, the one used here is 30 mm × 50 mm, and the photodetector array is able to be made smaller than 10 mm × 30 mm.

Practically, it is a waste of power to heat the whole AWG device because only the section of arrayed waveguides is of the interest in tuning the spectrum. A better solution is to employ a standard thin film heater fabrication process by depositing the heater directly on the surface of the arrayed waveguides. In this way, the power consumption can be reduced significantly, as the thermal mass is reduced to a large extent. The added advantage of this technique is that it significantly increases the interrogation speed to 500 Hz, as the response time is reduced to 2 ms for a silica waveguide-based device (51). This speed would be fast enough for any dynamic measurement application. In addition, it is found that smaller errors in the wavelength interrogation can be achieved by a smaller spectrum bandwidth of the AWG channel (52), which also works for the EDG-based interrogation units.

6.3.2.2 EDG Interrogator Based on the Thermal Spectrum Tuning Employing the same thermal spectrum tuning in the wavelength interrogation, we have prototyped miniaturized interrogation units using a monolithically packaged EDG device (53–56). The EDG device consists of a 1 × 32 EDG demultiplexer, a detector array, a thermal electric cooler (TEC), and an RTD. The whole device is made on an indium phosphide (InP) chip. In order to reduce the electronic complexity, we only connect the odd number EDG channels in our first prototype and use a DAQ device of 16 analog input channels. One analog input channel is assigned to detect the EDG temperature, and the others are assigned for the EDG output channels. The trade-off of this design is that the channel spacing of our first prototype becomes 1.6 nm rather than the 0.8 nm channel spacing of the EDG demultiplexer (57). This increases the temperature tuning range required and hence the electric power consumption. The detector array is integrated with the EDG multiplexer on the InP chip using semiconductor processing techniques (57). As the refractive index of the InP changes linearly with the chip temperature in the range of 10–60°C (58), the transmission wavelength of an EDG channel changes linearly with respect to the EDG temperature at a rate of approximately 90 pm/°C (53). Since this rate is much higher than the one of using an AWG, which is approximately 11 pm/°C (45), the temperature tuning range is significantly decreased in the EDG interrogator. Thus, in this design, a TEC is used to replace the film heater to control the temperature. By changing the electrical input to the TEC, the EDG temperature could be precisely controlled and tuned. The RTD performs the same function of providing the temperature feedback in the closed-loop temperature control.

Figure 6.16a shows the monolithically integrated EDG interrogator chip. It has a dimension of 20 mm × 7.3 mm. Figure 6.16b illustrates the packaged EDG device with 60 electrical pins. The whole packaged device has a weight of less than 60 g and a miniaturized dimension of 45 mm × 30 mm × 15 mm, making it ideal for the development of a micro FBG sensor system for air vehicle SHM.

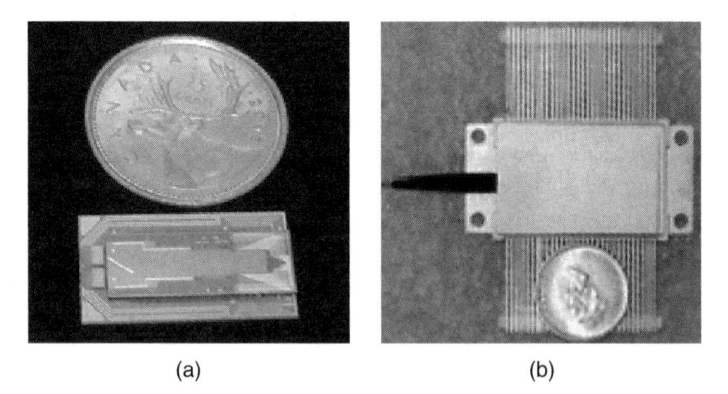

(a) (b)

FIGURE 6.16 Monolithically integrated EDG device. (a) Open box (chip dimension: 20 mm × 7.3 mm) and (b) closed box (box dimension: 45 mm × 30 mm × 15 mm.)

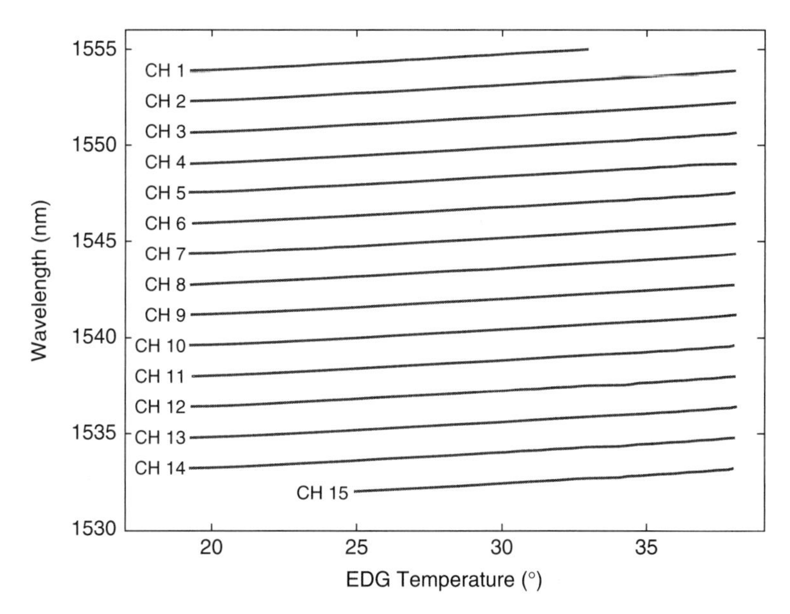

FIGURE 6.17 Relationship between the EDG temperature and the transmission wavelengths of the 15 EDG channels. *Source:* Reprinted, by permission, from Reference 54.

The channel spacing in our first prototype is 1.6 nm. As the wavelength tuning rate with respect to the EDG temperature is measured as approximately 90 pm/°C, a temperature tuning range of 20°C is enough for each EDG channel to cover the 1.6 nm. The relationship between the EDG temperature and the transmission wavelengths of the 15 EDG channels are tested and the results are shown in Figure 6.17.

It is seen that the transmission wavelength of all the EDG channels change linearly with the EDG temperature. All the EDG channels are cascaded to provide a measurement range of approximately 24 nm. The TEC used in this design can control the temperature with a resolution of 0.01°C, making the wavelength measurement resolution better than 1 pm.

Two laboratory prototypes are developed using this EDG device. The operation principle of the interrogation unit is shown in Figure 6.18, and the two laboratory prototypes are shown in Figure 6.19.

In Figure 6.18, an electric controller is used to perform the signal amplification, temperature control, A/D conversion, and DAQ. We use a DAQ card and a signal conditioning board from National Instrument as the electric controller in the first laboratory prototype as shown in Figure 6.19a. It has a dimension of 250 mm × 250 mm × 110 mm. In order to miniaturize the dimension, we develop the second laboratory prototype as shown in Figure 6.19b, which uses a microcontroller and a more sophisticated electric circuit to perform the electric controller, and the dimension is reduced to 125 mm × 73 mm × 50 mm. To demonstrate the performance of the prototypes, a variety of tests are implemented.

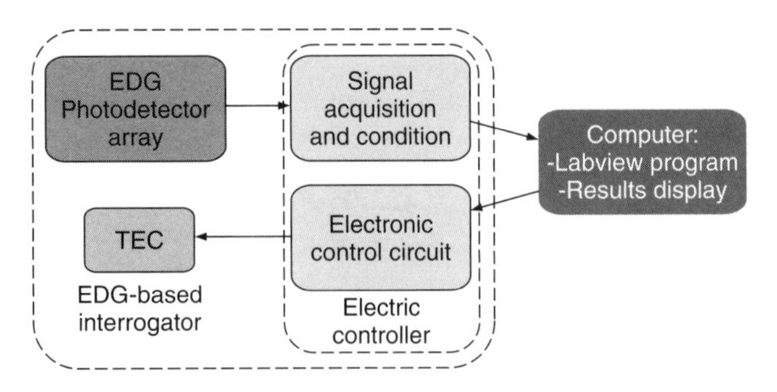

FIGURE 6.18 Illustration of the operation principle of the EDG-based interrogation unit.

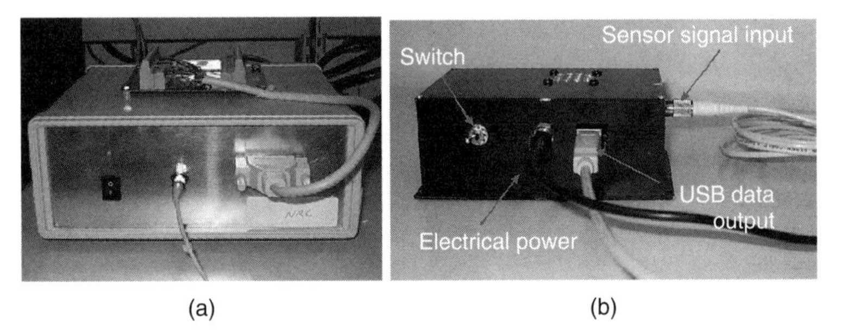

FIGURE 6.19 Prototypes of the EDG-based interrogation units. (a) Prototype based on NI DAQ system and (b) prototype based on microcontroller.

First, a remote temperature measurement is carried out (54). In this test, two FBG sensors are placed in a temperature-controlled oven. The distance between the sensors and our interrogation unit is 50 m. For temperature variations ranging from room temperature to approximately 80°C, the experimental results of interrogating FBG sensors by our prototyped EDG interrogator are shown in Figure 6.20.

The Bragg wavelengths of the two FBG sensors are 1543 nm and 1547 nm. During the test, both FBG sensors are fixed on a metal plate and placed on the shelf inside the oven to minimize any possible strain condition changes occurring on the sensors during the tests. Figure 6.20 shows the relationship between the interrogated wavelength of the two FBG sensors and the oven temperature. The results show that the interrogated wavelengths of the two FBG sensors change monotonically with respect to the ambient temperature. The rate difference between Figure 6.20a and b is believed to be attributed to the packaging of the two FBG sensors, which introduces different stress on the FBG sensors.

Second, mechanical sensing test is implemented (54). In this test, the EDG interrogator is used to monitor the load applied to a metal coupon by interrogating the wavelengths of the two FBG sensors with the test setup shown in Figure 6.21. The two FBG sensors are bonded at the mid-span of the aluminum testing coupon using the standard M-bond AE10 adhesive, where the strain distribution is known

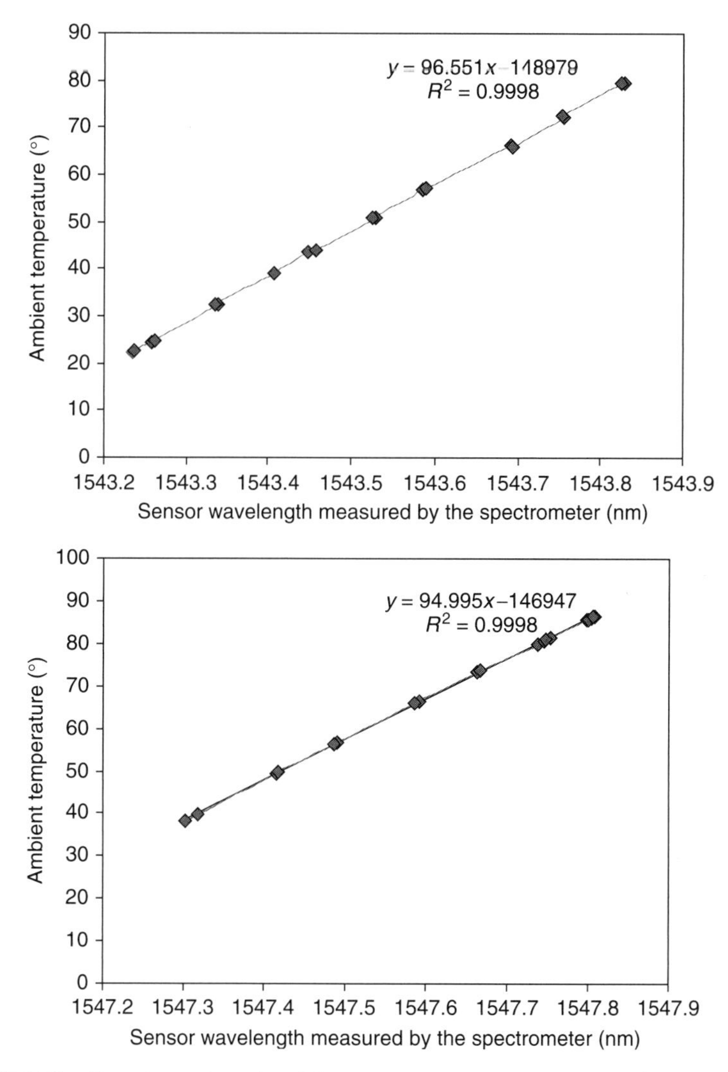

FIGURE 6.20 Experimental results of the remote temperature sensor monitoring. (a) 1543-nm FBG sensor and (b) 1547-nm FBG sensor. *Source:* Reprinted by permission from Reference 54.

to be uniform. The distance between the two FBG sensors is 10 mm. The coupon is then mounted on an MTS Load Frame for test.

Loads are applied to the coupon from 0 N to a maximum value of 71,528 N (~3200 μm of the coupon) at a step of 892.857 N with a 3-min holding period after each load increasing. Part of the loads profile is shown in Figure 6.22.

Figure 6.23 shows the relationship between the wavelengths measured by our EDG interrogation unit and the applied loads. In order to evaluate the performance

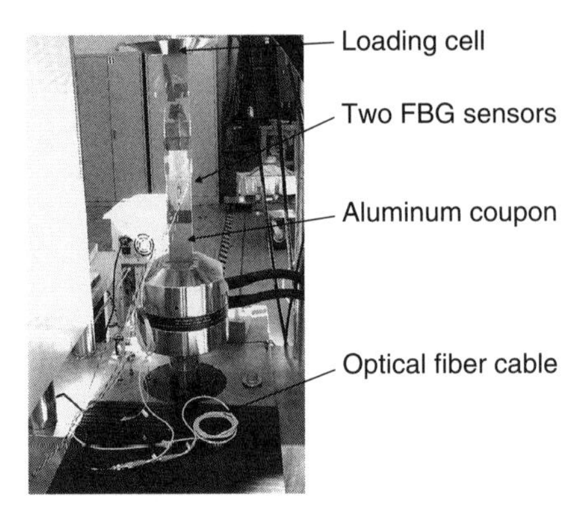

FIGURE 6.21 Setup for the mechanical sensing test. *Source:* Reprinted, by permission, from Reference 54.

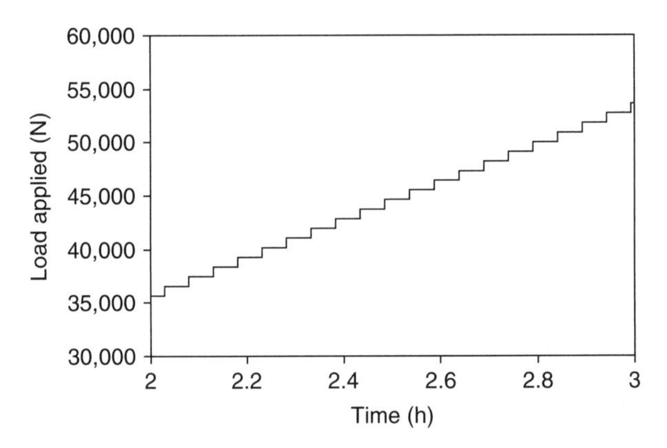

FIGURE 6.22 Illustration of the load applied to the coupon. *Source:* Reprinted by permission from Reference 54.

of our developed interrogation unit, the wavelength of the two FBG sensors is also measured by a multiwavelength meter (HP 86120C). The comparison between the two systems is shown in Figure 6.24. As the figure shows, a strong correlation between the two systems has been achieved, which confirms that the thermal-tuning-EDG-based interrogation unit can be used in the strain measurement and operational load monitoring.

Third, a simultaneous interrogation of a hybrid FBG/LPG sensor pair is performed using this interrogation unit, but employing a different thermal tuning scheme (55). In the previous approaches (53, 54), the EDG temperature is increased by steps. At each step, the light power from EDG channels and the EDG temperature are recorded. In order to reach a high resolution, EDG temperature is tuned with a 0.01°C step. This type of temperature tuning and DAQ significantly decrease the interrogation speed. To overcome this limitation, a different temperature control

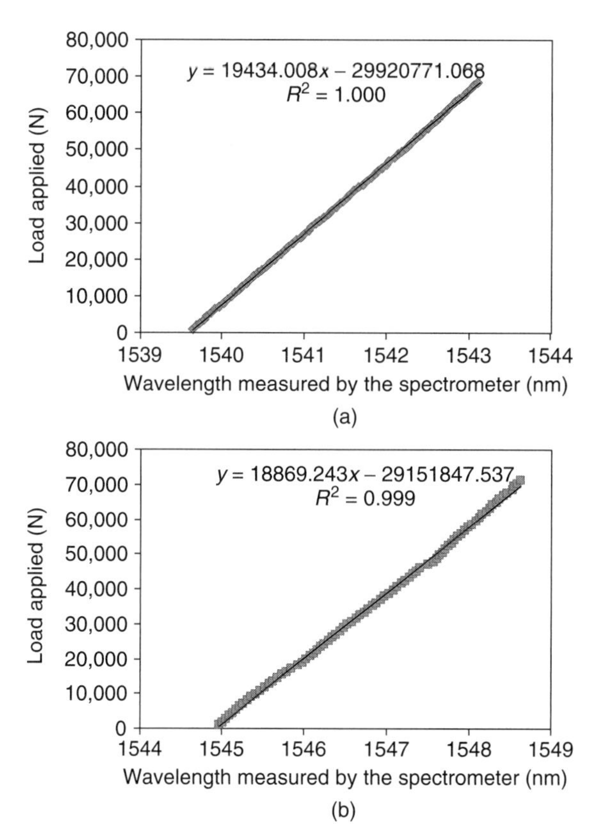

FIGURE 6.23 Bragg wavelengths measured by the thermal-tuning-based EDG interrogation unit with respect to the applied loads. (a) 1539 nm FBG sensor and (b) 1544 nm FBG sensor. *Source:* Reprinted by permission from Reference 54.

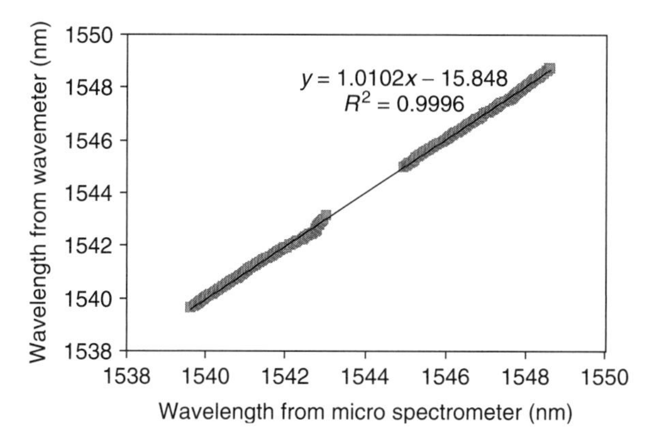

FIGURE 6.24 Correlation between the wavelengths measured by the thermal-tuning-based EDG interrogation unit and the HP multiwavelength meter. *Source:* Reprinted by permission from Reference 54.

method is adopted. The TEC is set directly from the lowest value to the highest value without introducing the tuning steps. Meanwhile, the light power from all the EDG channels and the temperature readings from the RTD are recorded with a DAQ card under the same trigger. In this case, all the data are recorded with respect to the sampling point. High resolution could still be achieved by setting a high sampling rate in the DAQ. The measurement speed is now limited by the inherent thermal characteristics of the EDG device rather than the implementation of a large number of temperature tuning steps. By this temperature tuning method and DAQ technique, the temporal cycle for each measurement is reduced from 30 to 10 s. Experimental results are shown in Figures 6.25–6.28.

The hybrid FBG/LPG sensor pair has a transmission spectrum as shown in Figure 6.25, which is measured by an OSA with a wavelength resolution of 10 pm. It is noted that the wavelength spacing between the two sensors is more than 4 nm and larger than the channel spacing of the EDG. Therefore, the measured spectra of the two sensors will be located in different EDG channels. In the measurement, the

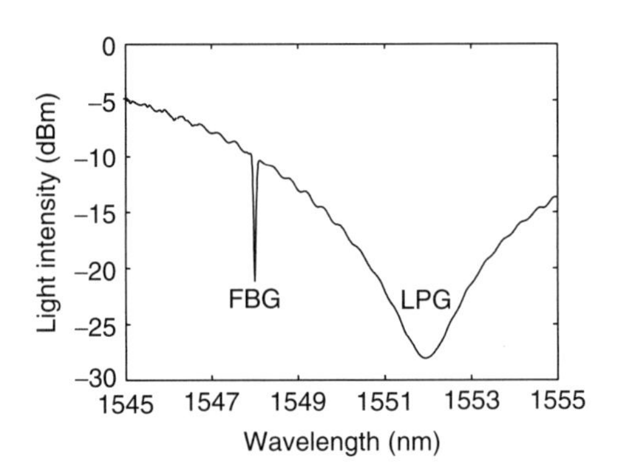

FIGURE 6.25 Spectrum of a hybrid FBG/LPG sensor pair measured by an OSA. *Source:* Reprinted by permission from Reference 55.

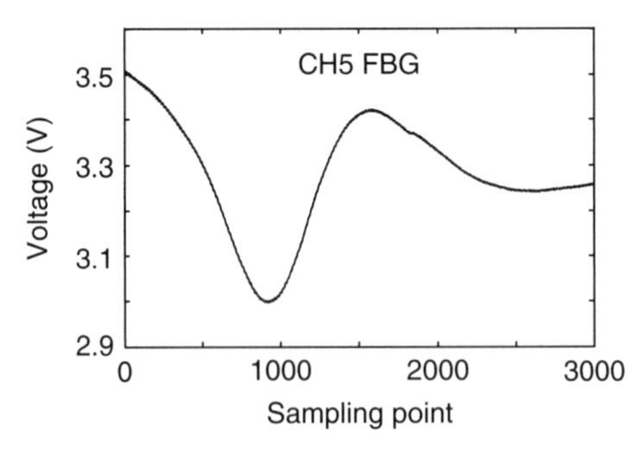

FIGURE 6.26 Measured light power for the FBG sensor with respect to the sampling points using channel 5 of the EDG-based interrogation unit. *Source:* Reprinted by permission from Reference 55.

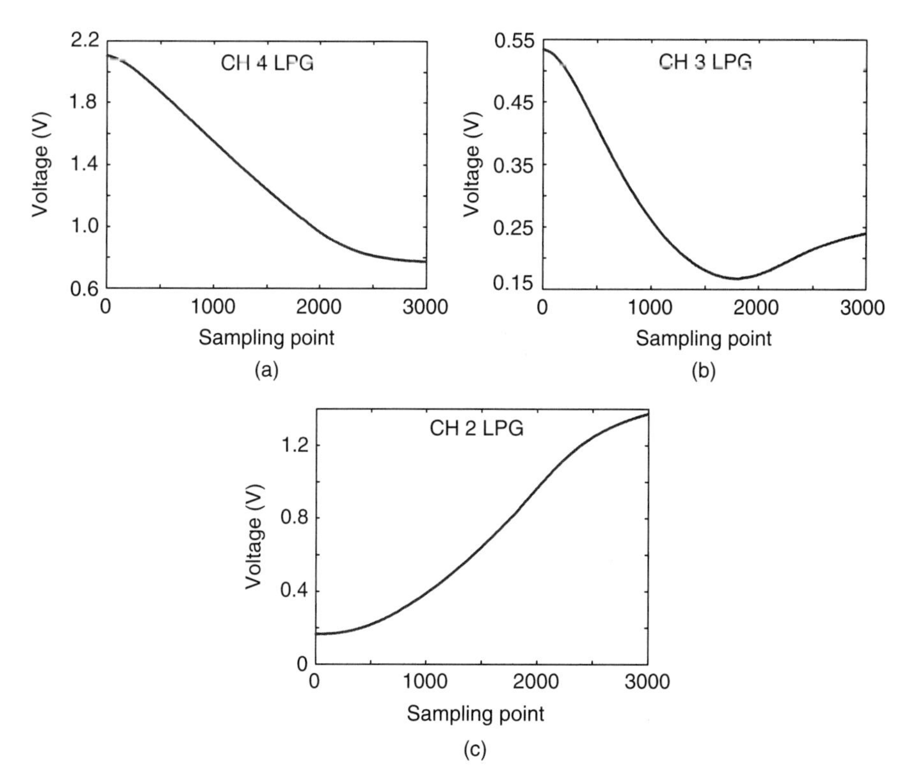

FIGURE 6.27 Measured light power for the LPG sensor with respect to the sampling points using channel 2–4 of the EDG-based interrogation unit. *Source:* Reprinted by permission from Reference 55.

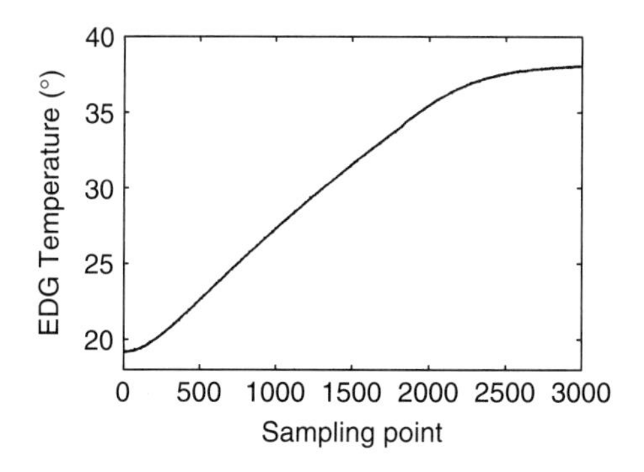

FIGURE 6.28 Measured EDG temperature with respect to the sampling points. *Source:* Reprinted by permission from Reference 55.

bandwidth of the FBG sensor is relatively small and one designated EDG channel (channel 5) is able to obtain its spectrum as shown in Figure 6.26. Although the LPG sensor has a broader spectrum, exceeding the channel spacing of the EDG device, the flexibility of the developed interrogation unit allows the use of multiple EDG channels to obtain its spectrum. In this test, three EDG channels (channel 2–4) are used to obtain the LPG spectrum as shown in Figure 6.27a–c, where the minimum light power is located at channel 3. Figure 6.28 shows the EDG temperature, which is measured by the RTD and recorded with the same time trigger as the one used for recording the light power.

For data processing, the sampling points representing the minimum value of the detected light power in Figures 6.26 and 6.27b are obtained for the FBG and LPG sensors, respectively. Then, the EDG temperature is calculated by introducing the above sampling points into the relationship shown in Figure 6.28. Finally, the wavelengths of the hybrid FBG/LPG sensor pair are calculated by correlating the EDG temperatures to the wavelengths using the linear relationship between the transmission wavelength of the EDG channel and the EDG temperature. Good agreement between the measured wavelength and the ones provided by the sensor manufacturer is achieved in this test.

Finally, the thermal-tuning-EDG-based interrogation unit is used to measure the microwave frequency with an experimental setup shown in Figure 6.29 (59).

In this application, the microwave signal is converted into an optical signal using a Mach–Zehnder intensity modulator. A single frequency light signal from the tunable laser source (TLS) is used as the optical carrier and modulated by the modulator with a signal of an unknown frequency. By properly selecting the DC bias of the modulator, the optical carrier can be suppressed, leaving two sidebands at the output of the modulator. One sideband is filtered out by an FBG and the other one is introduced into the developed interrogation unit for measuring its wavelength. Since the wavelength difference between the measured sideband and the optical carrier is equivalent to the unknown frequency of the input microwave signal, the developed interrogation unit can be used to measure the microwave frequency.

Figure 6.30 shows the measured sideband spectra with respect to the EDG temperature under different microwave frequencies applied on the modulator. It

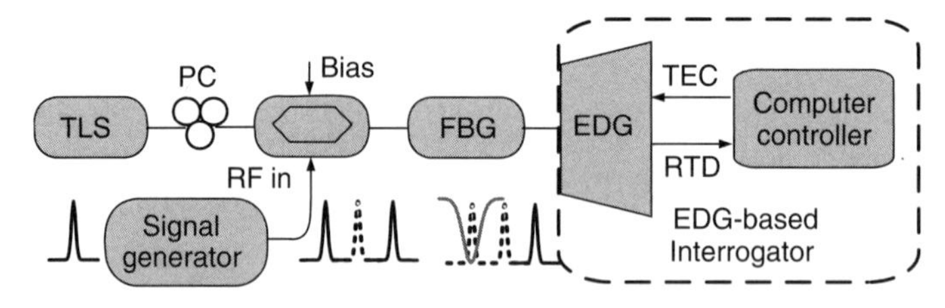

FIGURE 6.29 Experimental setup for the microwave frequency measurement using the thermal-tuning-EDG-based interrogation unit. *Source:* Reprinted by permission from Reference 59.

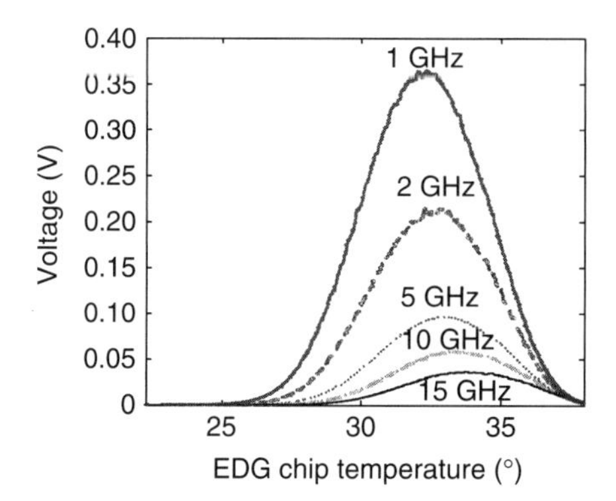

FIGURE 6.30 Samples of the sideband shift measured by the thermal-tuning-EDG-based interrogation unit. *Source:* Reprinted by permission from Reference 59.

is seen that the wavelength of the sideband is shifting toward larger wavelength when increasing the microwave frequency. The decreasing of the light power and the narrowing of the bandwidth of the measured sideband are due to the wavelength selection of the optical carrier.

By applying the above approach, the EDG temperature representing the peak light power and its corresponding wavelength of the remaining sideband under different microwave frequencies are obtained and shown in Figure 6.31a. Shown in Figure 6.31b is the comparison between the measured microwave frequencies and the actual microwave frequencies.

A resolution of 1 GHz in the frequency measurement is equal to 8 pm resolution in the spectrum, if the optical carrier is working around 1550 nm. From Figure 6.31b, it is seen that the measured frequency accords well with the actual

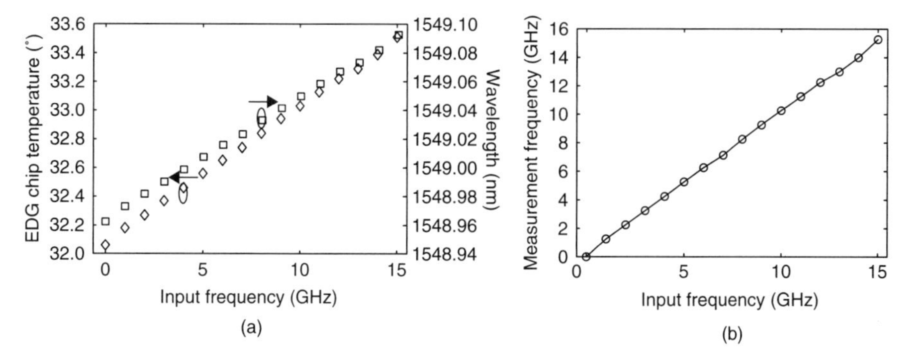

FIGURE 6.31 Experimental results of using the developed interrogation unit for the microwave frequency measurement. (a) Measured EDG temperature and the corresponding wavelengths with respect to the EDG temperature. (b) A comparison between the measured and actual microwave frequencies. *Source:* Reprinted by permission from Reference 59.

value with a resolution better than 1 GHz, which indicates that our EDG-based interrogation unit has a resolution better than 8 pm and is able to obtain accurate results in the microwave frequency measurement. Furthermore, our first prototype has 15 channels and each channel could be used for measuring one microwave signal. Therefore, a total of 15 microwave signals could be simultaneously characterized.

6.3.2.3 AWG Interrogator Based on Mechanical Spectrum Tuning Experimental tests using the AWG chip, shown in Figure 6.32, have been implemented to verify the performance. Initial results confirm the theoretical postulation that this technique can successfully perform the interrogation of FBG, LPG, and TFBG sensors (60–63). Compared with the thermal tuning methods, the mechanical tuning technique offers the advantages of a broader measurement range and a faster measurement speed.

It is noted that the AWG shown in Figure 6.32 does have neither the input waveguide nor the input fibers. The facet of the input coupler is right on the edge, facilitating the position scanning of an optical fiber along the side of the AWG chip. In addition, the Rowland Circle has been partly cutoff to minimize the insertion loss of the coupling between the optical fiber and the input coupler, as discussed and illustrated in Figure 6.8.

The space-to-wavelength mapping is first tested by measuring the relationship between the input light position along an AWG input coupler and the transmission wavelength of a designated AWG channel. The result is shown in Figure 6.33.

As shown in Figure 6.33a, the transmission wavelength of the designated AWG channel is shifted while the input light beam position changes. During the shifting, the spectrum profile remains unchanged, validating our assumption that the operation of the AWG is not affected by the structure change when slightly cutting the Rowland Circle. In addition, Figure 6.33b shows that the wavelength shifts at a constant rate, thus confirming that the space-to-wavelength mapping has a linear relationship, which agrees well with the theoretical simulation result shown in Figure 6.7 and Equation 6.8. The different power level of the shifted spectrum shown in Figure 6.33a is believed to be attributed to the nonuniform gain profile of the optical amplifier used in this test.

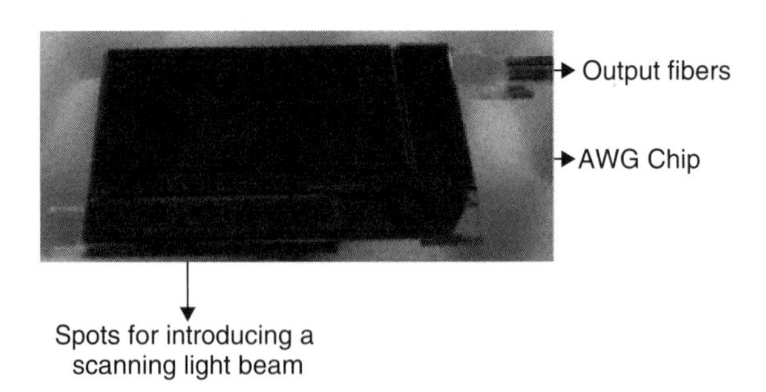

FIGURE 6.32 Illustration of the AWG chip used in the mechanical spectrum tuning technique.

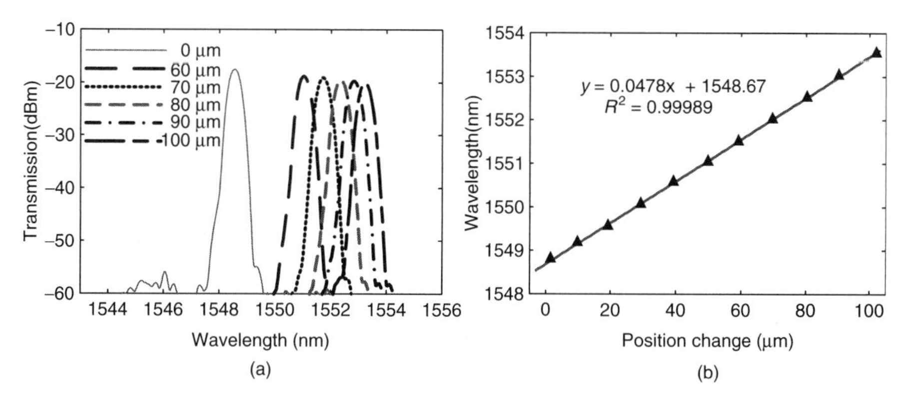

FIGURE 6.33 The space-to-wavelength mapping. (a) Shifted spectrum of one designated AWG channel with changes of the input light beam position. *Source:* Reprinted by permission from Reference 63. (b) Relationship between the transmission wavelength of the AWG channel and the input light position.

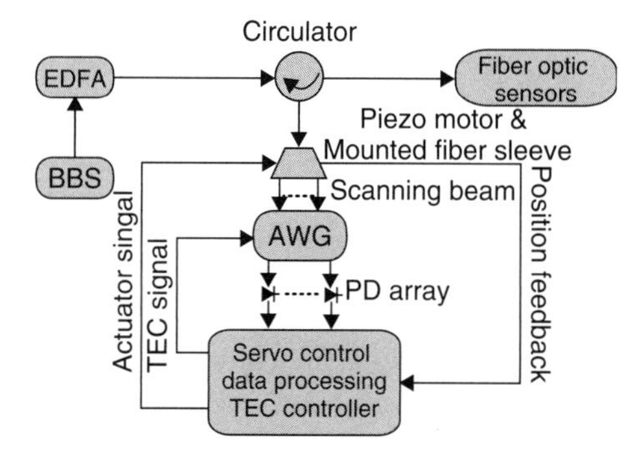

FIGURE 6.34 Experimental setup of the proposed interrogation technique based on the space-to-wavelength mapping. *Source:* Reprinted by permission from Reference 62.

Then, a single FBG sensor under a small ambient variation is tested using the mechanical tuning AWG. The experimental setup is shown in Figure 6.34.

In this setup, the position changes of the input light beam are achieved by moving an optical fiber that is mounted on the top surface of a closed-loop piezo motor. The optical fiber is fixed and protected by a fiber sleeve, which is then prealigned with the input coupler of the AWG using a positioning stage. A capacitive position sensor is embedded into the piezo motor to provide the absolute position of the scanning fiber tail and transfers the position to the servo control. Then, an actuator signal is properly set to drive the piezo motor to reach the specified position.

The small changes on the FBG sensor is introduced by changing the temperature applied to the FBG sensor, which is placed inside a temperature-controlled oven. Measurements are taken under four different temperatures (16, 19.5, 22, and

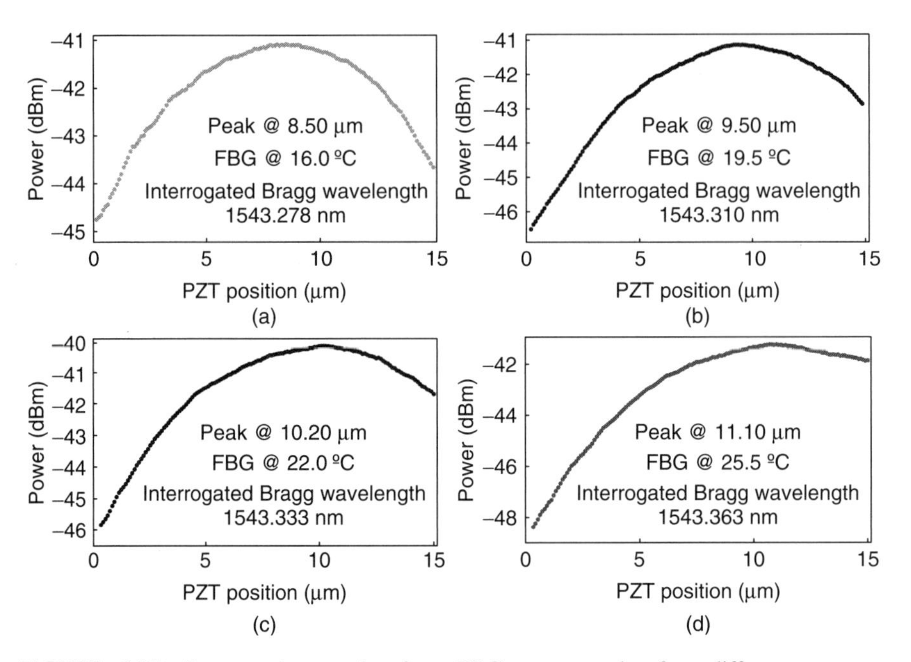

FIGURE 6.35 Interrogation result of an FBG sensor under four different temperatures, (a–d), as a function of the beam position. *Source:* Reprinted by permission from Reference 62.

25.5°C). At each temperature, the position of the input light beam changes from 0 to 15 μm with a step of 0.1 μm, resulting in a measurement resolution of approximately 5 pm. Experimental results of the output power as a function of the absolute position are shown in Figure 6.35.

The positions representing the peak light powers in Figure 6.35 are directly estimated by a standard programming module from the Labview program with the capability of searching the peak value. Then, the positions are introduced into the space-to-wavelength mapping to obtain the Bragg wavelength of the FBG sensor. Initial results show that the measured Bragg wavelengths are in good agreement with the actual values.

One significant feature of the AWG-based interrogation unit with mechanical spectrum tuning is that it has multichannel measurement capability. In the second application, an interrogation of four distributed FBG sensors using four designated AWG channels are implemented, where the temperature sensitivity of the four FBG sensors are measured with results shown in Figure 6.36.

Figure 6.36a shows the shifted spectrum of each AWG channel under four different input light beam positions (0, 5, 10, and 15 μm). It is seen that all the AWG channels have the same spectrum tuning capability. Therefore, it can be used for the multichannel measurement. The temperature sensitivity of the four FBG sensors are measured to be ~10 pm/°C near the wavelength of 1550 nm, which accords well with the results reported in Reference 16.

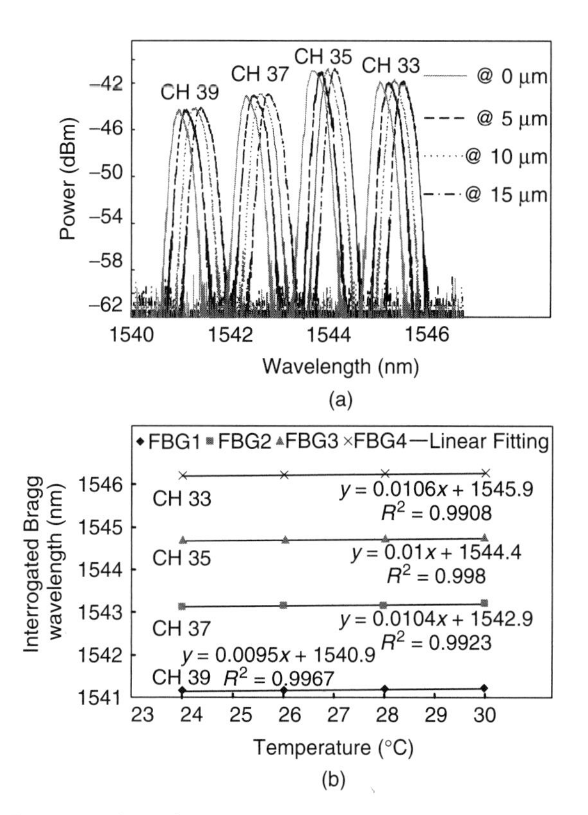

FIGURE 6.36 Demonstration of the multichannel measurement capability of the AWG interrogation unit based on space-to-wavelength mapping. (a) Shifted spectrum of each AWG channel under four different input light beam positions and (b) measured temperature sensitivity. *Source:* Reprinted by permission from Reference 62.

In the third application, the mechanical tuning AWG interrogation unit is used to monitor a TFBG-based refractive index sensor, which requires the covering of a large wavelength range. In the test, water–sugar solutions with different weight concentrations are used to provide a wide range of refractive indices. Reference of the refractive indices is obtained by an Abbe refractometer. The results show that the refractive indices vary from 1.35 to 1.40 with steps of 0.0083, corresponding to the weight concentration varying from 10% to 40% with steps of 5%. The transmission wavelength spectrum of the TFBG sensor immersed in the solutions is measured by the AWG interrogator based on space-to-wavelength mapping, and the result is shown in Figure 6.37.

In the transmission spectrum of a TFBG sensor, the Bragg wavelength is insensitive to the change of refractive index in the surrounding medium, as it is dominated by the core mode, and the core mode is only dependent on the refractive index in the fiber core region. Both cladding mode and core mode are sensitive to temperature variation and the impact induced by temperature to these modes are nearly the

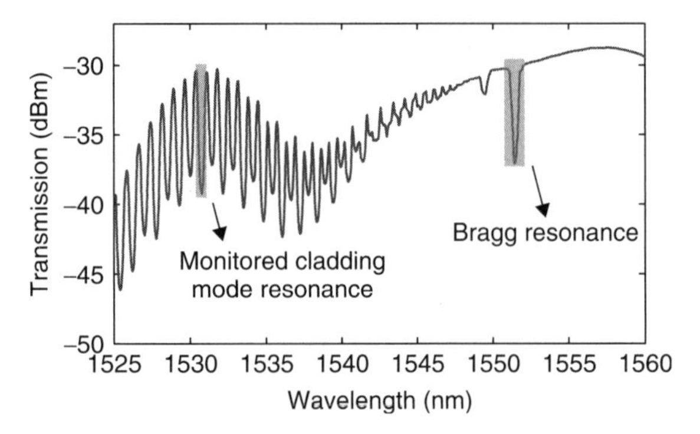

FIGURE 6.37 Measured transmission spectrum of the TFBG sensor using the space-to-wavelength mapping. *Source:* Reprinted by permission from (63).

same. Thus, by measuring the relative wavelength spacing between two resonances (usually one higher order cladding mode and the Bragg resonance), refractive index changes in the surrounding medium can be monitored disregarding the temperature changes (64). In our experiment, the wavelength spacing of the Bragg resonance (1551.506 nm as measured in the air) and a cladding mode resonance (1536.008 nm as measured in the air) are selected. The total travel range of the input light beam is 1 mm, which makes the measurement range up to 47.8 nm for a single AWG channel. However, only portion of this value is needed for the measurement of refractive index using a TFBG sensor according to its sensing principle as described above. The experimental result of the wavelength spacing as a function of the refractive index changes is shown in Figure 6.38.

Comparing this result with the one measured by an OSA, it is found that the deviations are within ±10 pm, thus confirming the applicability of the space-to-wavelength mapping technique to the large spectrum range measurement.

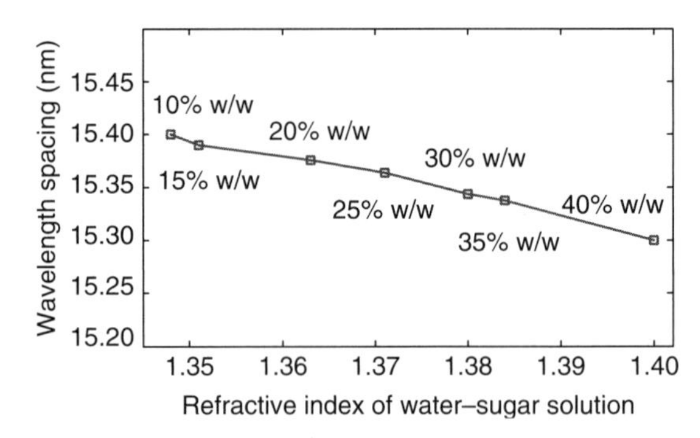

FIGURE 6.38 Experimental results of the wavelength spacing as a function of refractive index measured by the space-to-wavelength mapping. *Source:* Reprinted by permission from Reference 63.

In the studies presented above, a capacitive position sensor is used to provide the absolute position of the fiber along the AWG input coupler. Another solution is to introduce a wavelength reference device into the system. Therefore, the absolute position is not necessarily required, as the wavelength could be calculated by correlating the measured spectrum of the sensor and one of the wavelength reference device. An LPG sensor is successfully interrogated using this method (60).

Preliminary results show that (i) the measurement range is determined by the wavelength tunability, which is 47.8 nm in our design with this particular AWG, given that the travel range of the piezo motor is 1 mm and the space-to-wavelength mapping coefficient is 47.8 pm/μm. This range is broad enough for most sensor applications; (ii) the measurement resolution is determined by the spatial tuning step of the closed-loop piezo motor, which is 4.78 pm in this experiment by setting the spatial tuning step of 0.1 μm. Higher resolution can be reached by adopting a smaller spatial tuning step; (iii) the multichannel measurement capability is ensured by the characteristics of the AWG. All the AWG channels have the same wavelength tunability. The AWG used in this design has a total of 32 channels, making it possible to simultaneously interrogate 32 sensors; (iv) the interrogation speed is determined by the travel speed of the piezo motor. In our design, the piezo motor has a maximum speed of 500 mm/s, which makes the interrogation speed up to 1 kHz; (iv) since a commercial standard AWG is used in this experiment, there is no need to design and fabricate an AWG with new structures, which significantly decreases the total cost.

6.4 DUAL FUNCTION EDG-BASED INTERROGATION UNIT

FBG sensors have demonstrated capabilities in both operational load monitoring and impact damage inspection (33), it is preferable that the same wavelength interrogation unit will work for both types of air vechile SHM tasks, although the requirements are different. In addition, for air vehicle SHM, the interrogation unit should also have the features of small size, lightweight, low cost, and strong reliability (33). Our first prototype of EDG-based interrogation unit has been proved to serve in operational load monitoring with all the above features except for the capability of doing impact damage inspection (53, 54). In order to develop one interrogation unit for the two aspects of air vehicle SHM, we introduce two operation modes, that is, the sweeping mode and the parked mode, to the EDG-based interrogation unit we previously developed. The sweeping mode, in which the transmission wavelength of EDG channel is tuned by changing EDG temperature, can be used for the load monitoring. While the parked mode, in which the transmission wavelength of EDG channel is fixed, can be used for the damage detection, more specifically, the sweeping mode is used to apply the thermal spectrum tuning to an EDG for the wavelength interrogation (54). The parked mode is to use the fixed spectrum of an EDG for the wavelength interrogation with the same method used in the spectrum fixed AWG-based interrogation unit (21).

As discussed above, only odd number EDG channels are connected in our first prototype, resulting in a channel spacing of 1.6 nm. Considering the channel

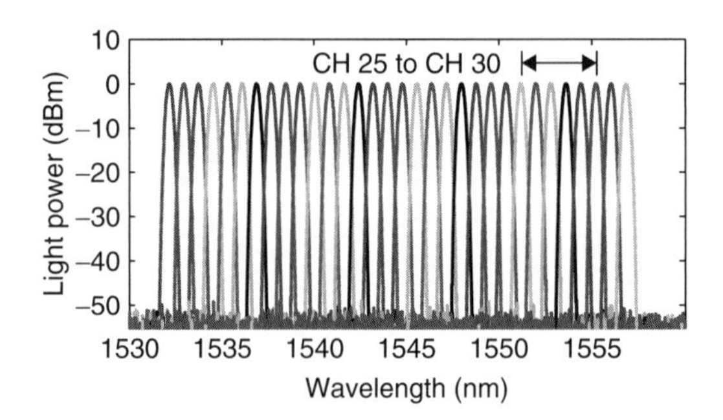

FIGURE 6.39 Full spectrum of the modified 32-channel EDG device.

bandwidth is 0.4 nm, two adjacent EDG channels have no overlapped region, making it unable to work in the parked mode. Therefore, the EDG-based interrogation unit is modified and all the 32 channels of the EDG are connected by two DAQ devices with the full spectrum shown in Figure 6.39.

The use of 32 channels makes the channel spacing reduced from 1.6 nm to 0.8 nm. Thus, in the sweeping mode, the EDG temperature tuning range required for one EDG channel to cover the channel spacing is reduced to half, resulting in a lower power consumption and an increase in the measurement repeatability. Also, in the parked mode, the reduction of the channel spacing makes the two neighboring EDG channels have an overlapped region. By the same method as described in Reference 32, the parked mode can reach an interrogation speed up to 300 kHz and be used to detect acoustic waves for impact damage inspection.

As the performance of the EDG-based interrogation unit in the strain measurement has already been demonstrated, only the parked mode is discussed in this section. In the parked mode, the EDG spectrum is fixed by keeping the EDG temperature constantly at 30°C, where the TEC is tested to have the best performance. The aim of testing the parked mode is to provide the proof-of-concept result and prepare initial data for the future development of detecting acoustic waves. The present achievement is to use the parked mode to monitor the movement of a piezo motor with the experimental setup shown in Figure 6.40.

An FBG sensor is mounted on a piezo motor. When the piezo motor moves, it applies strain on the FBG sensor. By knowing the wavelength of the FBG sensor, the movement of the piezo motor can be obtained. Selected EDG channels shown in Figure 6.39 are used to test the parked mode. First, two EDG channels are used with the result shown in Figure 6.41.

Figure 6.41a shows the light power from the two EDG channels while the piezo motor is moving, and Figure 6.41b shows the ratio of the light power, which could be used to reflect the wavelength shift of the FBG sensor and the movement of the piezo motor. On the basis of the data processing method in Reference 21, the wavelength shift is measured to be from 1553.810 to 1554.290 nm, equals to a strain of 480 $\mu\varepsilon$ applied on the FBG sensor when the piezo motor moves from one end to the other. Considering the length of 42 cm between the two ends, where the

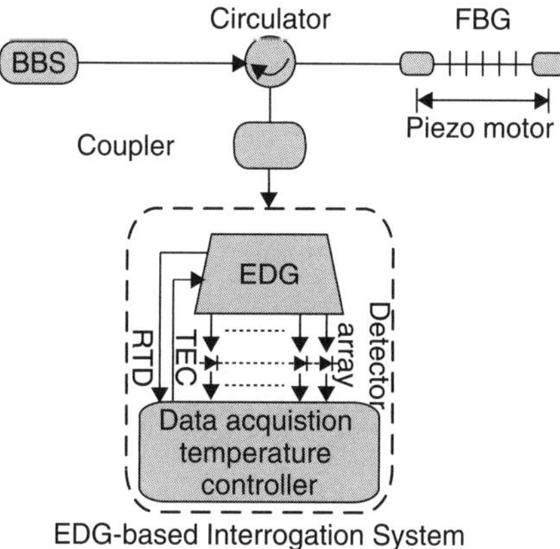

FIGURE 6.40 Experimental setup of testing the dual function EDG-based interrogation unit.

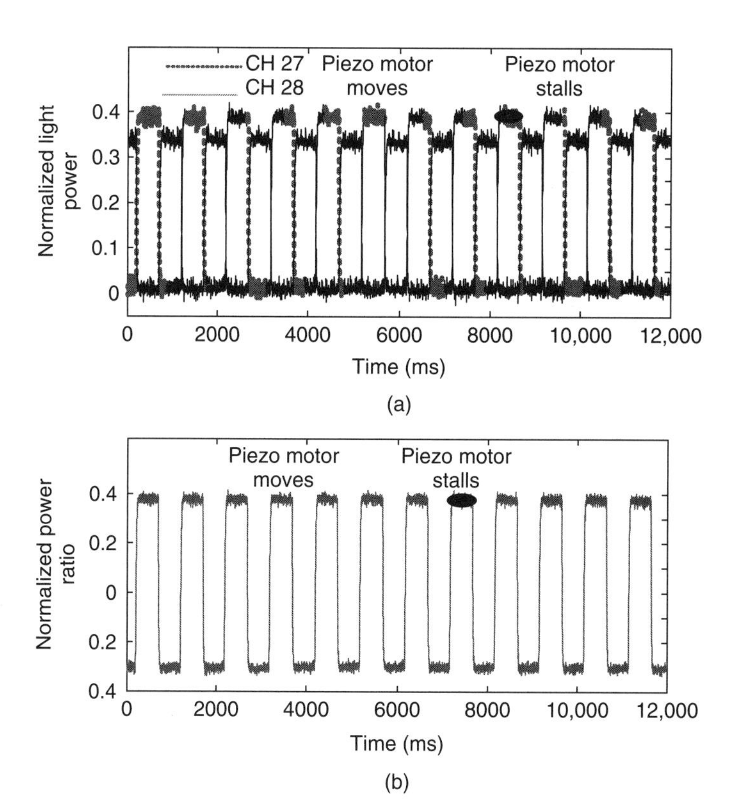

FIGURE 6.41 Experimental result of monitoring the movement of a piezo motor. (a) Light power from two designated EDG channels. (b) Illustration of the movement.

FBG sensor is fixed as shown in Figure 6.38, the travel range of the piezo motor is measured to be 200 μm. As shown in Figure 6.39, the piezo motor takes 10 ms to travel from one end to the other. The measured piezo motor movement accords well with the settings.

In previous study (21, 32, 41), the wavelength of the FBG sensor is required to be tuned at the middle between two adjacent channels before each measurement. This process introduces additional calibration work. Although it could be achieved by shifting the EDG spectrum, it is not feasible for the use of multiplexed FBG sensors in real implementation, because the calibration work for each sensor might not be the same. Considering that the parked mode is eventually applied to measure acoustic waves for the impact damage inspection, maximum of three EDG channels are able to provide sufficient measurement range, as the strain variation induced by acoustic waves is less than hundreds of micrometers (33). The use of three EDG channels is demonstrated with the experimental result shown in Figure 6.42.

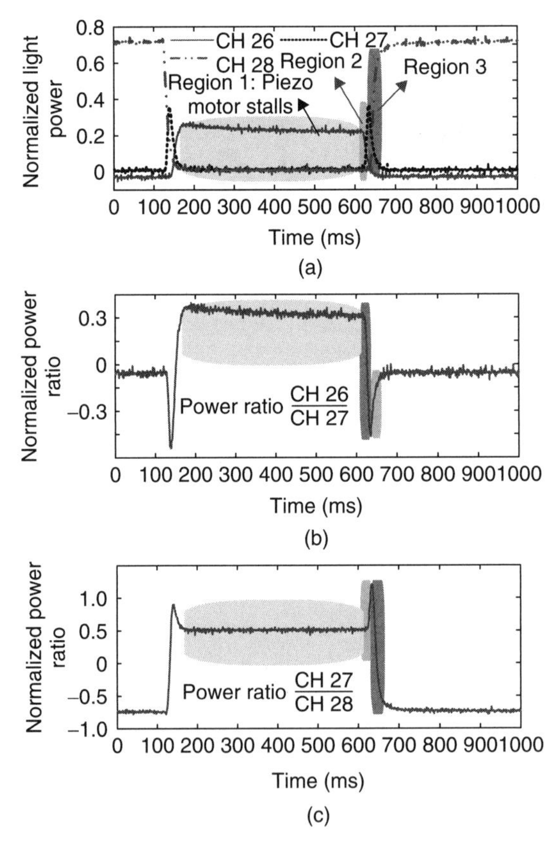

FIGURE 6.42 Experimental result of using three EDG channels. (a) Light power from the three channels, (b) light power ratio between CH 26 and CH27, and (c) light power ratio between CH 27 and CH 28.

Figure 6.42 shows the first 1000 measurement points in the test. In Figure 6.42a, light powcr from thrcc EDG channels, CII 26, CII 27, and CII 28, are illustrated. It is seen that Region 1 represents that the piezo motor stalls, and Region 2 and 3 represent that the piezo motor moves. More specifically, Region 2 indicates that the Bragg wavelength shifts between CH 26 and CH 27, and the wavelength of the FBG sensor should be interrogated by the power ratio in Figure 6.40b. Region 3 indicates that the Bragg wavelength shifts between CH 27 and CH 28, and the power ratio in Figure 6.42c is used to perform the wavelength interrogation. Similarly, the wavelength of the FBG sensor is measured to shift from 1553.135 to 1554.625 nm, and it represents that the travel range of the piezo motor is 626 μm, which accords well with the settings. Therefore, the use of three EDG channels in the parked mode has been verified, and no additional calibration work, such as tuning the EDG spectrum, is required before each measurement.

Broad bandwidth of the EDG channel and the power fluctuation could introduce minor error for the measurement. In the parked mode, the EDG spectrum bandwidth is tested to be proportional to the measurement resolution and error. Thus, better results could be obtained using an EDG with a smaller bandwidth. The effect induced by the power fluctuation on the measurement error mainly comes from the broadband light source. As discussed in Reference 53, a fast wavelength tuning or the use of one EDG channel as the reference will alleviate this challenge and provide a more accurate result.

Initial results have shown that the proposed sweeping mode and parked mode have a potential to be applied for operational load monitoring and impact damage inspection, respectively. In addition to the features of the EDG device, as described above, it also has the feature of low power consumption. The dual function EDG-based interrogation unit is powered by a +5 V power supply, which can be powered with battery. This feature allows the introduction of energy-harvesting techniques into the developed interrogation unit and makes it self-powered. Furthermore, a total of 32 EDG channels provide a capability to perform a multichannel interrogation for multiplexed FBG sensors, which is also a favorite feature for air vehicle SHM.

6.5 CONCLUSION

This chapter started with the introduction of the two aspects in air vehicle SHM, that is, operational load monitoring and impact damage detection. Owing to their unique advantages, most notably with lightweight, small size, and strong multiplexing capability, FBG sensors have been defined as one of the leading candidate and received considerable attention in the air vehicle SHM. However, the lack of suitable interrogation units for FBG sensors limited their deployment. We proposed and developed PLC-based interrogation units to overcome the limitations. Principles and applications of these interrogation units were investigated in detail. Furthermore, we proposed a dual function interrogation unit for both operational load monitoring and impact damage detection, which was achieved by the sweeping mode and parked mode, respectively. Results have shown that the developed interrogation unit could work for the two aspects of air vehicle SHM with the features of compact size, lightweight, broad measurement range, high measurement resolution, and multichannel measurement capability, making it ideal for air vehicle SHM.

ACKNOWLEDGMENTS

This work was supported in part by the Canadian Institute for Photonics Innovations, Defence Research and Development Canada of the Department of National Defence, and National Research Council (NRC) Canada.

REFERENCES

1. Prosser W. Development of structural health management technology for aerospace vehicles. Proceedings NASA LaRC, JANNAF 39th CS/27th APS/21st PSHS/3rd MSS Joint Subcommittee Meeting, Colorado Springs, CO, USA; 2003. p. 9.
2. Mrad N. Potential of Bragg grating sensors for aircraft health monitoring. Trans CSME 2007;31:1–17.
3. Brand C, Boller C. Identification of life cycle cost reduction in structures with self-diagnostic devices. Proceedings NATO RTO Symposium on Design for Low Cost Operation and Support. Ottawa, Canada; 1999. p. 17.
4. Boller C. Ways and options for aircraft structural health management. Smart Mater Struct 2001;10:432–439.
5. Glisic B, Inaudi D. *Fibre Optic Methods for Structural Health Monitoring*. 1st ed. Southern Gate, Chichester, England: John Wiley & Sons, Ltd.; 2007.
6. Boller C, Staszewski WJ. Aircraft structural health and usage monitoring. In: Staszewski W, Boller C, Tomlinson G, editors. *Health Monitoring of Aerospace Structures*. Munich: John Wiley and Sons; 2003. p. 29–73.
7. Def Stan 00–970 UK Ministry of Defence. Design and airworthiness requirements for service aircraft. *Defence Standard 00–970, Part1/2*, 2007. p. 5.
8. Department of Defense Standard Practice. Aircraft structural integrity program (ASIP). MIL-STD1530C (USAF); 2005.
9. Cracks blamed for 2002 China Airlines crash. CBC News; 2005. http://www.cbc.ca/news/world/story/2005/02/25/china-plane050225.html.
10. Tua P, Quek S, Wang Q. Detection of cracks in plates using piezo-actuated Lamb waves. Smart Mater Struct 2004;13:643–660.
11. Grattan S, Taylor S, Sun T, Basheer P, Grattan K. Monitoring of corrosion in structural reinforcing bars: performance comparison using in situ fiber-optic and electric wire strain gauge systems. IEEE Sens J 2009;9:1494–1502.
12. Udd E. Review of multi-parameter fiber Bragg sensors. Proc SPIE 2007;6770: 677002.
13. Majumder M, Gangopadhyay T, Chakraborty A, Dasgupta K, Bhattacharya D. Fibre Bragg gratings in structural health monitoring—Present status and applications. Sens Actuators A 2008;147:150–164.

14. Hill K, Meltz G. Fiber Bragg grating technology fundamentals and overview. IEEE/OSA J Lightw Technol 1997;15:1263–1276.
15. Betz D, Thursby G, Culshaw B, Staszewski W. Identification of structural damage using multifunctional Bragg grating sensors: I. Theory and implementation. Smart Mater Struct 2006;15:1305–1312.
16. Kersey A, Davis M, Patrick H, Leblane M, Koo K, Askins C, Putnam M, Prebele E. Fiber grating sensors. IEEE/OSA J lightw Technol 1997;15:1442–1463.
17. Duncan G. Sensor take the strain. Nat Photonics 2007;1:307–309.
18. Lee B, Jeong Y. Interrogation techniques for fiber grating sensors and the theory of fiber gratings. In: Rao Y, editor. *Optical Fiber Sensor Technology*. London: Chapman & Hall; 1998.
19. Mallinson S. Wavelength-selective filters for single-mode fiber WDM system using Fabry-Perot interferometers. Appl Opt 1987;26:430–436.
20. Kersey A, Berkoff T, Morey W. Multiplexed fiber Bragg grating strain-sensor system with a fiber Fabry-Perot wavelength filter. Opt Lett 1993;18:1370–1372.
21. Sano Y, Yoshino T. Fast optical wavelength interrogator employing arrayed waveguide grating for distributed fiber Bragg grating sensors. IEEE/OSA J Lightw Technol 2003;21:132–139.
22. Usbeck K, Ecke W, Hagemann V, Mueller R, Willsch R. Temperature referenced fiber Bragg grating refractometer sensor for on-line quality control of petrol products. Proceedings of 13th Optical Fiber Sensors Conference, Kyongju, Korea; 1999. p. 163–166.
23. Song M, Yin S, Ruffin P. Fiber Bragg grating strain sensor demodulation with quadrature sampling of Mach-Zehnder interferometer. Appl Opt 2000;39:1106–1111.
24. Davis M, Kersey A. All-fiber Bragg grating strain-sensor de-modulation technique using a wavelength division coupler. Electron Lett 1994;30:75–77.
25. Zhang L, Fallon R, Gloag A, Bennion I. Spatial and wavelength multiplexing architectures for extreme strain monitoring system using identical-chirped-grating-interrogation technique. Proceedings of 12th Optical Fiber Sensors Conferecne, Williamsburg, Virginia, USA; 1997, 452–455.
26. Zhao C, Demokan M, Jin W, Xiao L. A cheap and practical FBG temperature sensor utilizing a long-period grating in a photonic crystal fiber. Opt Commun 2007;276: 242–245.
27. Davis M, Kersey A. Application of a fiber Fourier transform spectrometer to the detection of wavelength-encoded signals from Bragg grating sensors. IEEE/OSA J Lightw Technol 1995;13:1289–1295.
28. Culshaw B, Thursby G, Betz D, Sorazu B. The detection of ultrasound using fiber-optic sensors. IEEE Sens J 2008;8:1360–1367.
29. Hongo A, Kojima S, Komatsuzaki S. Applications of fiber Bragg grating sensors and high-speed interrogation techniques. Struct Control Health Monit 2005;12:269–282.
30. Todd M, Nichols J, Trickey S, Seaver M, Nichols C, Vigin L. Bragg grating-based fibre optic sensors in structural health monitoring. Philos Trans R Soc Lond A 2007;365:317–343.
31. Frieden J, Cugnoni J, Botsis J, Gmur T, Coric D. High-speed internal strain measurements in composite structures under dynamic load using embedded FBG sensors. Compos Struct 2010;92:1905–1912.

32. Soejima H, Ogisu T, Yoneda H, Okabe Y, Takeda N, Koshioka Y. Demonstration of detectability of SHM system with FBG/PZT hybrid system in composite wing box structure. Proc SPIE 2008;6932:69322E.
33. Betz D, Staszewski J, Thursby G, Culshaw B. Multi-functional fibre Bragg grating sensors for fatigue crack detection in metallic structures. J Aerosp Eng 2006;220:453–461.
34. Cheben P. *Optical Waveguides: From Theory to Applied Technologies*. Boca Raton (FL): CRC Press; 2007.
35. Lopatin C, Mahmood S, Mendoza E, Moslehi B, Black R, Chau K, Oblea L. Progress in miniaturization of a multichannel optical fiber Bragg grating sensor interrogator. Proc SPIE 2007;6619:66193x.
36. Doerr C, Okamoto K. Advances in silica planar lightwave circuits. IEEE/OSA J Lightw Technol 2006;24:4763–4789.
37. Okamoto K. Recent progress of integrated optics planar lightwave circuits. Opt Quantum Electron 1999;31:107–129.
38. Sano Y, Hirayama N, Yoshino Y. Wavelength interrogator employing arrayed waveguide grating for distributed fiber Bragg grating. Proceedings of 14th Optical Fiber Sensors Conference, Venice, Italy; 2000. p. 788–791.
39. Norman D, Webb D, Pechstede R. Extended range interrogation of wavelength division multiplexed fiber Bragg grating sensors using arrayed waveguide grating. Electron Lett 2003;39:1714–1715.
40. Miewczas P, Willshire A, Dziuda L, McDonald J. Performance analysis of the fiber Bragg grating interrogation system based on an arrayed waveguide grating. IEEE Trans Instrum Meas 2003;53:1192–1196.
41. Cheben P, Post E, Janz S, Albert J, Laronche A, Schmid J, Xu D, Lamontagne B, Lapointe J, Delage A, Densmore A. Tilted fiber Bragg grating sensor interrogation system using a high-resolution silicon-on-insulator arrayed waveguide grating. Opt Lett 2008;33:2647–2649.
42. He J, Lamontagne B, Delage A, Erickson L, Davies M, Koteles E. Monolithic integrated wavelength demultiplexer based on a waveguide Roaland circle grating in InGaAsP/InP. IEEE/OSA J Lightw Technol 1998;16:631–638.
43. Ming Y, Tarter T, Weaver J, Xu H, Ho C. Integrated thin film heater and sensor with planar lightwave circuits. IEEE Trans Compon Pack Technol 2005;28:667–673.
44. Inoue Y, Kaneko A, Hanawa F, Takahashi H, Hattori K, Sumida S. Athermal silica-based arrayed-waveguide grating multiplexer. Electron Lett 1997;33:1945–1947.
45. Xiao G, Sun F, Lu Z, Zhang Z. A hybrid approach to dynamic-wavelength blocking/equalizing and optical-power monitoring. Microw Opt Technol Lett 2003;40:95–96.
46. Zhang Z, Sun F, Lin P, Xiao G, Wu Y. Heat control of an integrated silica VMUX using constant working power. IEEE J Sel Top Quantum Electron 2006;12:1064–1059.
47. Zhang Z, Xiao G, Liu J, Grover C. Coupling fibers to planar waveguides using a high-temperature epoxy. Fiber Integr Opt 2003;22:357–371.
48. Xiao G, Zhao P, Sun F, Lu Z, Zhang Z, Grover C. Interrogating fiber Bragg grating sensors by thermally scanning a demultiplexer based on arrayed waveguide grating. Opt Lett 2004;29:2222–2224.
49. Xiao G, Zhao P, Sun F, Lu Z, Zhang Z. Arrayed-waveguide-grating-based interrogator for wavelength-modulated multi-fiber-optic sensor applications. IEEE Photon Technol Lett 2005;17:1710–1712.

50. Guo H, Xiao G, Mrad N, Yao J. Interrogation of a long-period grating sensor by a thermally tunable arrayed waveguide grating. IEEE Photon Technol Lett 2008;20:1790–1792.

51. Miya T. Silica-based planar lightwave circuits: Passive and thermally active devices. IEEE J Sel Top Quantum Electron 2000;6:38–45.

52. Sun F, Xiao G, Zhang Z, Lu Z. Modeling of arrayed gratings for wavelength interrogation application. Opt Commun 2007;271:105–108.

53. Xiao G, Mrad N, Wu F, Zhang Z, Sun F. Miniaturized optical fiber sensor interrogation system employing echelle diffractive gratings demultiplexer for potential aerospace applications. IEEE Sensors J 2008;8:1202–1207.

54. Xiao G, Mrad N, Guo H, Zhang Z, Yao J. A planar lightwave circuit based micro interrogator and its applications to the interrogation of multiplexed optical fiber Bragg grating sensors. Opt Commun 2008;281:5659–5663.

55. Guo H, Xiao G, Mrad N, Yao J. Simultaneous interrogation of a hybrid FBG/LPG sensor pair using a monolithically integrated echelle diffractive grating. IEEE/OSA J Lightw Technol 2009;27:2100–2104.

56. Xiao G, Sun F, Zhang Z, Lu Z, Liu J, Wu F, Mrad N, Albert J. Miniaturized optical fiber Bragg grating sensor interrogator based on echelle diffractive grating. Microw Opt Technol Lett 2006;49:668–671.

57. Tolstikhin V, Densmore A, Pimenov K, Logvin Y, Wu F, Laframboise S, Gratchak S. Monolithically integrated optical channel monitor for DWDM transmission systems. IEEE/OSA J Lightw Technol 2004;22:146–153.

58. Gini E, Melchior H. The refractive index of InP and its temperature dependence in the wavelength range from 1.2 μm to 1.6 μm. Proceedings of the 8th International Conference on Indium Phosphide and Related Materials, Germany; 1996.

59. Guo H, Xiao G, Mrad N, Yao J. Measurement of microwave frequency using a monolithically integrated scannable echelle diffractive grating. IEEE Photon Technol Lett 2009;21:45–47.

60. Guo H, Dai Y, Xiao G, Mrad N, Yao J. Interrogation of a long-period grating using a mechanically scannable arrayed waveguide grating and a sampled chirped fiber Bragg grating. Opt Lett 2008;33:1635–1637.

61. Xiao G, Mrad N, Zhang Z. Wavelength tuning using an arrayed waveguide gratings demultiplexer. Microw Opt Technol Lett 2009;51:693–696.

62. Guo H, Xiao G, Mrad N, Albert J, Yao J. Wavelength interrogator based on closed-loop piezo-electrically scanned space-to-wavelength mapping of an arrayed waveguide grating. IEEE/OSA J Lightw Technol 2010;28:2654–2659.

63. Guo H, Xiao G, Mrad N, Shao L, Yao J. Wavelength interrogation of a tilted fiber Bragg grating sensor using space-to-wavelength mapping of an arrayed waveguide grating with closed-loop piezo-electrical control. IEEE Sensors Conference, Hawaii, USA; 2010.

64. Chan C, Chen C, Jafari A, Laronche A, Thomson D, Albert J. Optical fiber refractometer using narrowband cladding-mode resonance shifts. Appl Opt 2007;46:1142–1149.

CHAPTER SEVEN

Optical Coherence Tomography for Document Security and Biometrics

SHOUDE CHANG, YOUXIN MAO, and COSTEL FLUERARU
Institute for Microstructural Sciences, National Research Council Canada, Ottawa, ON, Canada

7.1 INTRODUCTION

Optical coherence tomography (OCT) is an emerging technology for high resolution cross-sectional imaging of 3D structures. The first OCT system was reported by Huang et al. (1) in 1991, and since then the OCT technology has been attracting the attention of researchers all around the world. Some good survey books and review articles are listed in References 2–5.

OCT relies on the interferometric measurement of coherent backscattering variation to sense the internal interface structures of test samples such as biological tissues or internal layered materials. It is analogous to ultrasound B mode imaging, except that it uses infrared light source rather than ultrasound (6–8). OCT takes advantage of the short temporal coherence length of the broadband light source that is used in the system to achieve precise optical sectioning in the depth dimension.

Advantages of OCT over other volume-sensing systems are as follows:

1. *High Resolution.* Compared to other systems: OCT, 5–20 μm; ultrasound, 150 μm; High resolution CT, 300 μm; MRI, 1000 μm. This feature enables greater visualization of details.

Photonic Sensing: Principles and Applications for Safety and Security Monitoring, First Edition.
Edited by Gaozhi Xiao and Wojtek J. Bock.

2. *Noninvasive and Noncontact.* This feature increases the safety and ease of use and extends the possibility for *in situ* and *in vivo* applications, which is important for biological applications such as biometrics.
3. *Fiber-Optics Delivery.* As fiber diameter is normally 125 μm, it allows OCT with an optic fiber probe to be used in very small space, such as tiny lumen, and intravascular.
4. *High Speed.* The new generation of OCT technology has no mechanical scanning procedures, which enables high resolution 3D sensing, particularly for the full-field optical coherence tomography (FF-OCT).
5. *Potential for Additional Information of the Testing Sample.* The new optical property of samples could be explored by functional OCT. For example, polarization contrast, Doppler effect, and spectroscopic information.
6. *Use of Nonharmful Radiation.* OCT systems work with visual and infrared band, unlike traditional CT working with X-ray and ultrasound relying on mechanical vibration.

Figure 7.1 shows the comparison of existing imaging technologies. OCT fills the gap between confocal microscopy and ultrasound methods. It has an excellent trade-off between resolution and penetration depth.

In the past decade, OCT systems have been developed mainly for medical and biomedical applications, especially for the diagnostics of ophthalmology, dermatology, dentistry, and cardiology (9–13). A few works have been reported for the nonmedical applications. References (14, 15) present works on materials analysis and data storage. Dunkers et al. (16) analyzed the applicability of OCT for nondestructive evaluation of highly scattering polymer matrix composites to estimate residual porosity, fiber architecture, and structural integrity. Fiber architecture and voids of glass-reinforced polymer composites have been successfully resolved. Another related application of OCT has been described by Bashkansky et al. (17). These authors detected and characterized the subsurface extent of the Hertzian

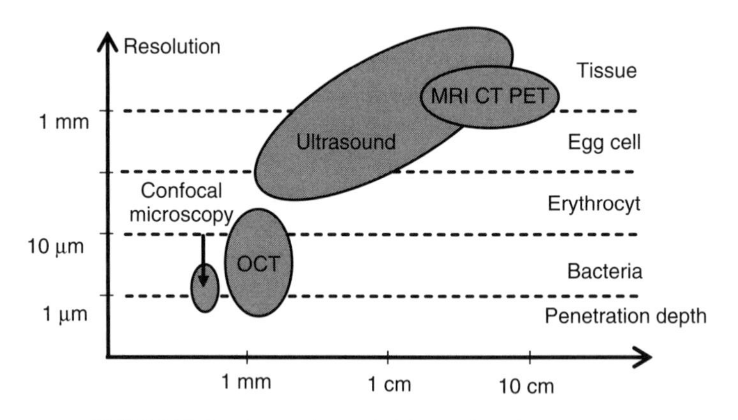

FIGURE 7.1 Comparison of deferent imaging systems.

crack on the surface of a silicon nitride ball to compare it with predictions of crack propagation theories. In Reference 18, Xu et al. used a free-space optics OCT system for the nondestructive evaluation of paints and coatings.

Reference 19 describes OCT diagnostics used for artworks, involving stratigraphic applications (20); varnish layer analysis (21, 22); and structural analysis and profilometric applications (23–25). In Reference 26, the use of different OCT systems for oil painting layer examination, varnish thickness determination, and environmental influence on paintings on canvas are described. In Reference 27, authors describe the use of OCT to monitor the subsurface morphologies of real and fake archaic jades. They focus their effort on observing the whitening area of jades. Archaeologically excavated archaic jades normally have different degrees of whitening due to the character of mineral composition, and variations in the underground environments of the jades. However, the artificially whitened jade probably was treated either by burning or by acidification. Authors reported that in naturally whitening area of jade, the backscattering intensity is weak, but for the fake one, the backscattering intensity is enhanced.

To explore the capabilities of an OCT system for probing the internal features of an object, we started our research in applying the OCT technology for information encoding and retrieving with a multiple-layer information carrier (28, 29). As OCT has the resolution of micrometer level and the ability of peeling cross-sectional images from the inside of an object, it has potential applications in documents security and object identification including biometrics.

For the document security/object identification, the most prevailing technology is the use of 1D or 2D barcode (30). As the barcode needs space to arrange the ordered data, either in 1D bar sequence or 2D image, the barcode reader has to scan the bar sequence or register the image. The visibility of the printed pattern of a barcode is vulnerable to counterfeiting. From the viewpoint of security and simplicity, multiple hidden information carriers of tiny sizes and are invisible to the human eye will be of great importance in the security applications. As the multiple-layer information carrier, namely, info-chip, can be made by low scattering clear material, the low signal/noise ratio caused by a scattering medium will no longer be a major problem in this circumstance. The confidential data is encoded layer by layer in the info-chip, which is then imbedded into a document or an object such as passport, ID card, and even fingernail. The information decoding and retrieval will be eventually performed by an OCT system.

Another important application of OCT technology is the new generation of biometrics. We call it "Internal Biometrics." The traditional biometric technologies used for security and person identification essentially deal with fingerprints, hand geometry, and face images. However, as all these technologies use external features of human body, they can be easily fooled and tampered by distorting, modifying, or counterfeiting the features. Therefore, to extract the internal features of human body that are unique to individual is becoming a new trend in biometrics. Except the well-known technologies for iris and retina recognition, other versatile technologies for internal biometrics have been currently developed.

Vein scan technology can automatically identify a person from the patterns of the blood vessels on the back of the hand. Vein patterns are distinctive between twins and even between a person's left and right hand. Developed before birth, they are highly stable and robust, changing only in the overall size throughout one's life (31).

Skin pattern recognition technology measures the characteristic of an individual's skin (32, 33). The exact composition of all the skin elements is unique to each person. For example, skin layers differ in thickness, the interfaces between the layers have different undulations, pigmentation differs, collagen fibers and other proteins differ in density, and the capillary beds have distinct densities and locations beneath the skin.

Since commencing operations in 1993, iris recognition has been introduced in biometric field (34). Iris recognition is based on the visible characteristics of the human iris, including rings, furrows, freckles, and the iris corona. Iridian's iris-recognition technology converts these visible characteristics into a template stored for future verification. From the 11-mm-diameter iris, each iris can have 266 unique spots—compared to 10–60 unique spots for traditional biometric technologies (35). Another eye-related internal biometrics is retinal scanning, which analyses the layer of blood vessels at the back of the eye (36).

Fingernail bed identification (37) is based on the distinct groove spatial distribution of the epidermal structure directly beneath the fingernail. This structure is mimicked in the ridges on the outer surface of the nail. When an interferometer is used to detect phase changes in back-scattered light shone on the fingernail, the distinct dimensions of the nail bed can be reconstructed and a one-dimensional map can be generated.

Human ear recognition is another interesting biometric field. Traditional Chinese medicine describes ear as the holograph of human, which has the shape of a human baby (Fig. 7.2a) and contains all the information of human organs and parts, as shown in Fig. 7.2b (38). Using ears for identifying people has been there for at least 100 years. Previous works (39–43) have shown that the ear is a promising candidate for biometric identification. In Reference 44, authors present a complete system for ear biometrics, including automated segmentation of the ear in a profile view image and 3D shape matching for recognition. They evaluated their system with the largest experimental study to date in ear biometrics, achieving a recognition rate of 97.6%. The algorithm they developed also shows good scalability of recognition rate with size of dataset.

However, all the internal structures of biometrics mentioned above cannot be explored by biometrics based on traditional external features. Features such as the micrometer-level accuracy, noncontact probing, and relatively cheap cost make OCT system the best candidate for internal biometric applications.

Summarizing the above statements, a technology that has the capability to probe some specific internal structures of human body or a pre-encoded information carrier is a powerful tool for the internal biometrics and document security. The OCT exhibits excellent capability to serve this purpose and therefore has a potential huge market for these applications.

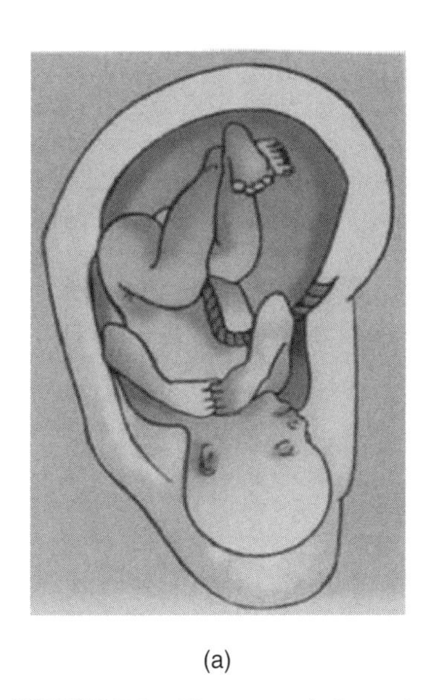

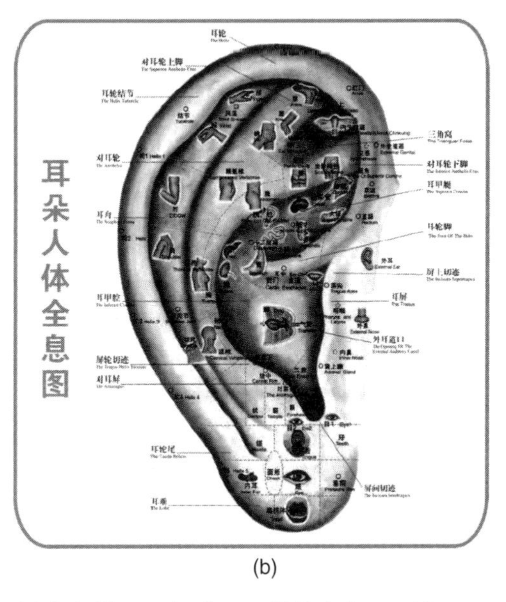

(a) (b)

FIGURE 7.2 Human ear holography. (a) Babylike main shape. (b) Relations with organs and body parts.

7.2 PRINCIPLE OF OCT

In direct imaging, such as ordinary photography, all the layers reflected from the underneath of an object are fused together. However, in OCT imaging, a coherence gate generated by an interferometer and broadband light source could be used to separate the fused layers and to extract the cross-sectional images from different depth.

7.2.1 Coherence Gate

The detection of light waves from an interferometer can be expressed as

$$I_{\mathrm{d}} = E_{\mathrm{s}}^2 + E_{\mathrm{r}}^2 + 2E_{\mathrm{s}}E_{\mathrm{r}} \cos\left(\frac{2\Delta L}{\lambda_0}\right), \tag{7.1}$$

where λ_0 is the central wavelength, ΔL is the optical path difference between two arms, and E_{s} and E_{r} are electromagnetic waves from sample arm and reference arm, respectively.

When a partial coherent light source with a Gaussian spectral distribution is used in the interferometer,

$$I_{\mathrm{d}} = E_{\mathrm{s}}^2 + E_{\mathrm{r}}^2 + 2E_{\mathrm{s}}E_{\mathrm{r}} \exp\left(\frac{-4\ln 2\Delta L}{L_{\mathrm{c}}^2}\right)\cos(\omega_0 \Delta L), \tag{7.2}$$

where $\exp(-4\ln 2\Delta L/L_c^2)\cos(\omega_0\tau)$ is called coherence function;L_c is the coherence length and is given by

$$L_c = \left(\frac{2\ln 2}{\pi}\right)\left(\frac{\lambda_0^2}{\Delta\lambda}\right) = 0.44\frac{\lambda_0^2}{\Delta\lambda}. \tag{7.3}$$

In OCT system, L_c acts as the coherence gate, which directly determines the depth resolution of an OCT system.

7.2.2 Time Domain and Fourier Domain OCT

In the earlier stages of OCT imaging, axial (depth) ranging is provided by linearly scanned low coherence interferometry (45, 46). This method of OCT, referred to as *time domain optical coherence tomography* (TD-OCT). TD-OCT system is based on a Michelson interferometer (47). There are two main configurations, free-space setup and fiber-based setup (48), shown in Figure 7.3a and b, respectively.

In TD-OCT systems, a broadband source is used in interferometer. The coherent gate is then created to separate the tomography at a certain layer. A mechanical scanning device is introduced to select different layers at different depths by moving the reference mirror. Figure 7.4 shows the basic concept. As the light source is broadband, when the mirror moves, the interferometer will produce a piece of moving interference fringes, called *coherence gate*, which scans through the sample. In ordinary camera imaging, all the reflected/scattered light from different layers, such as $L_1, L_2 \ldots L_n$, are collected together and eventually form a fused image. However, in the OCT imaging, only the layer whose optical length is the same as that in reference arm has been modulated by the interfering fringes, that is framed by the coherence gate. Using a specially designed algorithm, the image of this layer can be extracted from others. The broader the bandwidth the source possesses, the narrower the gate and the finer resolution the extracted cross-sectional image achieves.

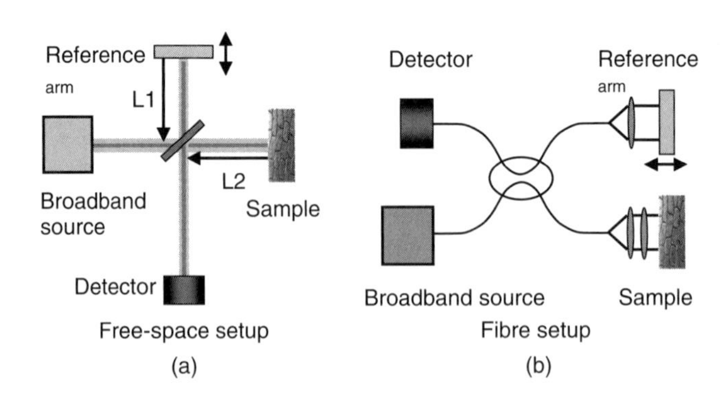

FIGURE 7.3 (a and b) Configurations of time domain OCT system.

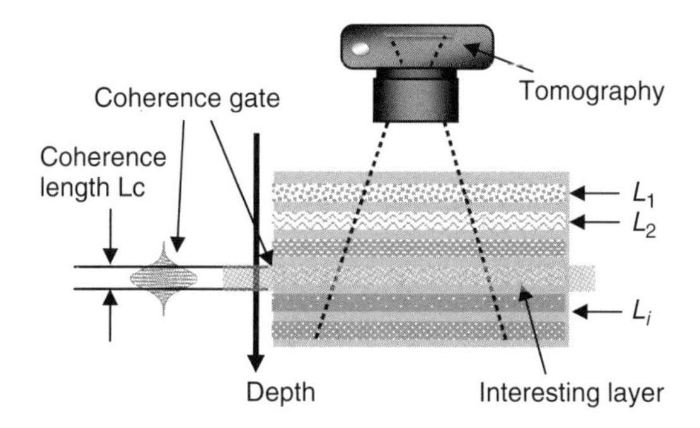

FIGURE 7.4 Coherence gate used for separating layers.

Considering the scanning procedure in TD-OCT, which is actually a procedure of convolution, Equation 7.1 can be expressed as

$$I_{\mathrm{d}} \propto E_{\mathrm{s}} \otimes E_{\mathrm{r}}, \tag{7.4}$$

its Fourier transform becomes

$$I_{\mathrm{d}}(\omega) \propto s_{\mathrm{s}}(\omega) \cdot s_{\mathrm{r}}(\omega), \tag{7.5}$$

where $s_{\mathrm{r}} = \frac{1}{2}s(\omega)$ and $s_{\mathrm{r}}(\omega)$ and $s_{\mathrm{s}}(\omega)$ are Fourier transform of E_{s} and E_{r}, respectively. $s(\omega)$ is the light source spectrum. Considering the interferometer structure, the signal detected by sensor is given by

$$I_{\mathrm{d}}(\omega) = |s_{\mathrm{r}}(\omega) + s_{\mathrm{s}}(\omega) \cdot s_{\mathrm{r}}(\omega)|^2 = S(\omega)[1 + s_{\mathrm{s}}(\omega)]^2, \tag{7.6}$$

where $S(\omega) = s^2(\omega)$.

Equation (7.6) is the foundation of Fourier domain optical coherence tomography (FD-OCT) (49–51). Actually, FD-OCT can be further divided into swept source optical coherence tomography (SS-OCT) and spectral domain optical coherence tomography (SD-OCT), which are illustrated in Figure 7.5.

In Figure 7.5a, a swept laser is used as the light source. When $s_{\mathrm{s}}(\omega_i)$ sends its ith wavelength, the detector receives the corresponding spectral signal $I_{\mathrm{d}}(\omega_i)$. After collecting all the wavelength response, an inverse Fourier transform is required to reconstruct the internal structures (52, 53). SD-OCT illustrated in Figure 7.5b extracts the spectral signal by means of a grating spectrometer and a linear detector array (54, 55). The reconstruction of the internal tomography is performed by an inverse Fourier transform of $I_{\mathrm{d}}(\omega)$. SD-OCT gets the full broadband $I_{\mathrm{d}}(\omega)$ in one shot but collects the signal in series from the linear detector array. However, SS-OCT collects the spectral signal in series by changing the wavelength of the light source. To build the internal construction, both need additional Fourier transform, implemented by either hardware or software.

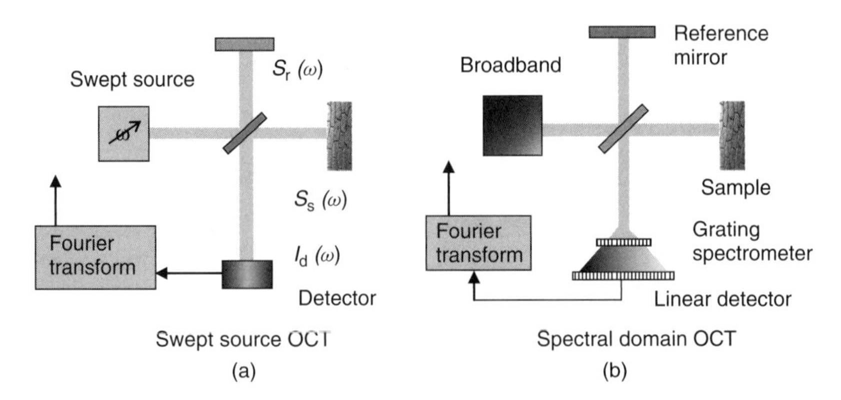

FIGURE 7.5 Fourier domain OCT: (a) SS-OCT and (b) SD-OCT.

SS-OCT and SD-OCT have several advantages over TD-OCT. Because of no mechanical scanning, the FD-OCT system is significantly faster, 50–100 times, than TD-OCT. In addition, both SD-OCT and SS-OCT have better sensitivity and signal-to-noise ratio (51).

7.2.3 Full-Field OCT (FF-OCT)

Most OCT systems use point-scanning based technology, especially the fiber-optics based interferometers. To get *an enface* image, the 2D scanning is a must. Depth scanning is achieved by the longitudinal translation of a reference mirror for TD-OCT, or wavelength scanning for FD-OCT. Such 3-axis scanning makes the system slow and cumbersome. To increase the acquisition speed and eliminate the need for lateral scanning, parallel detection schemes have been investigated. Parallel OCT system illuminate entire 2D target and collect light from all pixels simultaneously. These parallel OCT systems are usually called *FF-OCT systems*. A few OCT systems working directly on 2D full-field images were reported (28, 29, 56–58). Figures 7.3a and 7.6 show a set of Michelson-based FF-OCT. Figure 7.3a is the simplest configuration, which can be used for large area OCT imaging. Figure 7.6a shows a FF-OCT using two cameras, which can perform real-time video-rate OCT imaging (59, 60). Figure 7.6b is a typical full-field optical coherence microscopy (FF-OCM), in which two identical microscope objectives are used to image the tiny structures inside the sample (61, 62).

For the full-field interference, at a position where $\Delta L = 0$, Equation 7.2 can be expressed by a 2D function

$$(x, y) = I_0(x, y) + A_i(x, y)\cos[\phi(x, y)], \tag{7.7}$$

where $I_0(x, y) = {E_s}^2 + {E_r}^2$ is the background image, which is the intensity summation of all layer images; cos[] item represents the interference pattern, a function

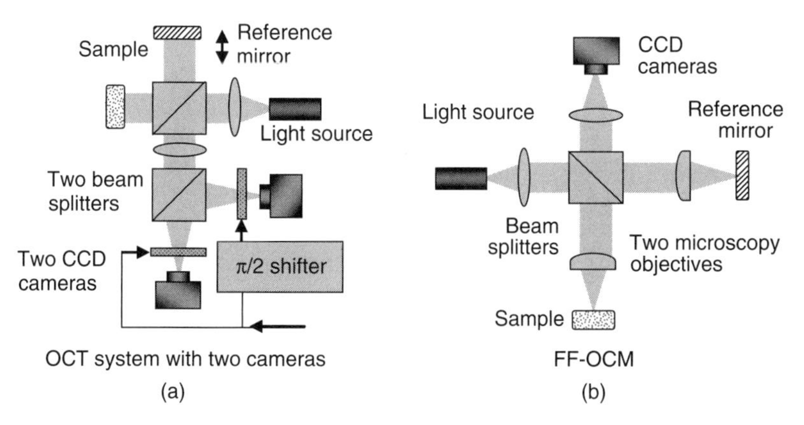

FIGURE 7.6 (a and b) Two configurations of FF-OCT system.

of variables x and y. $A_i(x, y) = 2E_s\, E_r$. As E_r can be taken as a constant, $A_i(x, y)$ is the tomographic image of ith layer.

Equation 7.3 indicates that the spectrum bandwidth dominates the resolution of an OCT system; the broader the band is, the finer the resolution will be. For a typical OCT light source, superluminescent diode (SLD) with central wavelength 830 nm and $\Delta\lambda$ of 15 nm, the system resolution is $L_C = 20\ \mu\text{m}$. To retrieve the information hidden in the interference images, the lateral spatial interfering fringes, represented by $\cos[\phi(x, y)]$, have to be removed through some processing. In Section 7.3.2, methods of removing fringes are described.

In FF-OCT, if the light source becomes swept source, it will eventually remove all the mechanical moving parts and therefore greatly speedup the acquisition and processing procedures. This system is called *swept source full-field optical coherence tomography*, (SS-FF-OCT).

7.3 OCT SYSTEMS: HARDWARE AND SOFTWARE

Generally, for all OCT systems, either TD-OCT and FD-OCT or fiber-based OCT and FF-OCT, the most critical part is the light source. Besides the requirement for the large bandwidth, the optical power also plays an important role in detecting the deep structure in the sample. In the hardware point of view, the photo sensor/camera with high sensitivity, large dynamic range, and corresponding data acquisition electronics is another critical part. For FD-OCT, the performance of Fourier transform board definitely affects the processing speed. In the view of software, algorithms for OCT image processing are needed.

7.3.1 OCT Systems and Components

In the past, more and more OCT systems were built as FD-OCT systems, including SD-OCT and SS-OCT. In this section, we first describe the details of a fibe-based

SS-OCT system and parts related to it. Then a time domain FF-OCT system and a SS-FF-OCT system are briefly introduced.

7.3.1.1 Full-Range Complex SS-OCT System In the traditional SS-OCT, because the input signal is a real signal, the reconstructed OCT images has a conjugate part, which yields a mirrored image. By introducing a 3×3 couplers and two deferential detectors, the quadrature detection can produce two outputs with a $\pi/2$ phase shift, which produces a complex signal, rather than a real signal, to feed the OCT system. The reconstructed image can eliminate the mirror image and compress the DC component and hence effectively improve the quality of reconstructed image (63). SS-OCT could also make possible for a quadrature interferometry based on multiport fiber couplers, for example, 3×3 quadrature interferometer (64, 65). Owing to its ability to have instantaneous complex signals with stable phase information, OCT with a 3×3 quadrature interferometer could suppress the complex conjugate artifact naturally, therefore to double the effective imaging depth. As it takes the full use of the Fourier domain for the useful signal, it is called *full-range SS-OCT*. By detecting the phase from the complex signals, it also could exploit additional information of the tissue to enhance image contrast, to obtain functional information, and to perform quantitative measurements (66, 67). In addition, SS-OCT could be made possible for an unbalanced input fiber interferometer and differential output detection by using a Mach–Zehnder interferometer (68). The unbalanced input could emit larger portion of the optical power from optical source to the tissue than that to the reference mirror for increasing sensitivity (69). The differential detection is used to reduce the excess intensity of noise to further sensitivity enhancement compared to SD-OCT (70). Figure 7.7 shows the experimental setup of the instantaneous complex conjugate resolved SS-OCT system that uses a 3×3 Mach–Zehnder

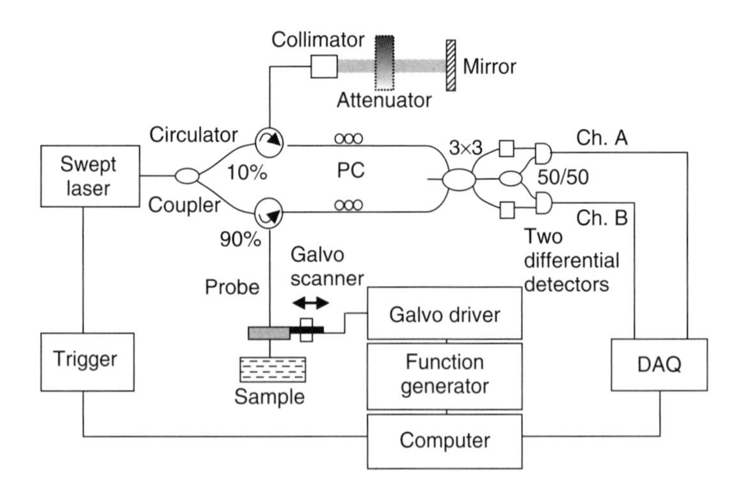

FIGURE 7.7 Full-range complex SS-OCT system.

interferometer topology with two differential detectors. The system consists of a 2×2 single-mode fiber coupler with 90/10 coupling ratio as the input coupler; a 3×3 fiber coupler with coupling ratios of 0.33/0.33/0.33, a 50/50 2×2 fiber couplers; and two variable fiber attenuators as the output couplers forming the dual-channel-balanced detection.

The swept source used in the system has a central wavelength of 1320 nm and a full scan wavelength range of 110 nm, which is sweeping linearly with optical frequency with a linearity of 0.2%. The average output power and coherence length of the swept source are 12 mW and 10 mm, respectively. A repetition scan rate of 20 kHz is used in our system, and the related duty cycle is 68%. The output light from the swept laser source is launched into the first coupler and then divided into the sample arm with 90% power and reference arm with 10% power by two fiber circulators. The reference arm is arranged with a fiber collimator and a mirror. A variable attenuator is inserted between the collimator and mirror for adjusting the optical power on reference arm to achieve the higher sensitivity. The light is illuminated on a sample through the lensed single-mode fiber probe with working distance (focus distance to lens surface in air) of 1.1 mm, depth of field (twice the Rayleigh range in air) of 1.1 mm, and $1/e^2$ spot diameter (transverse resolution) of 27 μm. A galvanometer scanner scans the fiber probe light transversely on the sample up to 4 mm at 20 Hz with 1000 transverse pixels. The total optical power illuminating on the sample is approximately 10 mW. Two polarization controllers (PCs) in both reference and sample arms are used for adjustment to match the polarization state of the two arms. The two-pair output signals from the output couplers are detected with two-pair photodiodes to obtain quadrature signals. Two differential photodetectors are used with adjustable bandwidth. A 3 dB bandwidth of 50 MHz is used in our system. The two detector outputs are digitized using a data acquisition card (DAQ) with 14-bit resolution and acquired at a sampling speed of 100 MS/s. The swept source generates a start trigger signal that is used to initiate the function generator for the galvo scanner and to initiate the data acquisition process for each A-scan (depth scan at a fixed point). As the swept source is linearly swept with wave number k, A-scans data with resolved complex conjugate artifact is obtained by a direct inverse Fourier transformation (IFT) from direct DAQ (data acquisition) sampling data without any resampling process. The maximum imaging rate of 20 f/s can be obtained when the lateral pixel number is 1000.

Figure 7.8 shows *in vivo* images of a human finger tip acquired by (i) a traditional SS-OCT system and (ii) full-range complex SS-OCT system. Figure 7.8b demonstrates the artefact-free OCT imaging and clearly shows the sweat glands, which is an important feature for distinguishing the real and fake fingers in biometrics.

7.3.1.2 Full-Field OCT Systems Large Area FF-OCT Using White Light Source. White light interferometry is certainly not new, however, by combining old white light interferometry techniques with modern electronics, computers, and software can produce powerful measurement tools (71, 72). White light interferometry method has already been established for the measurement of

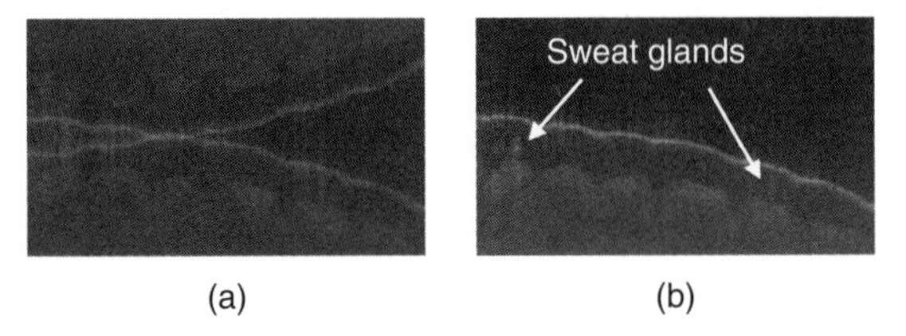

FIGURE 7.8 *In vivo* images of human finger tip. (a) OCT image mixed with mirror image. (b) OCT image generated by full-range complex SS-OCT system, in which the sweat glands can be seen clearly.

topographical features of sample surfaces (73). It can be modified to extract the cross-sectional image of an object. The basic configuration is shown in Figure 7.3a. In our system, a halogen lamp is used as the light source, which has a central wavelength 700 nm and bandwidth 200 nm. The depth resolution is 0.9 μm. Image grabbing rate is 30 frames per second, with a resolution of 1024 × 1024 pixels at 12-bit gray levels. Mechanical depth scanning accuracy is 37 nm. For such a high accurate system, the alignment and tuning up of the system is critical. A machine-vision-based technology was developed to perform this task. The working area is designed for 25 mm by 25 mm. As the halogen lamp has very strong power, the working area can be extended even larger. This system was built for applications in security and 3D imaging (74).

Swept Source Full-Field Optical Coherence Microscopy. This system is built based on a free-space interferometer Linnik microscope, with an imaging magnification of 60 times (75). Figure 7.6b shows basic configuration for Linnik FF-OCM. However, the light source used is a broadband-tunable laser. It has no mechanical scanning at all. Specifications are: wavelength 1300 nm; tuning range 110 nm; output power 40 mW; acquisition speed greater than 15 frames per second; and microscope working distance 20 mm. The lateral resolution is around 4 μm and depth resolution about 16 μm. This system can be used to fast generate the OCT images for the tiny internal structures of sample. Figure 7.9 shows the scheme of the SS-FF-OCT system.

7.3.2 Algorithms Used in OCT Signal/Image Processing

To retrieve the internal structural information contained in back reflected/scattered signal I_d, one has to solve the E_s or s_s from the following equations:

$$I_d = E_s^2 + E_r^2 + 2E_sE_r \exp\left(\frac{-4\ln 2\Delta L}{L_c^2}\right)\cos(\omega_0 \Delta L), \tag{7.2}$$

$$I_d(\omega) = |s_r(\omega) + s_s(\omega)s_r(\omega)|^2 = S(\omega)[1 + s_s(\omega)]^2, \tag{7.6}$$

$$I_d(x,y) = E_s(x,y)E_r(x,y) + 2E_s(x,y)E_r(x,y)\cos[\phi(x,y)]. \tag{7.7}$$

Equations 7.2, 7.6 and 7.7 are used for TD-OCT, FD-OCT, and FF-OCT, respectively.

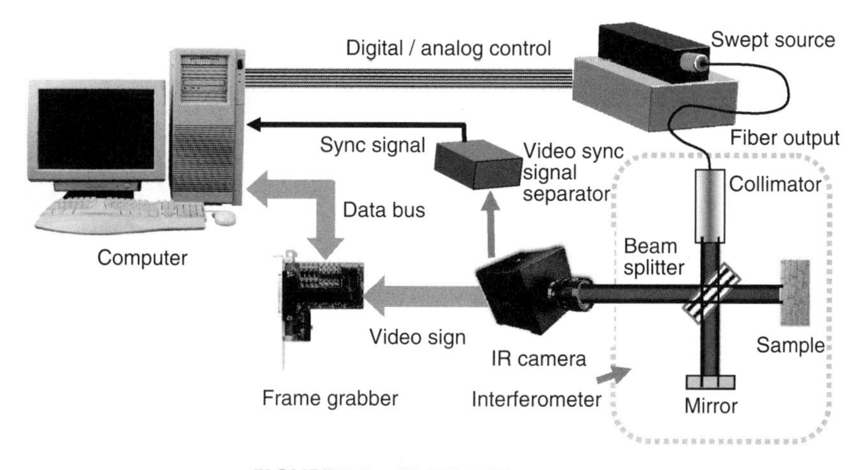

FIGURE 7.9 SS-FF-OCT system.

7.3.2.1 Algorithm for TD-OCT Considering E_r is a constant reference signal, Equation 7.2 can be expressed as

$$I_d = I_0 + AG(\Delta L)\cos(\omega_0 \Delta L), \tag{7.8}$$

where, I_0 is the background DC bias. A is the amplitude of back reflected/scattered signal; $G(\Delta L)$ is a Gaussian gate function; and cos() item represents the wavelength carrier. When $\Delta L = 0$, that is., the optical lengths are identical between two arms, OCT system yields maximum of the received signal.

To extract the tomographic signal hidden in background and interference fringes, the following two steps are needed:

1. Removing the DC part;
2. Extracting the maximal value of envelope of cos() item by a low pass filter. Normally, to enhance the visualization of the weak signal, a logarithm operation is used to yield a final signal:

$$I_{out} = \log_{10}(I_d). \tag{7.9}$$

7.3.2.2 Algorithm for FD-OCT Equation (7.6) shows the detected signal of FD-OCT:

$$\begin{aligned} I_d(\omega) &= |s_r(\omega) + s_s(\omega)s_r(\omega)|^2 \\ &= S(\omega)[1 + 2s_s(\omega) + s_s{}^2(\omega)]. \end{aligned} \tag{7.10}$$

After Fourier transform,

$$\begin{aligned} I_d(z) = \mathrm{FT}[I_d(\omega)] &= \mathrm{FT}[S(\omega)] + 2\mathrm{FT}[S(\omega)] \otimes \mathrm{FT}[s_s(\omega)] \\ &+ \mathrm{FT}[S(\omega)] \otimes \mathrm{FT}[s_s^2(\omega)]. \end{aligned} \tag{7.11}$$

where $\otimes$ represents a convolution. If the light source is broadband and smooth enough, $\mathrm{FT}[S(\omega)]$ could be considered as a constant k. Because $s_\mathrm{r}(\omega)$ represents the returned weak signal, the last term can be ignored. Therefore, Equation 7.11 could be expressed by

$$I_\mathrm{d}(z) = k + 2k\mathrm{FT}[s_\mathrm{s}(\omega)]. \tag{7.12}$$

By placing the reference mirror at an offset distance in front of the surface of the sample, the first term k can be easily removed from the Fourier transform result. In this case, the FD-OCT signal of the sample is extracted by the Fourier transform of the detected signal:

$$I_\mathrm{d}(z) \propto 2k\mathrm{FT}[s_\mathrm{s}(\omega)]. \tag{7.13}$$

Similarly, to get a better visualization result, a logarithmic operation, such as Equation 7.9, may be required.

In addition, before Fourier transformation, the N-points dataset was normally zero padded to $4 \times N$ points, in order to increase the resolution after Fourier transform.

7.3.2.3 Algorithm for FF-OCT For the FF-OCT, Equation 7.2 should be extended to x, y plane:

$$I_\mathrm{d}(x,y) = E_\mathrm{s}^2(x,y) + E_\mathrm{r}^2(x,y) + 2E_\mathrm{s}(x,y) \cdot E_\mathrm{r}(x,y)\cos\{\omega_0 \Delta L, [\phi(x,y)]\}. \tag{7.14}$$

In this case, two types of interferences occur. One is the temporal interference and another the spatial interference. The algorithm described in Section 7.2.3.1 only provides the means to remove the temporal, that is, depth interfering fringes. In FF-OCT, the lateral 2D interfering fringes also need to be removed.

7.3.2.4 Algorithms for Lateral Interfering Fringes Removal Let $\Delta L = 0$, the 2D raw OCT signal is described as

$$I_\mathrm{d}(x,y) = E_\mathrm{s}(x,y) \cdot E_\mathrm{r}(x,y) + 2E_\mathrm{s}(x,y) \cdot E_\mathrm{r}(x,y)\cos[\phi(x,y)]. \tag{7.15}$$

where cos() item represents the 2D lateral interfering fringes. When the light source is a plane light beam, $E_\mathrm{r}(x,y)$ can be considered as a constant, say, $E_\mathrm{r}(x,y) = 1$. Equation 7.15 can then be simplified as

$$I_\mathrm{d}(x,y) = I_0(x,y) + A_i(x,y)\cos[\phi_i(x,y)]. \tag{7.16}$$

where $A_i(x,y)$ is the cross-sectional image of the ith layer of the sample. To retrieve the $A_i(x,y)$, the interference fringe $\cos[\phi_i(x,y)]$, has to be removed. In Reference 60, a method of removing fringes by using two $\pi/2$ phase-shifted images $I_1(x,y)$,

$I_2(x,y)$, and a background image $I_0(x,y)$ is reported. Thus the tomography can be resolved by

$$A_i(x,y) = \{S_1(x,y)^2 + S_2(x,y)^2\}^{1/2}. \tag{7.17}$$

where $S_1(x,y) = I_1(x,y) - I_0(x,y)$ and $S_2(x,y) = I_2(x,y) - I_0(x,y)$.

The background image $I_0(x,y)$ can be obtained before or after the scanning. Two $\pi/2$ phase-shifted images $I_1(x,y)$, $I_2(x,y)$ can be acquired either by using two cameras, as illustrated in Figure 7.6a or by grabbing one after another with a phase shift in depth. When the phase shift is an arbitrary value, but not $\pi/2$, the following formula should be used (10):

$$A_i(x,y) = \left\{ S_1^2(x,y) + \left[\frac{(S_2(x,y) - S_1(x,y)\cos\varphi)}{\sin\varphi} \right]^2 \right\}^{\frac{1}{2}}. \tag{7.18}$$

However, both Equations 7.17 and 7.18 need to know the phase shift angle *a priori*. To solve the $A_i(x,y)$ without knowing the value of phase shift, more phase shift images have to be presented. An algorithm based on five-step phase shift, which was originally proposed by Hariharan equations (76–78), is given as

$$\begin{cases} I_1(x,y) = I_0(x,y) + A_i(x,y)\cos[\phi(x,y) - 2\varphi], \\ I_2(x,y) = I_0(x,y) + A_i(x,y)\cos[\phi(x,y) - \varphi], \\ I_3(x,y) = I_0(x,y) + A_i(x,y)\cos[\phi(x,y)], \\ I_4(x,y) = I_0(x,y) + A_i(x,y)\cos[\phi(x,y) + \varphi], \\ I_5(x,y) = I_0(x,y) + A_i(x,y)\cos[\phi(x,y) + 2\varphi]. \end{cases} \tag{7.19}$$

The tomography could be solved as

$$A_i(x,y) = \left[\left(\frac{I_2(x,y) - I_4(x,y)}{2\sin\varphi} \right) + \left(\frac{2I_3(x,y) - I_5(x,y) - I_1(x,y)}{4\sin^2\varphi} \right) \right]^{\frac{1}{2}} \tag{7.20}$$

where φ cannot be $n\pi$, n is an integer.

$$\tan\varphi = \frac{2[I_2(x,y) - I_4(x,y)]}{3I_3(x,y) - I_1(x,y)} \tag{7.21}$$

As five-step algorithm has to calculate the angle of phase step for each pixel in the captured interference images, the computation is relatively heavy, particularly when the image is large. Dubois et al. (79) proposed a four-integration buckets algorithm, in which four images with circular-shifted phase angles 0, $\pi/2$, π, and $3\pi/2$

are acquired. Analyzed by using Bessel functions of first kind, J_n, the tomography is obtained by

$$A_i(x,y) = k\{[I_1(x,y) + I_4(x,y) - I_2(x,y) - I_3(x,y)]^2 + [I_1(x,y) - I_4(x,y) - I_2(x,y) + I_3(x,y)]^2\}^{1/2}, \tag{7.22}$$

where k is a coefficient, a function of J_n

A more practical and fast algorithm was proposed by Chang et al. (80). On the basis of energy operator (EO) (81–83), the algorithm is defined by

$$A_i(x,y) = |[(I_3(x,y) - 2I_2(x,y - I_1(x,y)]^2 - \{I_4(x,y) - 3[I_3(x,y) - I_2(x,y)] - I_1(x,y)\}[I_2(x,y) - I_1(x,y)]|^{1/2}. \tag{7.23}$$

where $I_1(x,y)$, $I_2(x,y)$, $I_3(x,y)$, and $I_4(x,y)$ are four-phase-shifted images. The EO-based algorithm is about three times faster than other algorithms mentioned above.

7.3.2.5 Algorithms for Interlayer Demodulation Figure 7.10 schematically illustrates the imaging of a multilayer info carrier by a camera, in which R_i is the reflectance function and T_i the transmittance function of the ith layer. Assuming that there is no absorption in each layer, that is, ${R_i}^2 + T_i^2 = 1$, the complex amplitude R received by the camera would be given by References 28 and 29

$$R = R_1 + T_1R_2T_1 + T_1T_2R_3T_2T_1 + T_1T_2T_3R_4T_3T_2T_1 \ldots + M = \Sigma R_i \prod_{j=1}^{i-1} T_j^2 + M, \tag{7.24}$$

where M represents the stray lights resulting from multireflection among the layers. Concerning R_i is the cross-sectional image of the ith layer, a weighting factor of its intensity $\prod_{j=1}^{i-1} T_j^2$ represents the intensity modulations of the previous layers.

Besides the intensity modulation among layers, the phase modulation may also happen when the substrate and information pattern of the layer have different optical paths. Figure 7.11 shows this phenomenon. Figure 7.11a illustrates the structure

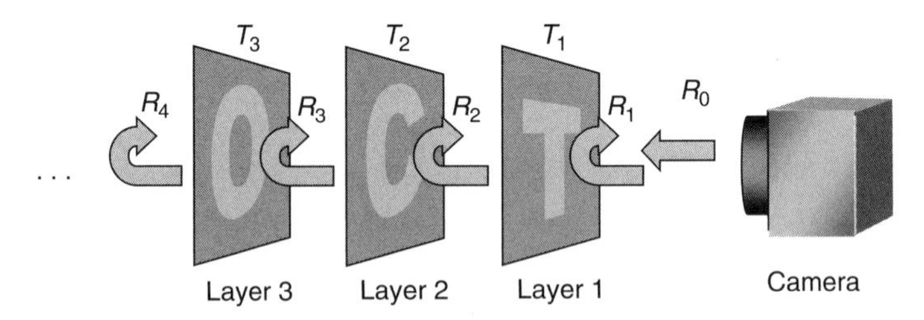

FIGURE 7.10 Images reflected from multiple layers. R_i: reflectance function of the ith layer; T_i: transmittance function of the ith layer.

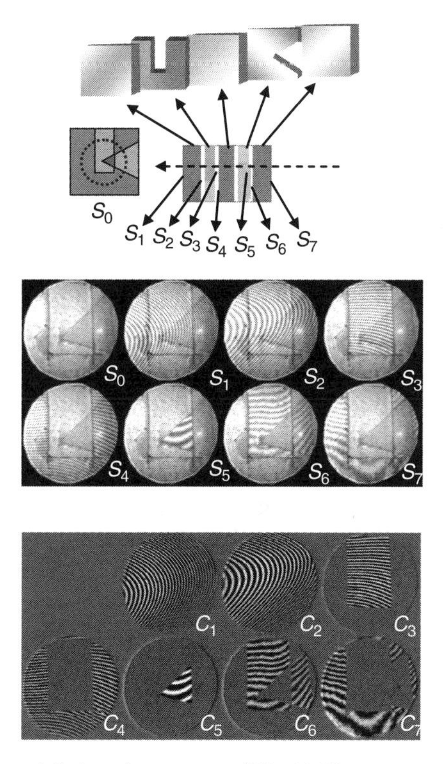

FIGURE 7.11 Phase modulation phenomenon (29). (a) The structure of a sample consisting of five layers, $T_1 - T_2$. (b) The images grabbed by a full-field OCT systems at different depths. (c) Cross-sectional images with the subtracted background.

of a sample consisting of five layers, T_1–T_5, each having a specific thickness. The patterns of layer T_2 and T_4 contain some 'empty' areas, meaning air-gaped between layer T_1 and T_3 and between T_3 and T_4 for these areas. Figure 7.11b shows the images grabbed by a FF-OCT system at different depths, where S_0 is the noninterfering background image, and S_1–S_7 are interference images for some selected surfaces of the layers, as shown in Figure 7.11a. Figure 7.11c provides seven cross-sectional images C_1–C_7, in which the background is subtracted from S_1–S_7. The shape of layer T_4 cannot be completely retrieved by any of the individual C_5, C_6, or C_7. This is due to the existence of the 'empty' area in the layers T_2 and T_4. The different refractive index in the air and glass produces different optical path lengths and make the phase variation across the surfaces of the sample. As the optical paths for information pattern and substrate are different, the cross phase variation caused by these layers, referred to as *phase modulation*, may eventually cause a segmented tomographic image.

When the information carrier is illuminated by a uniform partial coherent light and a reference R_e is introduced, the intensity image captured by the camera is given by

$$I_d = (R + R_e)^2 = R^2 + {R_e}^2 + R_e R^* + R{R_e}^*. \tag{7.25}$$

Then the interference part in Equation 7.25 becomes

$$R_e R^* + R{R_e}^* = R_e[R_i \prod_{j=1}^{i-1} T_j^2]^* + R_i \prod_{j=1}^{i-1} T_j^2 {R_e}^*. \tag{7.26}$$

Provided that the coherent length of illumination is shorter than the spacing between two layers, the interference images resulting from multireflection effect are negligible. Hence items in Equation 7.16) can be further written as

$$I_d(x, y) = I_0(x, y) + A_i(x, y) \cos \phi(x, y), \tag{7.16}$$

where

$$I_0(x, y) = R^2(x, y) + R_e^2(x, y), \tag{7.27}$$

$$A_i(x, y) = 2R_i(x, y) \prod_{j=1}^{i-1} T_j(x, y)^2 R_e(x, y). \tag{7.28}$$

A peeling algorithm can be obtained by using a recursive method to reduce the previous modulation layer by layer (75). Hence the demodulation algorithm for intensity is derived as:

$$R_i(x, y) = \frac{A_{\mathrm{i}}(x, y)}{\{\mathrm{Kr}[1 - R_j(x, y)]^2\}} \tag{7.29}$$

where $\mathrm{Kr} = 2R_e(x, y)$, a constant. However, this peeling algorithm defined by Equation 7.29 may not be effective if the information pattern on a layer is not perfectly planar and has some scattering effect to the probe beam. In this case, a practical algorithm is proposed to replace Equation 7.29:

$$R_i(x, y) = A_i(x, y) - w \times R_{i-1}(x, y). \tag{7.30}$$

where w is a weighting factor. Equation 7.30 only considers the intensity modulation of the previous layer, because the farther layers have little modulation effects due to the scattering. The weighting factor w is determined by experiments.

7.4 SENSING THROUGH VOLUME: APPLICATIONS

7.4.1 Security Data Storage and Retrieval

As OCT has the resolution of micrometer level and the ability of peeling cross-sectional images from the inside an object, it has potential applications in documents security and object identification (84). Assuming that the depth and lateral resolution of the OCT system are 10 μm and a single layer has 1024 × 1024 pixels, a 200-layer info-chip could have a volume size about 20 mm × 20 mm × 2 mm. If each pixel has 16 bits, this info-chip has a data content of 3.2 Gbit.

As the info-chip can be made by low scattering clear material, the low signal-to-noise ratio caused by scattering medium will no longer be a major problem in this circumstance. Therefore, the specific hardware, such as smart pixel device and lock-in detection apparatus, and complicated software designed to deal with the scattering effects in digital signal processing can be greatly simplified. In addition, as it does not require X–Y axis scanning, this 2D parallel OCT could be a simple and economic imaging system for applications of multiple-layer information extraction.

The multilayer information carrier can be constructed in such a way that it has a top layer, several information layers, and a base, as shown in Figure 7.12. The top layer has been covered with a band-pass coating, which allows only the known probing beam to pass through. All the information layers are of planar and transparent surfaces, on which the information, for example, text, image, or other 2D data, are encoded by applying a thin transparent coating, whose index of refraction is different from that of substrate. This difference makes the information observable. The base could be made of a solid blackened material to absorb all the incident lights and prevent reflections.

To reduce the effect of the phase modulation, the information carrier must be properly designed and fabricated, for example, by alternatively encoding the information as "negative" or "positive" images. The basic requirement for the design is that the overall optical path difference across the surfaces of all layers should be less than the depth resolution of the OCT system. Meanwhile, the phase modulation effect might be compensated by software, if prior knowledge of the materials used in information patterns and substrates is known.

An OCT system used for multilayer information extraction is of same configuration as that illustrated in Figure 7.3a, which is a Michelson interferometer with a vision and motion controller. The light source, a superluminescent diode (SLD), is a low coherence light source whose central wavelength is 830 nm and coherence

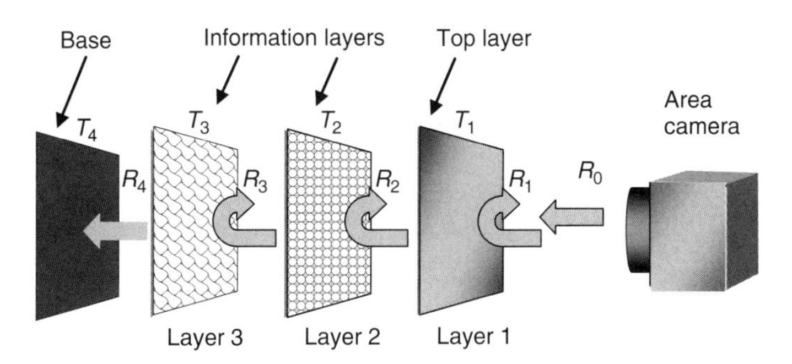

FIGURE 7.12 Info-chip: multilayer information carrier.

length is 20 μm. To reduce the interfering fringes, we have used the multiple phase-shifted images.

The system runs as follows

1. Starting scanning at a noninterference position;
2. Searching for the position of the nearest layer where the interference fringes become the strongest;
3. As soon as the layer is reached, grabbing the first interference image;
4. Moving one step, grabbing next interference image;
5. Repeating steps 3 and 4, until multiple interference images, depending on the algorithm, are grabbed;
6. Extracting the tomography of this layer by processing these images;
7. Repeating steps 2–6, until all the tomographic images of all layers are processed;
8. Starting the removing the interlayer modulations.

The methods of removing the interlayer modulations depend on the materials used in the sample. If all the layer substrate and information coatings are made of clear and flat materials, the demodulation algorithm for intensity could be Equation 7.29. In case that some scattering material is involved, for example, the fingerprint stamp, the more practical algorithm may be Equation 7.30.

Figure 7.13 shows a set of images resulting from multiple-layer information retrieval. The test sample is a four-layer info-chip. The first two layers are texts, and other two layers are fingerprints. Figure 7.13a illustrates the structure of the info-chip, and Figure 7.13b shows the directly reflected image of the chip, in which all the information are fused and the fingerprints are almost invisible. Figure 7.13c–f shows four extracted images of layers 1–4, respectively. Figure 7.13f shows the tomographic image of the last layer, which has passed the interlayer demodulation process by using Equation 7.29, where Kr is set to 0.3. The camera used in the experiment is an area CCD camera of PULNiX. Before performing tomography processing, the nonlinear intensity distortion of the camera has been calibrated by software.

7.4.2 Internal Biometrics for Fingerprint Recognition

Of the different biometric technologies, fingerprint is the most commonly used type for various applications, such as law enforcement, financial transactions, access control, and information security. Fingerprint recognition owns several merits, compared to other biometric identification techniques such as iris recognition, face recognition, and hand-geometry verification method, which made it the most popular biometric technology. First, fingerprint recognition has high permanence, as fingerprints form in fetal stage and remain structurally unchanged throughout the life. Secondly, it has high distinctiveness—even identical twins possess different fingerprints. In addition to high universality, a majority of the population have

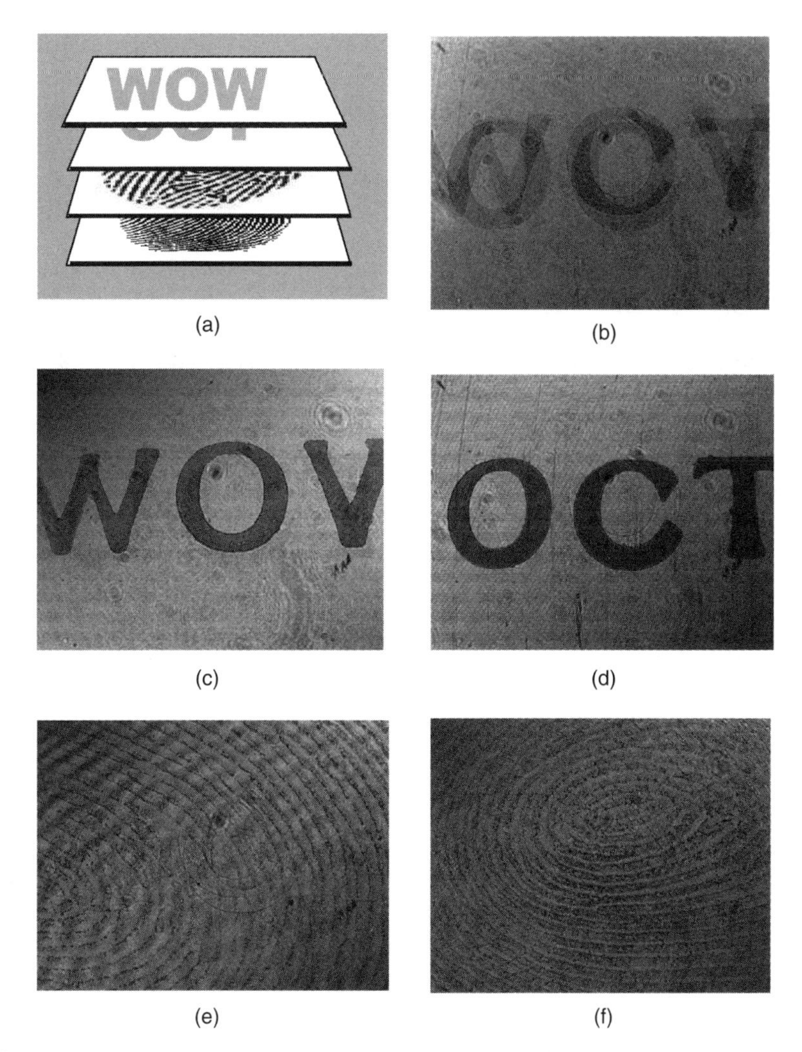

FIGURE 7.13 Four-layer info-chip and its OCT images of each layer (29). (a) Structure of four-layer info-chip; (b) Direct imaging of this info-chip; (c)–(f) Four OCT images of this info-chip.

legible fingerprints, which are more than the number of people who have license and passports. Last but not the least, fingerprint acquisition is nonintrusive, and it requires no training, which shows high acceptability. As every fingerprint is considered to be unique and immutable, people have a strong confidence in the fingerprint system. The advantages of using biometrics in identity verification over password or token have been broadly reported (85–87). However, it has been revealed that the fingerprint readers can be defeated either using a combination of low cunning, cheap kitchen supplies and a digital camera (88), or creating false thumbprint images (89). These attack methods are called *spoofing* because they are attempts to "fool" a biometric system by presenting a fake fingerprint trait to the sensor. With less than $10 worth of household supplies, artificial fingerprint gummies can be made and easily spoof the fingerprint system (88). Figure 7.14 shows an artificial human finger and a dummy fingerprint, on which all the visible external features are delicately created, which could successfully spoof a traditional fingerprint scanning system.

Different technologies have been proposed and tested both in hardware and software to defeat spoofing attacks as following:

- Analyzing skin details through very high resolution sensors (1000 dpi) to capture some details, such as sweat pores or coarseness of the skin texture, to distinguish the fake finger from the real one (90).
- Analyzing dynamic properties of the finger, such as pulse oximetry, blood pulsation, skin elasticity, and skin perspiration to detect a fake finger (91, 92).
- Analyzing static properties of the finger by adding hardware to capture information such as temperature, impedance or other electric measurements, and spectroscopy to recognize the fake finger (91, 93).
- Using multispectral imaging technology to measure the fingerprint characteristics that are at and beneath the surface of the skin to defeat prosthetic fingers (94).

Given the above factors, an enhanced fingerprint recognition system needs to be developed to deal with these fraudulent methods. During the past years, valuable improvements have been achieved by several scientific groups to enhance the surface-scanning-based fingerprint system. Making use of a fast fingerprint enhancement algorithm, which could improve the clarity of the fingerprint surface structures, and based on the estimated local ridge orientation and frequency, the verification accuracy and the false reject rate (FRR) could be improved (95). However, the improvement in the fingerprint recognition field was focused on obtaining a lower FRR and false accept rate (FAR), which made no contribution to detect whether artificial fingerprints were presented.

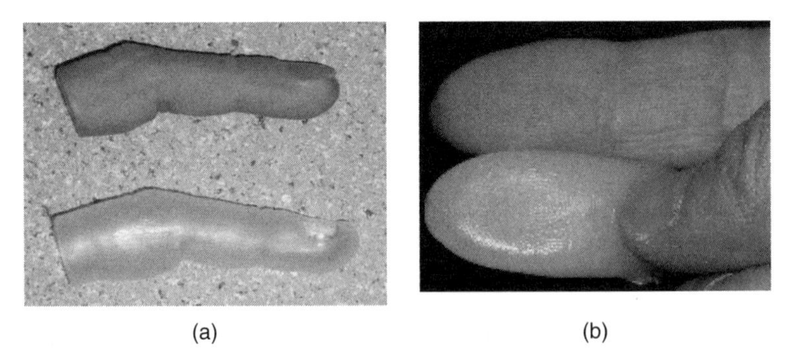

FIGURE 7.14 (a) Dummy finger before and after painting. (b) Dummy fingerprint.

As mentioned above, the OCT technique has unique ability of in-depth and lateral scanning to capture 2D images with resolution up to a few microns. In this way, by analyzing OCT images, additional artificial layers above the real finger could be identified. A study (96) has demonstrated that the OCT can be potentially applied to identify artificial fingerprints, which could be made of various artificial materials. In addition, artificial fingerprint dummy and real human tissue could be effectively distinguished by autocorrelation analysis of speckle noise in OCT images and signals (97). Therefore, artificial materials used to make fingerprints can be effectively detected by the new generation of OCT-enhanced fingerprint system. Unlike the traditional biometrics, which is mainly based on the external features of the object, the new generation of biometrics, termed *internal biometrics*, is deeply involved with the internal features of the object, which makes it more robust and discriminative.

Figure 7.15a shows typical cross-sectional 2D OCT image, which exposes artificial material layer placed over real human tissue. From this figure, one can see that characteristic layers of human skin (stratum corneum, epidermis, and dermis) are shifted down owing to the presence of an additional layer corresponding to the material. A photo of this artificial fingerprint dummy is shown in Figure 7.15b. The thickness of the dummy was about 150 μm. Although the ridges and valleys of the artificial fingerprint are tiny and transparent, all of them could be clearly detected in OCT 2D image Figure 7.15a. In addition, the artificial fingerprint layer can be further exposed and analysed from the related OCT signal curve shown in Figure 7.15b. Each pixel in Figure 7.15a could be converted to an equivalent intensity value, and hence the whole 2D image can be seen as a 2D intensity value matrix. The OCT signal curve, an A-scan signal, reflects the general profile of how the light goes through absorbing/scattering objects and how the photons reflect back. The two high peaks stand for the surfaces of the artificial and the

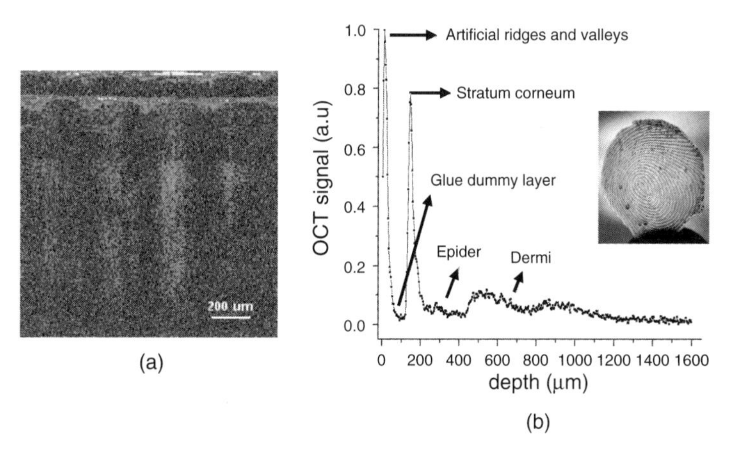

FIGURE 7.15 (a) OCT image obtained from dummy fingerprint placed over a real finger and (b) an A-scan line of corresponding OCT signal curve (98).

real fingerprint layers, respectively. After the second peak, which is the *stratum corneum* of the skin, typical OCT signals corresponding to epidermis and dermis can be seen. Thus both the 2D OCT image and corresponding signal curve clearly show the presence of artificial fingerprint layer placed above the real human finger.

The artificial fingerprint dummy can be easily made either by plasticene (Dixon Ticonderoga Company, Mexico) or by household cement (ITW Devcon Corp., Mass.). Testing experiments between a commercially available fingerprint scanning system (Microsoft Fingerprint Reader, Model: 1033, Redmond, WA) and the OCT system were carried out. First, the real fingerprint pattern from a volunteer's both hands, such as thumbs, forefingers, middle fingers, and ring fingers, were registered into the fingerprint scanning system using a computer. Secondly, the same fingers were used to prepare the corresponding artificial fingerprint dummies. Thirdly, the dummies were placed over another person's real finger and were tested by both the commercial fingerprint reader and the OCT system. This dummy spoofed the fingerprint scanning system mentioned above continuously for 10 times. However, at the same time the OCT system could recognize the artificial fraudulent dummy every time successfully and provided typical detection images as shown in Figures 7.16a and 7.15b.

As the artificial fingerprints are normally made by translucent materials, the FF-OCT becomes a powerful tool to fast and effectively recognize those dummies. FF-OCT can detect both surfaces of a dummy: fingerprint surface and nonprint surface, which does not exist in a real finger. In addition, a FF-OCT can explore the internal structure within these two surfaces, which is also different to a real finger (98). Figure 7.16 shows another set of cross-sectional images obtained from an artificial dummy by a FF-OCT system. During the depth scanning, two obvious surfaces of the dummy were found that exhibited different features. The outer surface shows a smooth 2D curve (Fig. 7.16a–f), which does not exist in the real fingerprint; however, the inner surface shows segmented fingerprints at different layers (Fig. 7.16g–l).

The summation of Figure 7.16a–f is given by Figure 7.17a, which is a bright area without any fingerprints in it. (As the image sampling separation is sort of too broad, 40 μm, the black fringes appear when they are overlapped.) Figure 7.17b shows the summation of these segmented fingerprints, as shown in Figure 7.16g–l. Figure 7.17c demonstrates the summation of all those tomograms, which completely destroys the fingerprint, and is totally different to the image captured by a common 2D camera used in a fingerprint recognition system (Fig. 7.17d).

Figure 7.18 provides another set of rotated 3D volume data. The threads existing inside the body of the dummy, which are totally different to the internal tissues of a real finger. The presence of those red threads and the patterns of those two surfaces prove that the object is an artificial fingerprint dummy.

After successfully distinguishing the artificial dummy fingerprint, the real fingerprint recognition can then start. The main flow chart of OCT-based fingerprint recognition system is illustrated in Figure 7.19.

This system gets the input from an OCT imaging system. The detection of the artificial dummy fingerprint is performed as the first step. After the input is

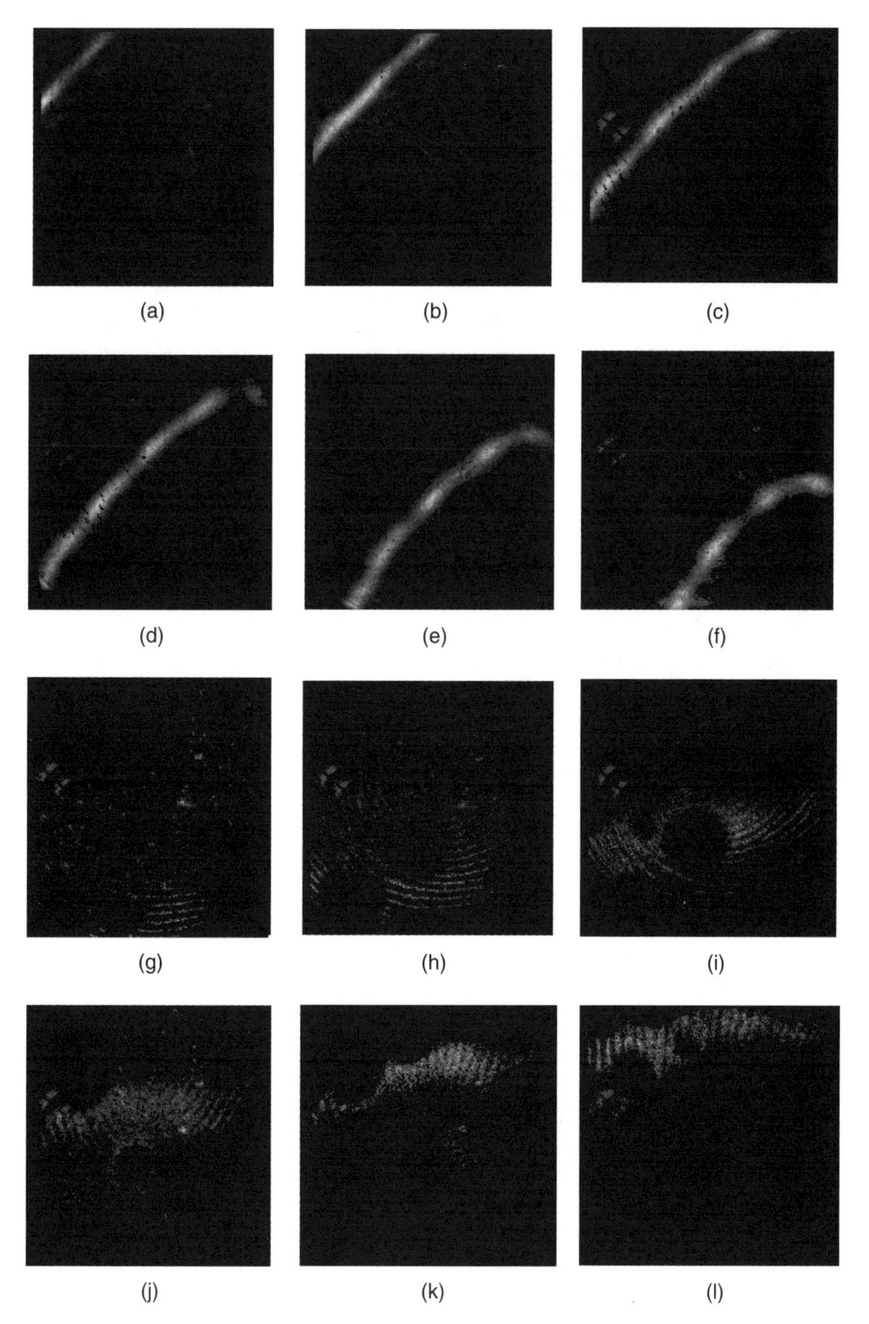

FIGURE 7.16 (a)–(f)OCT Images of a dummy fingerprint obtained by a full-field OCT system (98). (g)–(l)OCT images extracted from the outer surfaces. Layer distance: 50 μm. OCT images extracted from inner surfaces. Layer distance: 20 μm.

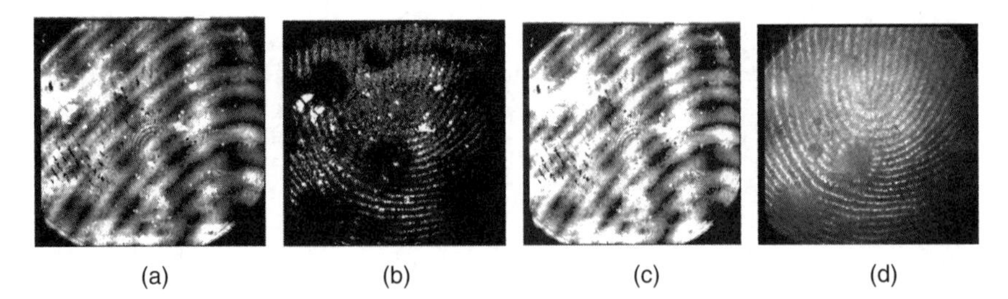

FIGURE 7.17 Images of an artificial fingerprint dummy (98). (a) 2D image of the outer surface of the dummy obtained from OCT images; (b) 2D image of the inner surface of the dummy obtained from OCT images; (c) Summation of above images (a) and (b); (d) Direct imaging of the dummy by a camera used in fingerprint recognition device.

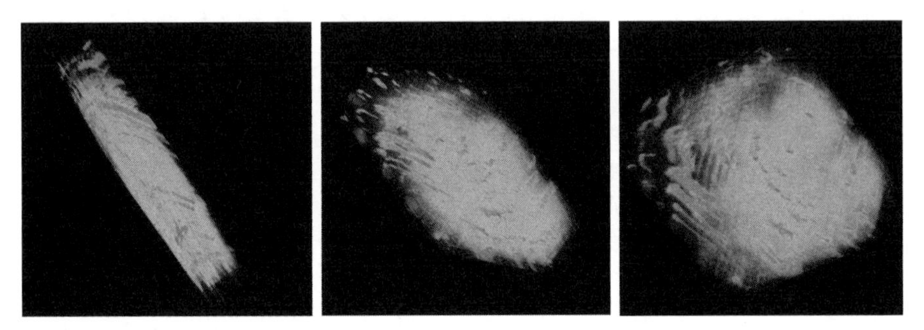

FIGURE 7.18 Three rotated images of the 3D volume data of a dummy fingerprint. The fingerprint can be seen only on the top surface of the dummy. The inside structure is totally different from a real finger (98).

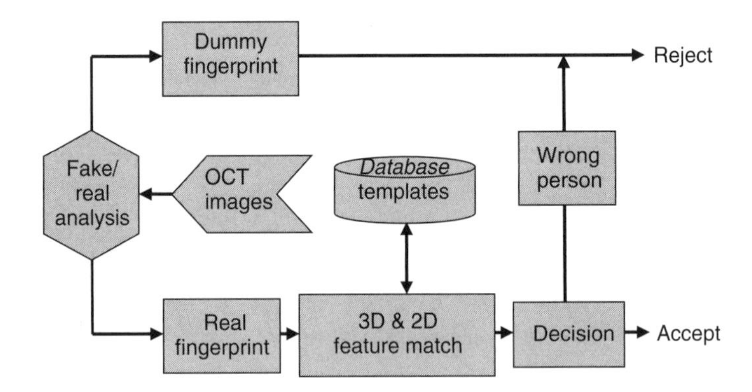

FIGURE 7.19 OCT-based fingerprint recognition system.

confirmed as a real one, the fingerprint identification procedure then starts. Differing from the traditional systems, which only process the 2D pattern of a fingerprint, this system extracts and analyses more features: 2D morphology features, 3D fingerprint profile pattern, and internal tissue structures.

In the traditional fingerprint reader, the finger has to press on a transparent flat surface to produce a 2D fingerprint pattern. There are some problems associated with this kind of devices. First, the motion of the finger may blur the imprinted image; secondly, to get a clear image the imprinting surface should be cleaned each time for a new user, which brings complex mechanical implement. Another important issue is that the 2D flat fingerprint pattern losses 3D profile feature of the fingerprint, which also provides unique ID information for individual. The OCT-based fingerprint recognition system integrates all the 2D morphologic features, 3D profile, and internal biology features, which has the higher discrimination ability, and is robust against the artificial dummy fingers or fingerprints.

Theoretically, OCT system can be extended to other biometric systems, for examples, retina, skin, nail bed, and so on. The OCT-based system can extract the internal structure/feature directly from object, which greatly increase the performance of the biometric system.

7.5 SUMMARY AND CONCLUSION

OCT paves a new avenue for the photonic sensing technologies. The feature of sensing through the object at a microlevel makes it unique from others. In this chapter, we have briefly presented the principle of different OCT systems and their applications in document security and biometrics. The main advantages of these technologies are as follows:

1. Capability to explore the internal structures of objects. It can be applied to most of the biologic samples, such as human skin, and materials transparent/translucent in the IR/visual band.
2. Higher depth resolution at micrometer level than other tomography imaging technology, such as Ultrasound, CT, and MRI (99).
3. Tiny sensing probe. On the basis of the fiber technology, its diameter can be as small as 125 μm, which is extremely important for the space limited applications, such as finger nail gap and tiny ID card spot/slot imaging.
4. Very high sensing speed. The latest imaging speed can reach 1000 frames per second; each frame has 500 × 400 pixels (100). That enables the *in vivo* sensing for objects in motion.
5. Extendable functionalities to serve different purposes. For example, Doppler OCT, polarization sensitive OCT, spectroscopic OCT, and so on can provide extra features for object identification (101).
6. No harmful radiation.

OCT technology has potential applications in many sensing-related fields, involving medical, security, environment, and industrial. In this chapter, we focus on applications to document security and biometrics. All the samples used in these applications should be transparent and translucent materials for the probing light source. However, even for the opaque materials, OCT can also produce accurate 3D profile image at micrometer level, which is important for the detection of the counterfeit valuable products. However, there are still many challenges to face. For examples, the weak signal enhancement in deep area, particularly for biosamples, owing to the multiple scattering effects; the speckle noise removal; ultra small probe control; and the huge data handling, particularly for the SS-FF-OCT. These will be our future targets.

REFERENCES

1. Huang EA, Swanson CP, Lin J, Schman S, Stinson WG, Chang W, Hee MR, Flotte T, Gregory K, Pullafito CA, Fujimoto JG. Optical coherence tomography. Science 1991; 254:1178–1181.
2. Drexler W, Fujimoto JG, editors. *Optical Coherence Tomography: Technology and Applications*. New York: Springer; 2008.
3. Fercher AF, Drexler W, Hitzenberger CK, Lasser T. Optical coherence tomography-principles and applications. Rep Prog Phys 2003;66:239–303.
4. Fercher AF, Hitzenberger CK. Optical coherence tomography. In: Wolf E, editor. *Progress in Optics*. Amsterdam: Elsevier Science B.V.; 2002.
5. Bouma BE, Tearney GJ. *Handbook of Optical Coherence Tomography*. New York: Marcel Dekker, Inc.; 2002, Chapter 4.
6. Kremkau FW. *Diagnostic Ultrasound: Principles and Instruments*. 5th ed. Philadelphia (PA): WB Saunders; 1998.
7. Hedric WR, Kykes DL, Starchman DE. *Ultrasound Physics and Instrumentation*. 4th ed. St Louis (MO): Elsevier Mosby; 2005.
8. Szabo TL. *Diagnostic Ultrasound Imaging: Inside Out*. Burlington (MA): Elsevier Academic Press; 2004.
9. Smolka G. Optical Coherence Tomography: Technology, Markets, and Applications 2008–2012. BioOptics World. Penn Well Corporation; 2008.
10. Yang VXD, Vitkin IA. Principles of Doppler optical coherence tomography. In: Regar E, Leeuwen TV, Serruys P, editors. *Handbook of Optical Coherence Tomography in Cardiology*. Oxford: Taylor and Francis Medical; 2006, Chapter 32.
11. Larin KV, Ghosn MG, Ivers SN, Tellez A, Granada JF. Quantification of glucose diffusion in arterial tissues by using optical coherence tomography. Laser Phys Lett 2007;4:312–317.
12. Podoleanu A, Richard GH, Rosen B. Combinations of techniques in imaging the retina with high resolution. Prog Retin Eye Res 2008;27:464–499.
13. Suzuki Y, Ikeno F, Koizumi T, Tio F, Yeung AC, Yock PG, Fitzgerald PJ, Fearon WF. *In vivo* comparison between optical coherence tomography and intravascular ultrasound for detecting small degrees of in-stent neointima after stent implantation. J Am Coll Cardiol Intv 2008;1:168–173.

14. Wiesauer K, Pircher M, Götzinger E, Bauer S, Engelke R, Ahrens G, Grützner G, Hitzenberger CK, Stifter D. En face scanning optical coherence tomography with ultra-high resolution for material investigation. Opt Express 2005;13(3):1017–1024.
15. Chinn SR, Swanson EA. Multilayer optical storage by low-coherence reflectometry. Opt Lett 1996;21(12):899–901.
16. Dunkers JP, Parnas RS, Zimba CG, Peterson RC, Flynn KM, Fujimoto JG, Bouma BE. Optical coherence tomography of glass reinforced polumer composites. Composites 1999;30A:139–145.
17. Bashkansky M, Lewis D, Pujari V, Reintjes J, Yu HY. Subsurface detection and characterization of hertzian cracks in Si3N4 balls using optical coherence tomography. NDT&E Int 2001;34:547–555.
18. Xu F, Pudavar HE, Prasad PN. Confocal enhanced optical coherence tomography for non-destructive evaluation of paints and coatings. Opt Lett 1999;24:1808–1810.
19. Targowski P, Gora M, Wojtkowski M. Optical coherence tomography for artwork diagnostics. Laser Chem 2006. DOI: 10.1155/2006/ 35373.
20. Szkulmowska A, Góra M, Targowska M, Rouba B, Stifter D, Breuer E, Targowski P. In: Nimmrichter J, Kautek W, Schreiner M, editors. The Applicability of Optical Coherence Tomography at 1.55 um to the Examination of Oil Paintings. In: Lasers in the Conservation of Artworks, LACONA VI Proceedings, Vienna, Austria, Sept. 21–25, 2005. Berlin-Heidelberg-New York: Springer Verlag; 2007, pp. 487–492.
21. Liang H, Cid M, Cucu R, Dobre G, Podoleanu A, Pedro J, Saunders D. En-face optical coherence tomography—a novel application of non-invasive imaging to art conservation. Opt Express 2005;13:6133–6144.
22. Rie ER. The influence of varnishes on the appearance of paintings. Stud Conserv 1987;32(1):1–13.
23. Spring M, Liang H, Peric B, Saunders D, Podoleanu A. Optical coherence tomography—a tool for high resolution non-invasive 3D-imaging of the subsurface structure of paintings. In: International Council of Museums (ICOM) Committee for Conservation Triennial Conference, Preprints Vol. II, New Delhi, 22–26 September 2008. New Delhi: Allied Publishers; 2008, pp. 916–923.
24. Targowski P, Rouba B, Wojtkowski M, Kowalczyk A. The application of optical coherence tomography to nondestructive examination of museum objects. Stud Conserv 2004;49(2):107–114.
25. Targowski P, Góra M, Bajraszewski T, Szkulmowski M, Rouba B, Łękawa-Wysłouch P, Tymińska-Widmer L. Optical coherence tomography for tracking canvas deformation. Laser Chem 2006:8. (doi:10.1155/2006/93658).
26. Szkulmowska A, Gora M, Targowska M, Rouba B, Stifter D, Breuer E, Targowski P. Volume 116, *Lasers in the Conservation of Artworks*, Springer Proceedings in Physics, Madrid, Spain; September 17–21, 2007. ISSN: 0930–8989.
27. Yang ML, Lu CW, Hsu IJ, Yang CC. The use of optical coherence tomography for monitoring the subsurface morphologies of archaic jades. Archaeometry 2004;46(2): 171–182.
28. Chang S, Liu X, Cai X, Grover CP. Full-field optical coherence tomography and its application to multiple-layer 2D information retrieving. Opt Commun 2005;246:579–585.

29. Chang S, Cai X, Flueraru C. Image enhancement for multilayer information retrieval using full-field optical coherence tomography. Appl Opt 2006;45:5967–5975.
30. Barcode from Wikipedia. http://en.wikipedia.org/wiki/Barcode.
31. Wilson C. *Vein Pattern Recognition: A Privacy-enhancing Biometrics*. Richardson (TX): Mullaney Publishing Group, L.L.P.; 2008.
32. Denyer PB. Skin-pattern recognition method and device. US patent 4805223; 1989.
33. Kamei T. Skin pattern and fingerprint classification system. US patent 5901239; 1999.
34. Flom L, Safir A. Iris recognition system. US patent 4,641,349, Patent and Trademark Office, Washington (DC); 1987.
35. Daugman JD. High confidence visual recognition of personals by a test of statistical independence. IEEE Trans Pattern Anal Mach Intell 1993;15:1148–1160.
36. Retinal scanning biometrics. http://www.retina-scan.com/.
37. Topping A, et al. Method and apparatus for the automated identification of individuals by the nail beds of their fingernails. US patent 5751835 1998
38. Human ear holography. http://f1.net4p.com/gate/gb/www.relativehumanity.com.tw/cm/ear.htm.
39. Lammi HK. Ear Biometrics. http://www2.it.lut.fi/kurssit/03-04/010970000/seminars/Lammi.pdf.
40. Burge M, Burger W. Ear biometrics in computer vision. 15th International Conference of Pattern Recognition, Barcelona, Spain, Volume 2; September 3–7, 2000. p 822–826.
41. Chang K, Bowyer K, Barnabas V. Comparison and combination of ear and face images in appearance based biometrics. IEEE Trans Pattern Anal Mach Intell 2003;25:1160–1165.
42. Chen H, Bhanu B. Human ear detection from side face range images. 17th International Conference on Pattern Recognition, Cambridge, England, UK; 2004. p 574–577.
43. Chen H, Bhanu B. Contour matching for 3D ear recognition. 7th IEEE Workshop on Application of Computer Vision Breckenridge, Colorado, USA; 2005. p 123–128.
44. Yan P, Bowyer K. Biometric recognition using three-dimensional ear shape. http://www.cse.nd.edu/Reports/2006/TR-2006-01.pdf.
45. Youngquist R, Carr S, Davies DEN. Optical coherence-domain reflectometry: a new optical evaluation technique. Opt Lett 1987;12:158–160. ISSN: 0146–9592.
46. Takada K, Yokohoma I, Chida K, Noda J. New measurement system for fault location in optical waveguide devices based on an interferometric technique. Appl Opt 1987;26:1603–1606. ISSN: 0003–6935.
47. Michelson interferometer from Wikipedia. http://en.wikipedia.org/wiki/Michelson_interferometer.
48. Gilgen HH, Novak RP, Salathe RP, Hodel W, Beaud P. Submillimeter optical reflectometry. IEEE J Lightw Technol 1989;7:1225–1233.
49. Leitgeb R, Hitzenberger CK, Fercher AF. Performance of Fourier domain vs. time domain optical coherence tomography. Opt Express 2003;11:889–894.
50. Choma M, Sarunic M, Yang C, Izatt J. Sensitivity advantage of swept source and Fourier domain optical coherence tomography. Opt. Exp 2003;11:2183–2189. ISSN:1094–4087.

51. de Boer JF, Cense B, Park BH, Pierce MC, Tearney GJ, Bouma BE. Improved signal to noise ratio in spectral-domain comared with time-domain optical coherence tomography. Opt Lett 2003;28:2067–2069.
52. Chinn SR, Swanson EA, Fujimoto JE. Optical coherence tomography using a frequency-tunable optical source. Opt Lett 1997;22:340–342.
53. Yun SH, Tearney GJ, de Boer JF, Iftimia N, Bouma BE. High-speed optical frequency-domain imaging. Opt Express 2003;11:2953–2963.
54. Fercher F, Hitzenberger C, Kamp G, El-Zaiat S. Measurement of intraocular distances by backscattering spectral interferometry. Opt Commun 1995;117:443–448. ISSN:0030–4018.
55. Hausler G, Lindner MW. Coherence Radar and Spectral Radar—new tools for dermatological diagnosis. J Biomed Opt 1998;3:21–31.
56. Beaurepaire E, Boccara AC, Lebec M, Blanchot L, Saint-Jalmes H. Full-field optical coherence microscopy. Opt Lett 1998;23(4):244–246.
57. Na J, Choi W, Choi E, Seon Young Ryu S, Lee B. Image restoration method based on two fringe images for a high-speed full-field optical coherence tomography. Appl Opt 2008;47(3):459–466.
58. Dubois A, Vabre L, Boccara AC, Beaurepaire E. Appl Opt 2002;41:805–812.
59. Akiba M, Chan KO. *In vivo* video-rate cellular-level full-field optical coherence tomography. J Biomed Opt 2007;12:064024.
60. Akiba M, Chan KP, Tanno N. Full-field optical coherence tomography by two-dimensional heterodyne detection with a pair of CCD cameras. Opt Lett 2003;28:816–818.
61. Moneron G, Boccara AC, Dubois A. Stroboscopic ultrahigh-resolution full-field optical coherence tomography. Opt Lett 2005;30:1351–1353.
62. Dubois A, Vabre L, Boccara AC, Beaurepaire E. High-resolution full-field optical coherence tomography with a Linnik microscope. Appl Opt 2002;41(4):805–812.
63. Flueraru C, Kumazaki H, Sherif SS, Chang S, Mao Y. Quadrature Mach–Zehnder interferometer with application in optical coherence tomography. J Opt A: Pure Appl Opt 2007;9:5–8.
64. Choma MA, Yang C, Izatt JA. Instantaneous quadrature low-coherence interferometry with 3x3 fiber-optic couplers. Opt Lett 2003;28:9672162–9672164.
65. Mao Y, Sherif S, Flueraru C, Chang S. 3x3 Mach-Zehnder interferometer with unbalanced differential detection for full range swept-source optical coherence tomography. Appl Opt 2008;47:2004–2010.
66. Sticker M, Hitzenberger CK, Leitgeb R, Fercher AF. Quantitative differential phase measurement and imaging in transparent and turbid media by optical coherence tomography. Opt Lett 2001;26:114–116.
67. Zhao Y, Chen Z, Saxer C, Xiang S, de Boer JF, Stuart Nelson J. Phase-resolved optical coherence tomography and optical Doppler tomography for imaging blood flow in human skin with fast scanning speed and high velocity sensitivity. Opt Lett 2000;25:114–116.
68. Mach-Zehnder interferometer from Wikipedia. http://en.wikipedia.org/wiki/Mach–Zehnder_interferometer.
69. Rollins AM, Izatt JA. Optimal interferometer designs for optical coherence tomography. Opt Lett 1999;24:1484–1486.

70. Podoleanu A. Unbalanced versus balanced operation in an optical coherence tomography system. Appl Opt 2000;39:173–182.
71. Denisyuk YN. On the reproduction of the optical properties of an object by the wave field of its scattered radiation. Pt. II. Opt Spectrosc (USSR) 1965;18:152–156.
72. Leith EN, Swanson GJ. Achromatic interferometers for white light optical processing and holography. Appl Opt 1980;19:638–644.
73. Dresel T, Häusler G, Venzke H. Three-dimensional sensing of rough surfaces by coherence radar. Appl Opt 1992;31:919–925.
74. Chang S, Sherif S, Mao Y, Flueraru C. Large area full-field optical coherence tomography and its applications. Open Opt J 2008;2:10–20.
75. Chang S, Sherif S, Mao Y, Fluerauru C. Swept-source full-field Optical Coherence Microscopy. Invited paper. SPIE Photonics North 2009. Proceedings of SPIE Volume 7386, Quebec, Canada, 738604. DOI: 10.1117/12.838213.
76. Greivenkamp JE, Bruning JH. Optical shop testing. In: Malacara D, editor. *Phase Shift Interferometer*. 2nd ed. New York: John Wiley and Sons; 1992, Chapter 14.
77. Carré P. Installation et utilisation du comparateur photo-electrique et interferential du bureau international des poids de mesures. Metrologia 1966;2:13–23.
78. Novak J. Five-step phase-shifting algorithms with unknown values of phase shift. Optik-Int J Light Electron Opt 2003;114(2):63–68.
79. Dubois A. Phase-map measurements by interferometry with sinusoidal phase modulation and four integrating buckets. J Opt Soc Am A 2001;18:1972–1979.
80. Chang S, Cai X, Flueraru C. An efficient algorithm used for full-field optical coherence tomography. Opt Lasers Eng 2007;45:1170–1176.
81. Maragos P, Kaiser JF, Quatieri TF. On amplitude and frequency demodulation using energy operators. IEEE Trans Signal Process 1993;41:1532–1550.
82. Santhanam B, Maragos P. Energy demodulation of two-component AM-FM signal mixtures. IEEE Signal Process Lett 1996;3:294–298.
83. Larkin KG. Efficient nonlinear algorithm for envelope detection in white light interferometry. J Opt Soc Am A 1996;13:832–843.
84. Chang S, Mao Y, Sherif S, Flueraru C. Full-field optical coherence tomography used for security and document identity. Europe Security & Defence Symposium; 2006 Sept; Stockholm, Sweden, Proceedings of SPIE Volume 6402 64020Q1-9.
85. Boulgouris NV, Plataniotis KN, Micheli-Tzanakou E. *Biometrics: Theory, Methods, and Applications*. Hoboken (NJ): Wiley-IEEE Press; 2009. p 745.
86. Cavoukian A, Stoianov A. Biometric encryption: a positive sum technology that achieves strong authentication, security and privacy. White Paper, Office of the Information and Privacy Commissioner of Ontario; 2007.
87. Jain AK, Ross A, Prabhakar S. An introduction to biometric systems. IEEE Trans Circ Syst Video Technol 2004;14(1):4–20.
88. Matsumoto T, Matsumoto H, Yamada K, Hoshino S. Impact of artificial gummy fingers on fingerprint systems. Proceedings of SPIE, Optical Security and Counterfeit Deterrence Techniques IV, Yokohama, Japan, Volume 4677; 2002. p 275–289.
89. Chirillo J, Blaul S. *Implementing Biometric Security*. Indianapolis (IN): Wiley Publishing; 2003.

90. Maltoni D, Maio D, Jain AK, Prabhakar S. *Handbook of Fingerprint Recognition*. New York: Springer; 2003.
91. Osten D, Carim HM, Arneson MR, Blan BL. Biometric, personal authentication system. US patent 5719950; 1998.
92. Parthasaradhi STV, Derakhshani R, Hornak LA, Schuckers SAC. Time-series detection of perspiration as a liveness test in fingerprint devices. IEEE Trans Syst Man Cybern C Appl Rev 2005;35(3):335–343.
93. Nixon KA, Rowe RK, Allen J, Corcoran S, Fang L, Gabel D, Gonzales D, Harbour R, Love S, McCaskill R. Novel spectroscopy-based technology for biometric and liveness verification. Proc SPIE 2004;5404:287–295.
94. Rowe RK, Nixon KA, Corcoran SP. Multispectral fingerprint biometrics. Proceedinigs of the 6th IEEE Systems, Man and Cybernetics Information Assurance Workshop, West Point, NY; 2005. p 14–20.
95. Suzuki H, Yamaguchi M, Yachida M, Ohyama N, Tashima H, Obi T. Experimental evaluation of fingerprint verification system based on double random phase encoding. Opt Express 2006;14:1755–1766.
96. Manapuram RK, Ghosn M, Larin KV. Asian J Phys 2006;15:15–27.
97. Cheng Y, Larin KV. Artificial fingerprint recognition by using optical coherence tomography with autocorrelation analysis. Appl Opt 2006;45:9238–9245.
98. Chang S, Cheng Y, Larin KV, Mao Y, Sherif S, Flueraru C. Optical coherence tomography used for security and fingerprint sensing applications. IET Image Process 2008;1:48–58.
99. Smolka G. editor. *Optical Coherence Tomography: Technology, Markets, and Applications*. Nashua, NH: PennWell; 2008.
100. Gora M, Karnowski K, Szkulmowski M, Kaluzny BJ, Huber R, Kowalczyk A, Wojtkowski M. Ultra high-speed swept source OCT imaging of the anterior segment of human eye at 200kHz with adjustable imaging range. Opt Express 2009;17(17):14880–14894.
101. Drexler W, Fujimoto JG, editors. *Optical Coherence Tomography: Technologies and Applications*. New York: Springer; 2008.

CHAPTER EIGHT

Photonics-Assisted Instantaneous Frequency Measurement

SHILONG PAN and JIANPING YAO
Microwave Photonics Research Laboratory, University of Ottawa, Ottawa, ON, Canada

8.1 INTRODUCTION

In the field of electronic warfare (EW), it is of critical importance to analyze an intercepted radio frequency (RF) signal from a hostile radar or communication system. The EW environment is totally different from that of commercial wireless communications, where the carrier frequency, modulation format, and bandwidth of the signal are known. But in an EW environment, the information about the signal is not known, and also the signal is usually specifically designed to avoid being detected by an RF signal receiver. Typically, the frequency of a radar or communication signal can be varied in a range from several hundreds of megahertz to several hundreds of gigahertz, but an RF signal receiver has to be operated in a very narrow frequency band to minimize the noise and interference. An instantaneous frequency measurement (IFM) module is thus required to instantaneously identify the carrier frequency of a unknown RF signal before passing it into a specialized receiver for processing (1–3). IFM is also of great importance in an EW system to jam an enemy receiver. Typically, a signal emitted by an enemy transmitter would vary in frequency in a predetermined manner. In order to effectively jam such a

Photonic Sensing: Principles and Applications for Safety and Security Monitoring, First Edition.
Edited by Gaozhi Xiao and Wojtek J. Bock.

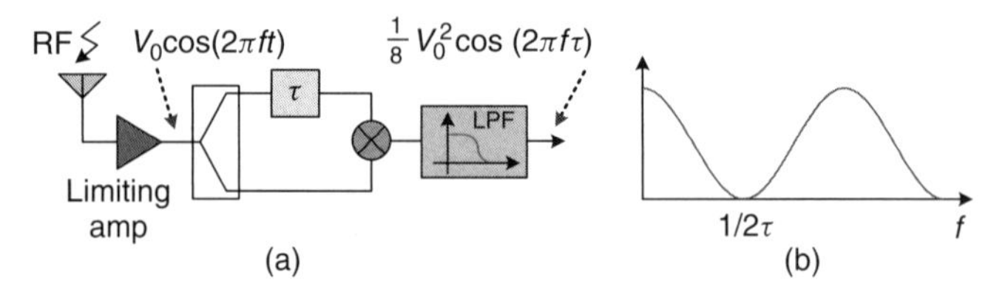

FIGURE 8.1 A typical electrical approach to IFM. (a) The configuration and (b) the received DC power versus the microwave frequency.

signal, the frequency of the signal from the enemy transmitter must be accurately and quickly determined. Knowing the frequency, a jammer can be easily configured to concentrate its energy in the desired frequency range (2).

The development of IFM can date back as early as 1948 (4). In Reference 4, Earp proposed the first electrical IFM receiver, in which the frequency information was displayed in polar coordinates on a scope. In his IFM receiver, an RF signal with the frequency to be measured was split into two paths; one path was delayed by a constant time with respect to the other one. A phase difference was resulted owing to the constant time delay. The relative phase angle is frequency dependent. By measuring this phase angle, the frequency information was obtained. The technique is simple but effective. Most of the modern electrical IFM receivers are still based on the same principle (5–20). Figure 8.1a shows the block diagram of a typical electrical IFM system (21). The received RF signal is first passed through a limiting amplifier to remove the amplitude variation. The amplified signal is then divided into two equal portions by an electrical splitter. One portion is delayed relative to the other by a time of τ. The two signals are then mixed at an electrical mixer. After passing through a low pass filter, a DC signal with the level that is a function of the microwave frequency is obtained. The relationship between the output DC power and the microwave frequency is shown in Figure 8.1b. By monitoring the DC power, the microwave frequency can be measured. To avoid the measurement ambiguity, a band-pass filter may be incorporated at the input of the receiver, and the measurement range is limited to $0-1/2\tau$.

Thanks to the development in the last few decades, the current electrical IFM systems can measure not only the microwave frequency but also the microwave amplitude, the pulse width (PW), and the time of arrival (TOA). The typical frequency measurement range is 0.5–18 GHz with a measurement resolution of about 3 MHz. The detectable signal level can be from −50 to +10 dBm, and the operating temperature is from −50 to +85°C (22).

The electrical solutions can achieve high measurement resolution and large dynamic range, but the frequency measurement range is limited owing to the electronic bottleneck. The electrical solutions also suffer from other limitations such as high power consumption, vulnerability to electromagnetic interference, and bulky size. To overcome these limitations, photonics-assisted microwave frequency measurement has been proposed (23). Thanks to the broad bandwidth provided by optics, the photonics-assisted IFM can potentially achieve a measurement range up to hundreds of gigahertz. In addition, photonics-assisted IFM also offers advantages such as low loss, light weight, and immunity to electromagnetic interference.

Generally, photonics-assisted IFM can be classified into two main categories. In the first category, an optical channelizer (24–27) is employed to spectrally divide the optical microwave signal (a microwave signal modulated on an optical carrier) into many contiguous parallel channels. The channelized optical signals are then detected by an array of low speed photodetectors (PDs). The optical spectrum of the optical microwave signal is thus obtained, which includes the information of the microwave carrier frequency. In the second category, the frequency of a microwave signal is obtained by power monitoring. The "power" can be either a microwave power or an optical power depending on the configuration of the system. The fundamental concept of the techniques in this category is that a fixed relationship between the microwave frequency and the microwave or optical power is established. By monitoring the microwave (28–38) or optical power (39–41), the microwave frequency is obtained.

In this chapter, a comprehensive review of the techniques for photonics-assisted IFM is performed, with an emphasis on the frequency measurement based on an optical channelizer and optical or microwave power monitoring. Other techniques for photonics-assisted IFM, including frequency measurement based on an optical scanning receiver (42), a photonic Hilbert transformer (43), a monolithically integrated echelle diffractive grating (EDG) (44), and frequency-to-time mapper (45), are also briefly reviewed. The advantages and limitations of these techniques are discussed. The challenges in implementing these techniques for practical IFM are also discussed.

8.2 FREQUENCY MEASUREMENT USING AN OPTICAL CHANNELIZER

An optical channelizer is a device that was originally developed to instantaneously analyze the spectrum of an arbitrary optical signal. As the carrier frequency of an RF signal can be retrieved from its spectrum, optical channelizers can also be used for microwave frequency measurement. Figure 8.2 shows a conceptual diagram of microwave frequency measurement using an optical channelizer. A microwave signal with its frequency to be measured is first converted into an optical signal at an intensity modulator, to which an optical carrier emitted from a laser diode (LD) is applied. The optical microwave signal is then split into many contiguous parallel channels using the optical channelizer. Each channel corresponds to a particular microwave frequency band. The split optical signals are then detected by an array of low speed PDs. The converted electrical signals are sent to a low speed electrical processing unit, where the spectral information of the microwave signal is obtained.

The key device in this system is the optical channelizer, which can be seen as a group of parallel optical narrow-band filters with a fixed spacing between adjacent center frequencies. A wavelength-dependent splitter or a wavelength division demultiplexer is actually an optical channelizer. In general, a wavelength division multiplexer (WDM) can be used as a wavelength division demultiplexer by

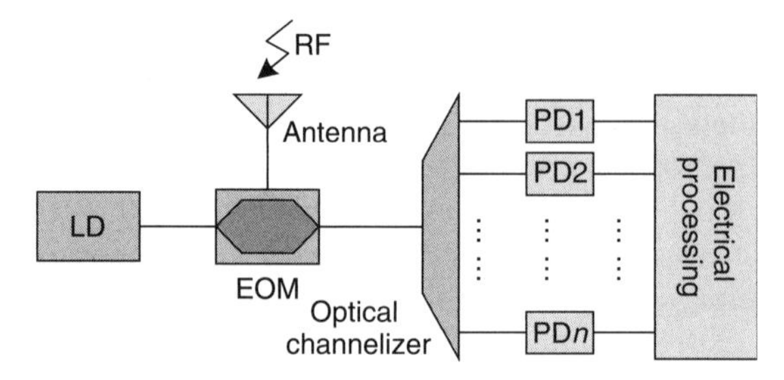

FIGURE 8.2 Microwave frequency measurement based on an optical channelizer. LD, laser diode; RF, radio frequency; EOM, electro-optical modulator; PD, photodetector.

propagating a light wave in a reversed direction. As a WDM is a key component in a wavelength-division-multiplexed optical communication system, it was well developed in the past two decades (46). The channel spacing determines the measurement resolution. For microwave frequency measurement, a small channel spacing ($\leq$1 GHz) is usually required. Typically, the channel spacing of a WDM for communication applications is tens of gigahertz. Thus special designs are needed to reduce the channel spacing. An optical channelizer could be realized using an integrated optical phased array consisting of multiple electro-optic waveguide delay lines (24), a high resolution free-space diffraction grating (25), an array of phase-shifted gratings (26), or an integrated Bragg grating with a Fresnel lens (27).

The key advantage of an IFM receiver implemented based on an optical channelizer is its capability to measure the frequencies of multiple signals simultaneously. The major difficulty that limits the use of an optical channelizer for practical applications is the limited channel number and small channel spacing. To increase the channel number and to reduce the channel spacing, the fabrication cost would be significantly increased. At the current stage, an optical channelizer with a channel number of 16 would have channel spacing larger than 1 GHz. In addition, the currently available optical channelizers are also complicated, bulky, unstable, and expensive. In the following sections, a review of optical channelizers is presented.

8.2.1 Optical Phased Array WDM

An integrated optical passive phased array WDM is also known as an *arrayed waveguide grating* (*AWG*). In an AWG, an incoming optical signal is coupled to an array of planar waveguides. The optical signal in each waveguide will experience a different phase shift owing to the different path lengths of the waveguides. The phase shifts are also wavelength dependent because of the frequency dependence of the mode-propagation constant. As a result, the incoming light components at different wavelengths will be deflected at different angles and then coupled again to different output waveguides. To make the channel spacing as small as 1 GHz, the maximum difference in time delay between two adjacent waveguides should as large as 1 ns, which corresponds to a maximum delay length difference of 7.7 cm.

Such a long length is difficult to implement when laying the waveguides out on a single chip. A possible solution to this problem is to add electro-optic waveguide delay lines to an optical phased array WDM. As the phase of each waveguide delay line is electrically controlled, the phase shift between different waveguides can be increased by introducing different bias voltages to the waveguide delay lines. Such a device with a channel number of 16 was fabricated by Heaton et al. (24), by using 16 high confinement GaAs/AlGaAs optical waveguides in combination with a multimode interference (MMI) splitter, 128 compact low loss corners, and 16 electro-optic phase shifters. The layout of the device is shown in Figure 8.3. As can be seen, an input light wave is resolved to 16 channels. The channel spacing was 1 GHz. The total chip size is as small as 26 mm × 0.7 mm, which comprises (i) a straight input guide; (ii) a l-to-16 MMI splitter (96 μm in width and 1884 μm in length); (iii) 16 electro-optic phase shifters with 5-mm-long electrodes; (iv) 16 delay lines; and (v) output array with parallel guides on a 6 μm pitch. The input, electro-optic, and delay-line guides are all 3.0 μm wide, and the electrodes are 1.2 μm wide. The delay lines are laid out to achieve a large spread in delay length and to have the same number of corners and S-bends. All of the corners and bends are of same size and shape, to ensure each delay-line guide to have the same insertion loss.

The key limitation related to the microwave frequency measurement based on this type of optical channelizers is that the measurement resolution and measurement range are limited to 1 GHz and 1 ~ 16 GHz, respectively. Doubling the frequency range to 32 GHz or fining the resolution to 500 MHz might significantly increase the device width and the excess insertion losses.

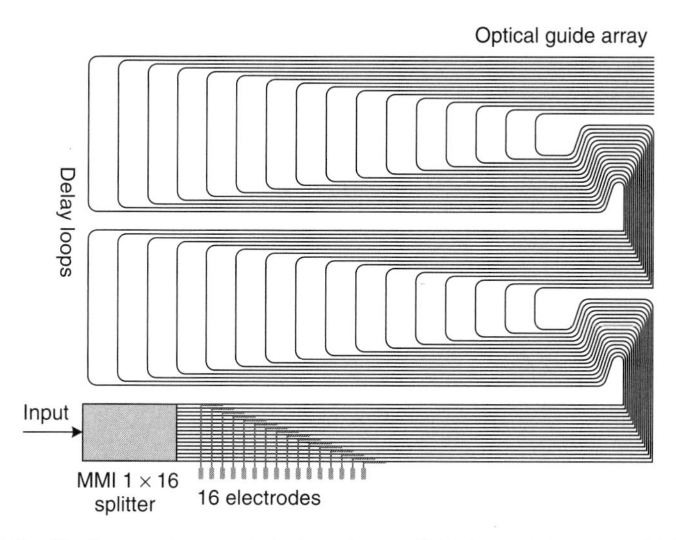

FIGURE 8.3 The layout of an optical phased array WDM consisting of multiple electro-optic waveguide delay lines.

8.2.2 Free-Space Diffraction Grating

To increase the operating frequency range, optical channelizers based on a high resolution free-space diffraction grating would be a potential solution. For instance, an instantaneous measurement bandwidth of over 100 GHz with a resolution of 1 GHz is possible by use of a free-space diffraction grating with coherent optical heterodyne detection (25). Figure 8.4 illustrates the concept of a coherent optical channelizer based on a free-space diffraction grating. The input light wave carrying a microwave signal with unknown frequency is collimated and incident to the diffraction grating. Different frequency components in the incident signal are dispersed with different angles and subsequently imaged on a detector array. The position of the detector corresponds to a particular microwave frequency, and the spatial separation of the detectors determines the measurement resolution. When the detected electrical signals are sent to an electrical processing unit, the spectral information of the microwave signal is derived. As the most advanced technology available currently for postprocessing is based on digital electronics, and its bandwidth is restricted by the limited speed of the state-of-the-art analog-to-digital (A/D) converters, which is about 1 GHz. Therefore, it is essential to down convert the central frequencies of all the channel outputs to common intermediate frequency (IF) that can be handled by a low speed digital signal processing circuit. Coherent optical heterodyne detection technology is usually applied to perform the down conversion. A local oscillator (LO) beam, with its spectrum comprising an optical frequency comb with a frequency spacing equal to the channel spacing, is incident on the grating at the same offset angle with respect to the optical microwave signal

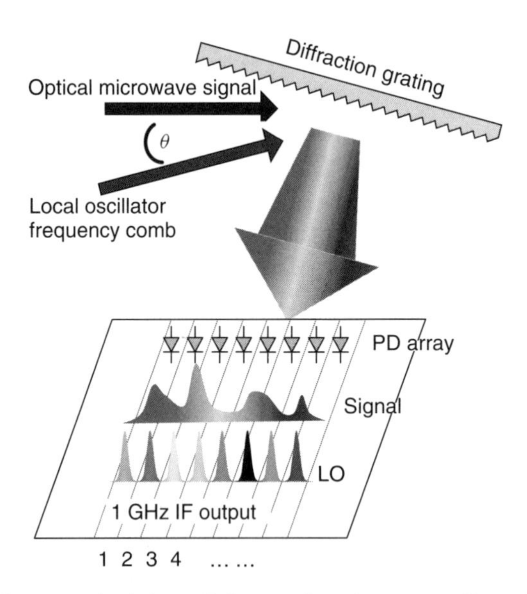

FIGURE 8.4 Coherent optical channelizing receiver that uses a dispersion grating to map the optically carried RF signal and a local oscillator light beam on a photodetector array.

beam. Such spatial offset will result in a constant frequency difference between the spatially overlapped optical microwavc signal bcam and thc LO comb lines. When the two beams are combined at the PDs, a heterodyne beat tone will be generated at the output of each PD. In this way, every portion of the signal spectrum is translated to the same IF band so that each channel may use the same postprocessing electronics. A high degree of coherence between the optical microwave signal and the LO comb is required to minimize the phase noise in the heterodyne RF signal, which can be obtained by using a continuous-wave (CW) light wave as the optical carrier and using a mode-locked laser that is injection-locked by the same CW light wave as the LO comb.

8.2.3 Phase-Shifted Chirped Fiber Bragg Grating Arrays

The major limitation of the phased array WDM and the optical channelizers based on free-space diffraction grating is the large insertion loss. For example, the total insertion loss for a signal path in a coherent optical RF channelizer based on a diffraction grating is about 28 dB (25). Therefore, to ensure an acceptable signal-to-noise ratio (SNR) at the output of the channelizer, the input signal should have a high power or the frequency measurement error would be large. The high insertion loss can be significantly reduced if an all-fiber optical channelizer is employed. Hunter et al. (26) reported an optical channelizer based on an array of phase-shifted fiber gratings. The schematic representation is shown in Figure 8.5. The optical microwave signal is split and applied to a grating array via a coupler tree. The grating array consists of a number of phase-shifted chirped fiber Bragg gratings (FBGs), which have a transmission notch within the reflection spectrum. The center wavelength and the width of the transmission notch can be controlled by adjusting the spatial position of the phase shift along the length of the grating. If the spatial position of the phase shift in each grating is carefully controlled, the transmission notch will be located at a position that corresponds to one of the modulation sidebands. Therefore, the optical energy transmission through any of

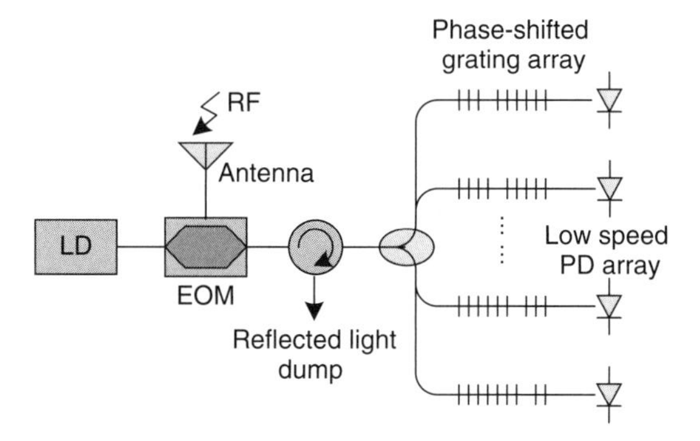

FIGURE 8.5 The schematic representation of an optical channelizer based on an array of phase-shifted fiber gratings.

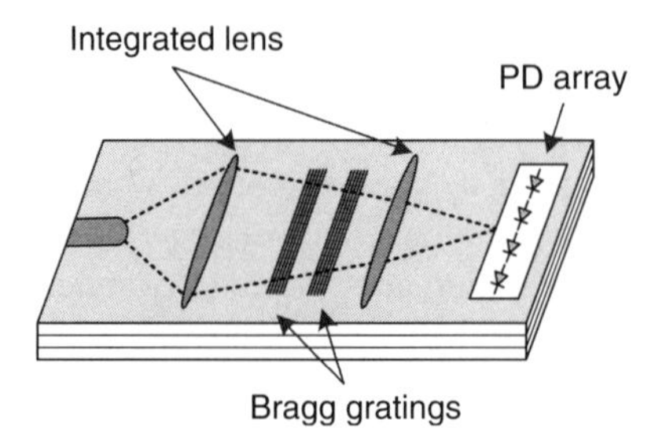

FIGURE 8.6 The schematic representation of an optical channelizer based on an integrated optical Bragg grating Fabry–Perot etalon.

the fixed notches is associated with a particular RF frequency band, which would channelize the intercepted microwave signal. In an experimental demonstration, a 1×16 optical channelizer with an average insertion loss for each channel of $\sim$15 dB and a resolution of 2 GHz was realized. The use of the channelizer for the channelization of a 2–18 GHz microwave signal was successfully realized.

8.2.4 Integrated Optical Bragg Grating Fabry–Perot Etalon

An optical channelizer can also be realized based on an integrated optical Bragg grating Fabry–Perot etalon and integrated hybrid Fresnel lens system. The architecture of such a channelizer is depicted in Figure 8.6 (27). In the device, the optical phase front of the optical microwave signal is expanded, and the integrated optical lens system would cause this signal to diverge laterally. A Fabry–Perot etalon formed with two Bragg gratings acts on this wave front to provide microwave channelization, because a resonance with periodic peaks that is a function of the angle and wavelength of the incident light is resulted by the two Bragg gratings. The device is compact, but the insertion loss is very large and dependent on the wavelength of the incident light. The accuracy of the system is severely limited by the finesse of the Fabry–Perot etalon. A demonstration performed by Winnall et al. (27) showed that the channelizer can provide a measurement range of 1–23 GHz with a measurement resolution of 2 GHz.

8.3 FREQUENCY MEASUREMENT BASED ON POWER MONITORING

As frequency measurement based on an optical channelizer has a poor resolution and a relatively small measurement range, it may not meet the requirements for applications where a wide operating frequency range and high resolution are required. For IFM of a single-channel signal, we need not to know the detailed spectrum of the microwave signal. As a result, the measurement can be significantly simplified, while the measurement resolution and measurement range are evidently improved. One of such techniques is the frequency measurement based on power monitoring. For this type of IFM, if multiple signals are received with one of the signals that has a much higher power, the frequency measurement would give an estimated frequency corresponding to the highest power signal. There are numerous applications that require the frequency information of an RF signal with the highest power. For example, in an adaptive antijamming receiver, the jamming signal always has the highest power. If the IFM can obtain the frequency information of the jamming signal, it can be used to control a tunable notch filter to remove the undesirable jamming frequency. Given that the techniques for IFM of multiple signals are not mature, one can use cascaded IFMs based on power monitoring and add/drop multiplexers to receive multiple signals. In each stage of the receiver, the frequency of the RF signal with the highest power is measured. The frequency information is then used to drive an add/drop multiplexer to pass the signal to a specialized receiver.

A typical scheme for microwave frequency measurement based on power monitoring is shown in Figure 8.7. As usual, the microwave signal is first converted to an optical microwave signal at an electro-optic modulator. Then, the optical microwave signal is split into two channels to undergo different frequency-dependent power penalties. In the electrical processing unit, the frequency-dependent powers are monitored and used to retrieve the frequency information of the received microwave signal. The frequency-dependent power functions can be realized using a dispersive element that would introduce chromatic-dispersion-induced (CD) microwave power fading, a photonic microwave filter or an optical filter. In addition, the microwave power penalty functions are dependent on the modulation schemes.

8.3.1 Chromatic-Dispersion-Induced Microwave Power Penalty

For conventional double sideband + carrier (DSB+C) modulation, a MZM (Mach-Zehnder modulator) is usually employed, which is biased at the quadrature point, thus providing one optical carrier and two optical sidebands. When the optical DSB+C signal is propagating in a dispersive fiber, the optical carrier and sidebands would experience different phase shifts owing to the CD. The beating of the optical carrier with the upper sideband and the lower sideband at a PD would generate a microwave signal with its power being a function of the CD and the microwave

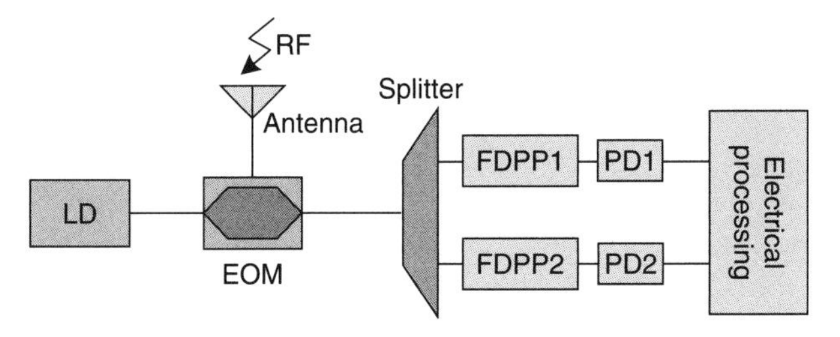

FIGURE 8.7 A diagram to show microwave frequency measurement based on power monitoring. FDPP, frequency-dependent power penalty.

frequency. Mathematically, the microwave power at the output of a PD is given in Reference 47

$$P = R\cos^2\left(\frac{\pi DL\lambda^2 f^2}{c} + \arctan\alpha\right), \tag{8.1}$$

where D is the dispersion parameter, L the length of the dispersive fiber, λ the carrier wavelength, f the microwave frequency, c the light velocity in vacuum, α the residual chirp parameter of the modulator, and R a constant, which is related to the optical power, the phase modulation index at the modulator, the insertion losses of the optical devices, and the responsivity of the PD.

As can be seen from Equation 8.1, the microwave power has a direct relationship with the microwave frequency. For a given measurement system, D, L, and λ are fixed and can be considered insensitive to environmental variations. But R is dependent on the microwave power of the unknown microwave signal, which is usually not known. In addition, R is also sensitive to the losses of the optical path, which may be affected by environmental variations. On the basis of this consideration, two different carriers (denoted as λ_1 and λ_2) can be selected to obtain two different power fading functions. The ratio of the two power fading functions, referred to as an *amplitude comparison function* (*ACF*), is independent of the microwave power. If the ACF is denoted as γ, we have

$$\gamma = \frac{P_1}{P_2} = \frac{R_1\cos^2\left(\frac{\pi D_1 L_1 \lambda_1^2 f^2}{c} + \arctan\alpha\right)}{R_2\cos^2\left(\frac{\pi D_2 L_2 \lambda_2^2 f^2}{c} + \arctan\alpha\right)} \tag{8.2}$$

R_1 and R_2 can be made identical by adjusting the optical powers of the two optical carriers. Thus by measuring the ACF, the microwave frequency can be estimated.

Figure 8.8 shows two power fading functions and the corresponding ACF for a frequency measurement system with two dispersion values of 185 and 325 ps/nm.

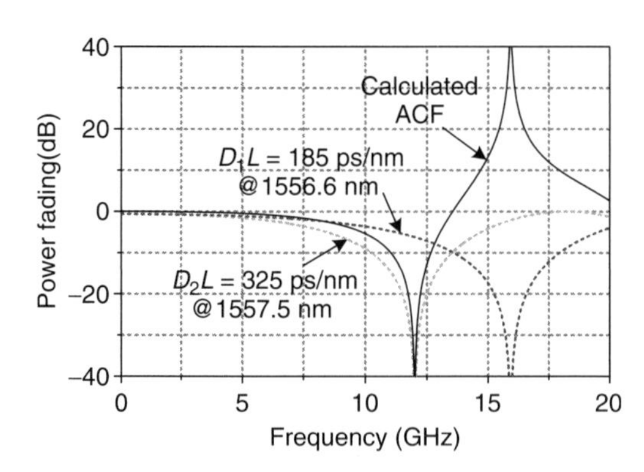

FIGURE 8.8 Typical power fading functions and the corresponding ACF function. $\lambda_1 =$ 1556.6 nm; $\lambda_2 =$ 1557.5 nm; $D_1L =$ 185 ps/nm; $D_2L =$ 325 ps/nm; and $\alpha = 0.4$.

The residual chirp of the MZM is 0.4. As can be seen, the ACF is monotonically decreasing for $0<f<12$ GHz and monotonically increasing for $12<f<16$ GHz. Owing to the nonmonotone nature of the ACF, frequency measurement ambiguities exist. The problem can be solved by applying a band-pass filter with its passband set to cover the frequency range of interest. Usually, the system is designed to make the first notch as the maximum frequency to be measured, so the bandwidth of the frequency measurement system is from DC to the first notch of the ACF, which is given by Nguyen and Hunter (28)

$$\mathrm{BW} = \sqrt{\frac{c}{2(DL\lambda^2)_{\max}}\left(1 - \frac{2}{\pi}\arctan\alpha\right)}, \tag{8.3}$$

where $(DL\lambda^2)_{\max}$ is equal to either $D_1L_1\lambda_1^2$ or $D_2L_2\lambda_2^2$, whichever is greater. For practical applications, the lowest measurable frequency cannot reach zero even if the MZM is chirp free, as the slope of the ACF is very small at low frequencies. The system noise would cause a large measurement error at the low frequency band.

Frequency measurement based on dispersion-induced microwave power penalty can be implemented based on a system shown in Figure 8.9 (28). Two optical carriers from two LDs (1556.71 nm and 1559.13 nm) are multiplexed and modulated by a microwave signal with its frequency to be measured at an electro-optic modulator. A multichannel chirped fiber grating (MCFG) is employed as the dispersive element. The MCFG is designed to have a value of dispersion of 186 ps/nm at 1556.71 nm and 714 ps/nm at 1559.13 nm. The optical signals from the MCFG are demultiplexed and then converted to microwave signals at the PDs. Two detector logarithmic video amplifiers (DLVAs) are used to detect the microwave powers. The outputs from the DLVAs are digitized by an A/D conversion card. The receiver is controlled through a LabView interface with the frequency information extracted through a simple subtraction operation of the resultant logarithms of the power fading functions. A measurement range of 4–12 GHz with 100-MHz accuracy is achieved (28). The performance of this IFM receiver is dependent on a number of factors. The frequency resolution is dependent on the resolution of the A/D

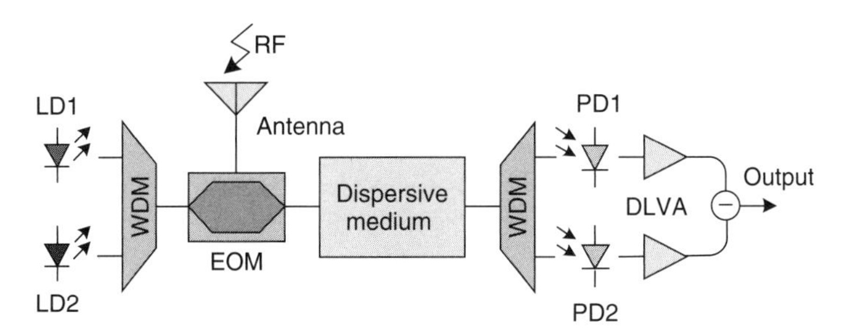

FIGURE 8.9 A microwave photonic IFM receiver based on dispersion-induced microwave power penalty. DLVA, detector logarithmic video amplifier.

converter. The accuracy is dominated by the ripples present in the ACF, which is attributed to two parts: the RF impedance mismatches between the PDs and the DLVAs, and the group delay ripples present in the MCFG.

The work in Reference 28 can be used for frequency measurement for a fixed measurement range and resolution. To achieve frequency measurement with an adjustable measurement range, an optical approach was recently proposed and demonstrated (29). The key difference between References 28, 29 is that a dispersive fiber was employed to replace the MCFG. Compared with the MCFG, the spacing of the two wavelengths used to achieve the different power fading functions could be tuned to get different dispersion values, resulting in an adjustable frequency measurement range and resolution. In addition, the measurement accuracy is improved, as a dispersive fiber has less dispersion ripples than an MCFG.

In Reference 29, a 20-km single-mode fiber (SMF) was employed as the dispersive fiber. Two tunable laser sources (TLSs) are used to provide the two wavelengths with tunable wavelength spacing. Six different wavelengths, 1470, 1500, 1520, 1550, 1580, and 1600 nm, are chosen to get different power penalty functions. The corresponding dispersion coefficients are 234.5, 273, 298, 335, 370, and 393 ps/nm. The power fading functions at these wavelengths are shown in Figure 8.10a. It is seen that the location of the first notch of the power fading function is determined by the total dispersion and the wavelength. The measurement range and resolution can thus be tuned by setting the spacing of the two wavelengths. If the two wavelengths are first set at 1470 and 1600 nm, the measurement range is from 5 to 12.1 GHz for an error within ±0.1 GHz, as shown in Figure 8.10b. If the longer wavelength of 1600 nm is tuned to 1520 nm keeping the shorter wavelength of 1470 nm unchanged, a new measurement range of 9–14.7 GHz is achieved. Meanwhile, as reported, the resolution can be improved by using smaller wavelength spacing. But a trade-off between the measurement range and resolution, that is, a higher resolution is achieved at the cost of a relatively smaller measurement range, has to be considered. To overcome this limitation, one can add a third wavelength. The use of three wavelengths would divide the measurement range into two parts, each having a higher resolution. For example, if three wavelengths at 1470, 1550, and 1600 nm are used, the use of 1470 and 1600 nm would provide a range of 5–11.5 GHz, while the use of 1470 and 1550 nm would offer another range of 11.5–15.6 GHz. Thus a total measurement range of 5–15.6 GHz is achieved with an error less than 0.1 GHz.

In References 28 and 29, the decreasing region of the ACF is used for frequency measurement. In that region, the slope of the ACF is small. The system noise would significantly impact on the measurement accuracy. To improve the measurement accuracy without complicating the system, in Reference 30, the increasing region of the ACF is utilized. A new frequency measurement range that exhibits fast variation in the ACF is obtained, leading to an improved measurement range and resolution. With the same experimental setup as used in Reference 29, two wavelengths of 1470 and 1600 nm are used to obtain the ACF shown in Figure 8.11. The two regions of the ACF to estimate the microwave frequency are evaluated. As can be seen, the variation of the ACF is very low in region A, which results in poor

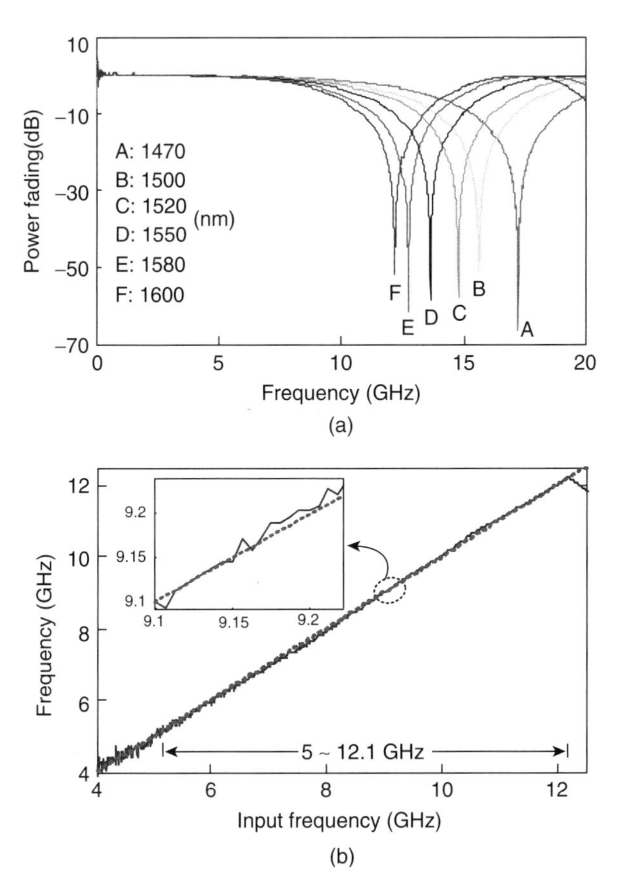

FIGURE 8.10 (a) Measured power fading functions for six different wavelengths. (b) Frequency measurement range for the system operating at 1470 and 1600 nm.

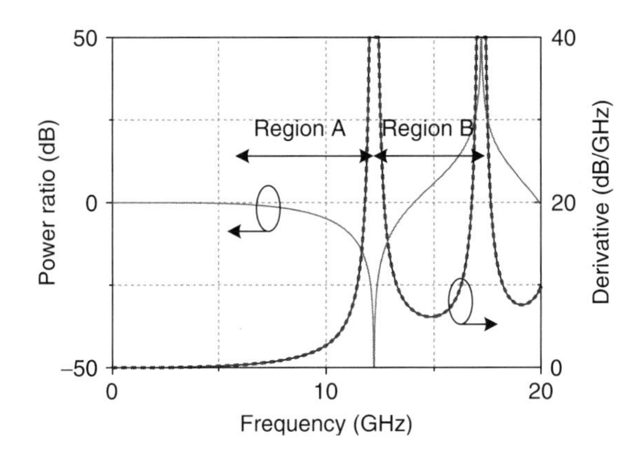

FIGURE 8.11 Simulated ACF with the wavelengths of 1470 and 1600 nm, and the absolute value of the first-order derivative of the ACF.

measurement resolution, especially when the system has large noise (28, 29). But region B has a much faster variation because the absolute values of the first-order derivatives in region B are larger than those in region A. Hence, the use of region B would offer an improved measurement resolution. Furthermore, for a given variation rate, region B would also provide a larger measurement range than region A. For example, the measurement range in region B is 12.3–17.1 GHz with a minimum absolute first-order derivative of 6.14 dB/GHz, while the measurement in region A is 9.9–12.2 GHz with a minimum absolute first-order derivative of 2.5 dB/GHz.

The frequency-dependent microwave power penalty in the above studies is achieved in an intensity-modulation-based system. The major limitation associated with intensity modulation for the frequency measurement is the bias drifting problem, which makes the system unstable. A sophisticated circuit is thus needed to control the bias voltage, which would increase the system complexity and cost. By using a phase modulator to replace the intensity modulator, the bias drifting problem would be eliminated because a phase modulator needs no bias, which leads to an improved system stability. This feature is highly desirable in defense systems, where the system stability is one of the major concerns. In addition, the use of an optical phase modulator would make the system simpler with lower insertion loss.

When a phase-modulated microwave signal is transmitting in a dispersive medium and then detected by a PD, the output electrical power is given by Zeng and Yao (48)

$$P = R\sin^2\left(\frac{\pi DL\lambda^2 f^2}{c}\right). \tag{8.4}$$

Similar to the technique based on intensity modulation, two different microwave power penalty functions can be obtained by using two different wavelengths. The ratio between the two penalty functions can be used to calculate the microwave frequency. Figure 8.12 shows the simulated results of the RF power distribution as a function of the microwave frequency and the power ratio distribution against the microwave frequency, that is, ACF. To obtain Figure 8.12, the two optical wavelengths are set at 1520 and 1630 nm; the corresponding values of dispersion are 362 and 512 ps/nm, which correspond to the values of dispersion of a 25-km standard SMF at 1520 and 1630 nm, respectively. To estimate the microwave frequency based on the ACF without ambiguity, one monotone interval should be chosen. The upper frequency bound is dependent on the larger dispersion and its corresponding wavelength, which is given by

$$\mathrm{BW} = \sqrt{\frac{c}{(DL\lambda^2)_{\max}}}. \tag{8.5}$$

The lower bandwidth bound is limited by the predetermined measurement error tolerance because large errors would be resulted in the low frequency range owing to a relatively slow varying ACF, making the measurement more sensitive to the system noise.

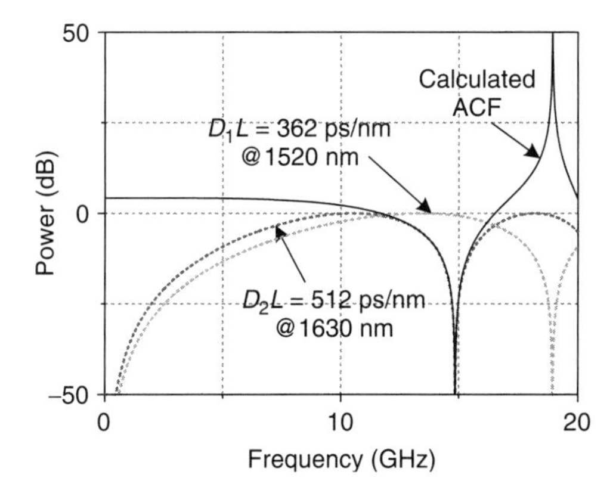

FIGURE 8.12 Typical power fading functions and the corresponding ACF function in the technique based on phase modulation. $\lambda_1 = 1520$ nm; $\lambda_2 = 1630$ nm; $D_1L = 362$ ps/nm; $D_2L = 512$ ps/nm.

In a proof-of-concept experiment (31), two wavelengths at 1520 and 1620 nm are employed, which provide a microwave frequency measurement range of 7.3–15.05 GHz with a measurement accuracy of ±0.5 GHz. If the two wavelengths are set at 1520 and 1540 nm, the measurement range is 11.75–17.95 GHz for an error within ±0.5 GHz. Again, there is a trade-off between the measurement range and the measurement accuracy. Higher measurement resolution results in a smaller measurement range, while a lower measurement resolution leads to a relatively wider measurement range.

8.3.2 Break the Lower Frequency Bound

The calculated ACF based on dispersion-induced microwave power penalty functions at two different wavelengths could only provide IFM with a limited measurement range, typically less than 10 GHz. This measurement range may be not broad enough for practical applications such as EW. There are two possible solutions to extend the frequency measurement range. The first solution is to increase the frequency of the first notch in the ACF (i.e., the upper frequency bound), but this will sacrifice the measurement accuracy. The second solution is to break the lower frequency bound. The ACF at low frequencies has a slow increasing slope, which makes the lower frequency bound at a relatively high frequency. If the lower frequency bound can be smaller, the entire measurement range can then be increased.

A method to increase the slope of the ACF in the low frequency band is to introduce dispersion-induced power penalties to two frequencies, that is, f and $f_{LO} - f$, where f_{LO} is a fixed LO frequency, to generate two complementary power fading functions. Figure 8.13a shows the diagram of such an IFM scheme. The unknown signal is divided into two paths, one is mixed in a broadband mixer with a fixed LO frequency. The LO frequency is selected to be just beyond the frequency range of operation required by the receiver. The mixing will produce two spectral

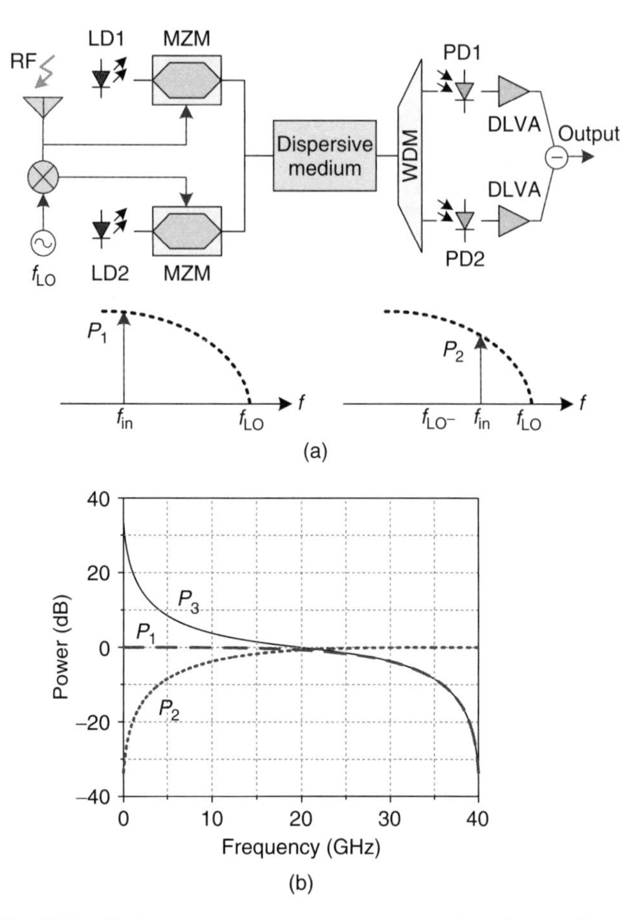

FIGURE 8.13 IFM with increased measurement range by introducing dispersion-induced power penalties to two frequencies, to generate two complementary power fading functions. (a) Schematic representation of the IFM system, insets: detected RF frequencies from the two optical paths, the dotted lines show the dispersion-induced microwave power penalty functions. (b) Simulated ACF with a 40-GHz operational bandwidth, $DL = 25.5$ ps/nm. P_1 shows the detected RF power for the upper optical signal path, P_2 the RF power for the lower optical signal path, and P_3 the ACF.

components at $f_{\text{LO}} - f$ and $f_{\text{LO}} + f$. As the $f_{\text{LO}} + f$ component lies well outside the bandwidth of interest, only the $f_{\text{LO}} - f$ component will be considered. The RF signals from both paths will modulate two independent optical carriers at two MZMs, both biased at the quadrature point, thus providing two optical DSB+C signals. These signals are then combined and passed through a dispersive element such as an optical fiber or a chirped FBG. The dispersed signals are demultiplexed, detected by two PDs and amplified using DLVAs in order to measure the power of the two signals. The dispersion will introduce a microwave power penalty on each of the signals owing to the DSB+C nature of the signals. The ACF can be given by

$$\gamma = \frac{P_1}{P_2} = \frac{R_1 \cos^2\left(\frac{\pi D_1 L_1 \lambda_1^2 f^2}{c} + \arctan\alpha\right)}{R_2 \cos^2\left(\frac{\pi D_2 L_2 \lambda_2^2 (f_{\mathrm{LO}} - f)^2}{c} + \arctan\alpha\right)}. \tag{8.6}$$

By adjusting the powers of the two optical carriers or the microwave powers to the two MZMs, R_1 and R_2 can be made identical. Figure 8.13b shows the typical power fading functions and the corresponding ACF when $f_{LO} = 40$ GHz, $D_1 L_1 \approx D_2 L_2 = 38.5$ ps/nm, $\lambda_1 = 1549.2$ nm, $\lambda_2 = 1550$ nm, and $\alpha = 0$. The curve named as P_3 in Figure 8.13b, that is, ACF, shows a monotonically decreasing function in which each frequency associated with a unique amplitude in a 40-GHz operational bandwidth. This response, when compared with a calibrated response, will allow the determination of the frequency of the unknown input signal.

An experiment based on the setup in Figure 8.13 was performed. A mixer operating from 4 to 40 GHz was employed. The LO frequency was 22 GHz. The total dispersion of the dispersive element was ~184 ps/nm. Two DLVAs with a bandwidth of 19 GHz were connected at the outputs of the two PDs (32). An extended frequency measurement range from 4 to 19 GHz was achieved.

As a mixer always has a lower operation frequency limit, the IFM system still has lower frequency bound. In addition, the input RF signal is split and processed in two different paths; the nonlinearity and uneven frequency responses of the electrical devices and the MZMs would inevitably introduce additional frequency measurement errors.

The measurement range can be further extended by reducing the lower bound to DC. For a DSB+C optical signal transmitting in a dispersive medium, the microwave frequency response of the system is low pass, while for a phase-modulated optical signal transmitting in a dispersive medium, the frequency response of the system is band pass. Owing to the complementary nature of the two frequency responses, the ratio of the two frequency responses, that is, ACF, is monotonically increasing or decreasing from DC to the frequency of the first notch of the power fading function of the intensity-modulated signal. The use of intensity modulation and phase modulation for microwave frequency measurement was demonstrated in Reference 33. The approach offers a wide measurement range without the need for the tuning of the optical carrier wavelength. Figure 8.14a shows the experimental setup. As can be seen only a laser source is employed. Compared with the technique in Reference 32, the complexity and cost are considerably reduced. Figure 8.14b shows the experimentally measured and theoretically calculated ACFs. A frequency measurement range of 0–13.8 GHz is obtained with a measurement accuracy of better than ±0.3 GHz. The measurement error mainly originates from the length difference of the SMFs in the upper and the lower arms. As the SMF lengths can be controlled precisely using a commercially available optical time domain reflectometer, the measurement error

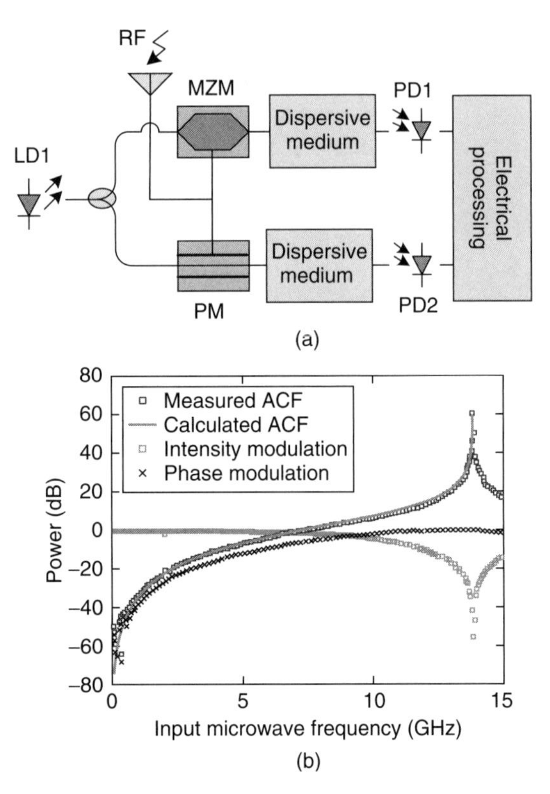

FIGURE 8.14 A microwave frequency measurement system based on intensity and phase modulation. (a) Schematic representation of the system, (b) measured and simulated ACF, where $DL = 340$ ps/nm.

should be greatly reduced. The measurement range can be further extended by reducing the dispersion of the dispersive medium. As two modulators are used, it is difficult to make the ratio of the modulation indices of the two modulators identical in the whole frequency measurement range, which may introduce additional measurement errors.

The above method using an intensity modulator and a phase modulator can be implemented using a much simple scheme based on a polarization modulator (PolM) (34). The schematic representation of system using a PolM is shown in Figure 8.15. In the system, the microwave signal with its frequency to be measured is applied to the PolM via the RF port to modulate two linearly polarized optical light waves at different wavelengths from two laser sources. The PolM is a special phase modulator that supports both transverse electric (TE) and transverse magnetic (TM) modes, but with opposite phase modulation indices (49). The polarization direction of one light wave is aligned with one principal axis of the PolM. Phase modulation is thus imposed on this light wave. The polarization direction of the other light wave is oriented with an angle of 45° with respect to the principal

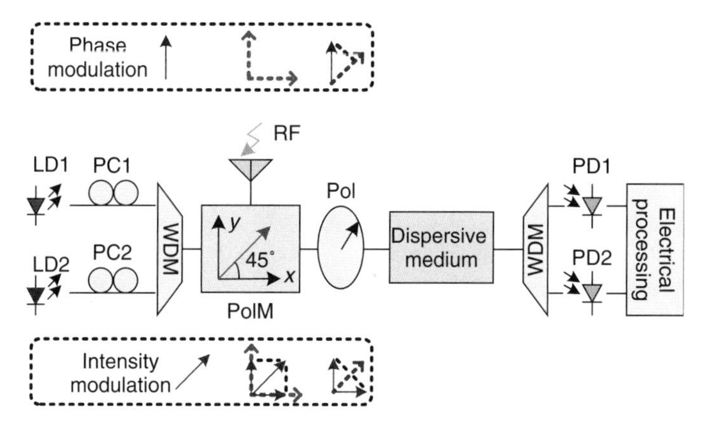

FIGURE 8.15 A microwave frequency measurement system using a polarization modulator. PolM, polarization modulator.

axis of the PolM. As an optical polarizer is connected with its transmission axis oriented with an angle of 45° to the same principal axis of the PolM, the phase-modulated signals along the two principal axes are projected to the transmission axis of the optical polarizer, leading to the generation of an intensity-modulated signal. Both the intensity-modulated and the phase-modulated optical signals are then sent to a dispersive element. Similar to the technique in Reference 33, the frequency responses corresponding to the two light waves along the two directions would be a low pass and band pass. Thus an ACF similar to that shown in Figure 8.14 is obtained. It was demonstrated that the system could perform frequency measurement with a measurement ranging from 3 to 18 GHz with a measurement accuracy of ±0.2 GHz. The major advantage of this approach over the method in Reference 33 is that only one modulator and one dispersive element are required, which would reduce the complexity and cost of the system.

8.3.3 IFM Based on Photonic Microwave Filters with Complementary Frequency Responses

It is seen from Figure 8.13 and 8.14 that the broadband frequency measurement range can be achieved if the ACF is calculated using two complementary frequency responses, such as a low pass frequency response and a band-pass frequency response. As a photonic microwave filter can be designed to have a low pass or band-pass frequency response, it is thus possible to perform microwave frequency measurement with extended frequency measurement range using two photonic microwave filters with complementary frequency responses.

Photonic microwave filters are usually implemented in the optical domain based on a delay-line structure, in which a microwave input signal is modulated on one or multiple optical carriers at an optical modulator; the modulated light waves are then sent to a time delay device to introduce different time delays; the

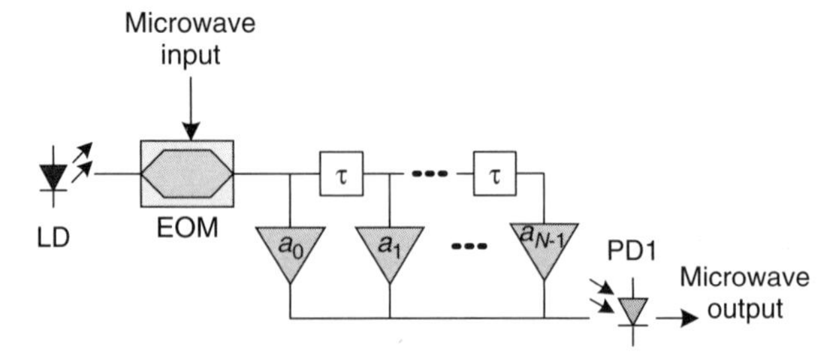

FIGURE 8.16 Schematic representation of an N-tap photonic microwave delay-line filter.

time-delayed microwave signals are then detected at a PD (50–55). The schematic representation of an N-tap photonic microwave delay-line filter is shown in Figure 8.16. There are two key parameters that determine the frequency response of the filter: the time delay difference τ, which is inversely proportional to the free spectral range (FSR), and the coefficients a_n, which determine the shape of the filter response. Mathematically, the frequency response of an N-tap microwave delay-line filter is given as (53)

$$H(f) = \sum_{n=0}^{N-1} a_n e^{-j2n\pi f\tau}. \tag{8.7}$$

For a two-tap filter with coefficients of (1, 1) or (1, −1), the frequency response is given as

$$|H_{(1,1)}(f)|^2 = 2\cos^2(\pi f\tau), \tag{8.8}$$

$$|H_{(1,-1)}(f)|^2 = 2\sin^2(\pi f\tau). \tag{8.9}$$

As seen in Equations 8.8 and 8.9, the two frequency responses are complementary, corresponding to the frequency responses of a low pass and a band-pass filter, respectively. The ratio between the two power spectral responses, again referred to as the *ACF*, is given by

$$\mathrm{ACF} = \frac{|H_{(1,1)}(f)|^2}{|H_{(1,-1)}(f)|^2} = \frac{1}{\tan^2(\pi f\tau)}. \tag{8.10}$$

From Equation 8.10, we can see that the ACF is monotonically decreasing over a frequency band from DC to $1/2\tau$. Therefore, based on the value of the ACF, we can calculate the frequency of the input microwave signal by the following equation:

$$f = \frac{1}{\pi\tau\tan^{-1}\sqrt{\mathrm{ACF}}}. \tag{8.11}$$

The maximum measurement range is determined by τ. To have a larger measurement range, we may design the filter pair with a greater FSR. As the photonic microwave filter pair is generally implemented in the incoherent regime using an incoherent light source (54), the IFM scheme would have a low requirement for the quality of the laser source. The system would also have improved system stability because the wavelength drift of the laser source would not affect the performance of the filter pair owing to the incoherent operation of the filter pair. This feature is highly desirable in defense systems.

Figure 8.17 shows the schematic representation of the microwave frequency measurement system using photonic microwave filters with complementary frequency response. It consists of an LD, a polarization controller, a PolM, a polarizer, two sections of polarization-maintaining fiber (PMF), two PDs, and an electrical processing module. A linearly polarized CW light wave from the LD is fiber coupled to the PolM with its polarization direction oriented at an angle of 45° to a principal axis of the PolM by the polarization controller. The microwave signal with its frequency to be measured is applied to the PolM via its RF port. In the

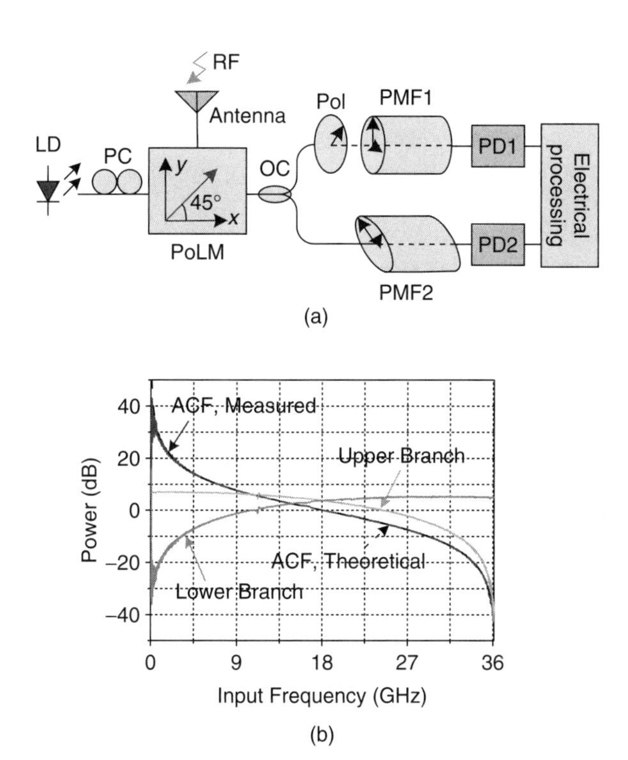

FIGURE 8.17 (a) Schematic representation of the microwave frequency measurement system using photonic microwave filters with complementary frequency response. (b) Measured and calculated frequency responses of the two photonic microwave filters and the ACF based on the measured frequency responses.

PolM, the light waves would experience complementary phase modulations along the two principal axes (x and y), or complementary intensity modulations along the axes that have an angle of 45° to the principal axes. The signal at the output of the PolM is then split into two branches by a polarization-maintaining coupler. In the upper branch, an optical polarizer with its transmission axis oriented at an angle of 45° to one principal axis of the PolM is incorporated to convert the phase-modulated signal to an intensity-modulated signal. A section of PMF (PMF1) with its fast axis aligned to one principal axis of the PolM is connected after the polarizer. The intensity-modulated signal after the polarizer is thus split equally along the fast and slow axes of PMF1. Owing to the differential group delay of the PMF, a time delay difference determined by the birefringence and the length of the PMF is produced between the two signals along the two principal axes. The time-delayed signals are then detected at a PD. The entire operation corresponds to a two-tap photonic microwave filter with two coefficients of (1, 1). In the lower branch, the phase-modulated signals are directly sent to a second section of PMF (PMF2) with its fast axis oriented at an angle of 45° to one principal axis of the PolM. The phase-modulated signals are then converted to two complementary intensity-modulated signals along the fast and slow axes. After experiencing a time delay difference in PMF2, the two signals are detected at a second PD. Owing to the complementary nature of the two time-delayed signals, a two-tap photonic microwave filter with coefficients of (1, −1) is implemented (52).

Figure 8.17b shows the experimental results. As expected, the two photonic microwave filters have complementary frequency responses, one as a low pass filter and the other as a band-pass microwave filter. The ratio of the two frequency response is calculated. Owing to the complementary nature of the frequency responses, a monotonically decreasing ACF in a range from 1 to 36 GHz is obtained, which agrees well with the theoretically calculated ACF. A microwave frequency measurement range as large as 36 GHz with a measurement error less than ±0.2 GHz is demonstrated. The wavelength of the LD is tuned from 1535 to 1570 nm; no significant increase in measurement error is observed, demonstrating the wavelength-independent feature of the scheme.

A similar scheme is also demonstrated using a two-tap photonic microwave filter with two coefficients of (1, −1) to produce a frequency response that is complementary to that produced based on dispersion-induced microwave power penalty in a phase-modulation-based system. The ratio of the two frequency responses leads to an ACF that varies monotonically from negative infinity to positive infinity (36). Frequency measurement based on the ACF is performed in the range from 1.7 to 12.2 GHz with a measurement error within ±0.07 GHz.

8.3.4 First-Order Photonic Microwave Differentiator

As the dynamic range (the ratio of the maximum and the minimum detectable powers) of an IFM receiver cannot be infinite, to obtain a maximized measurement range with almost a uniform measurement resolution, it is highly desirable that the ACF is linear. A linear ACF would also allow an accurate measurement of the center frequency of a pulsed microwave signal. On the other hand, an ACF is also dependent on other parameters of the system, such as the wavelengths of the laser sources, the responsivity of the PDs, and the insertion losses of the transmission paths. Practically, a calibration should be performed before the measurement to guarantee high measurement accuracy. If the ACF is directly proportional to the microwave frequency, the calibration only requires a reference microwave signal at a fixed frequency. To obtain such an ACF, a first-order photonic microwave differentiator may be employed.

The first-order photonic microwave differentiation can be implemented by optical phase modulation and phase modulation to intensity modulation conversion in an optical band-pass filter (OBPF) (38). Mathematically, the optical field of a phase-modulated signal can be expressed as

$$E_{\mathrm{PM}}(t) = E_0 \exp(j\omega_c t + j\beta \sin \Omega t), \tag{8.12}$$

where E_0 and ω_c are the amplitude and the angular frequency of the input optical field, β is the phase modulation index, and Ω is the angular frequency of the microwave signal.

Assume that the frequency response of an OBPF has two linear slopes and the carrier of the phase-modulated signal is located at one of the two slopes, say, the left slope, as shown in Figure 8.18. At the output of the OBPF, we have

$$\begin{aligned} E'(t) = E_0 e^{j\omega_c t}\{&K(\omega_c - \omega_1) \\ &+ K(\omega_c + \Omega - \omega_1) J_1(\beta) e^{j\Omega t} - K(\omega_c - \Omega - \omega_1) J_1(\beta) e^{-j\Omega t}\}, \end{aligned} \tag{8.13}$$

where K is the slope ($K > 0$) of the OBPF frequency response, ω_1 the left zero transmission point of the filter, and $J_1(\beta)$ the first-order Bessel function of the first kind, and β the phase modulation index. When this signal is detected by a PD, the AC term of the photocurrent is given by

$$I_1 = R_1 |E_0|^2 K^2 J_1(\beta)(\omega_c - \omega_1)\Omega \sin \Omega t, \tag{8.14}$$

where R_1 is a parameter related to the optical loss of the optical path and the responsivity of the PD. From Equation 8.14, we can see that the output photocurrent is dependent on the modulation index. To remove this dependence, we may apply intensity modulation in the other branch. If the intensity-modulated signal is sent to a PD for square law detection, the AC term of the photocurrent is

$$I_2 \approx R_2 |E_0|^2 J_1(\beta) \sin \Omega t. \tag{8.15}$$

On the basis of Equations 8.14 and 8.15, we have the ACF, given by

$$\mathrm{ACF}(f) = \frac{|I_1|}{|I_2|} = \frac{R_1}{R_2} K^2 (\omega_c - \omega_1)\Omega. \tag{8.16}$$

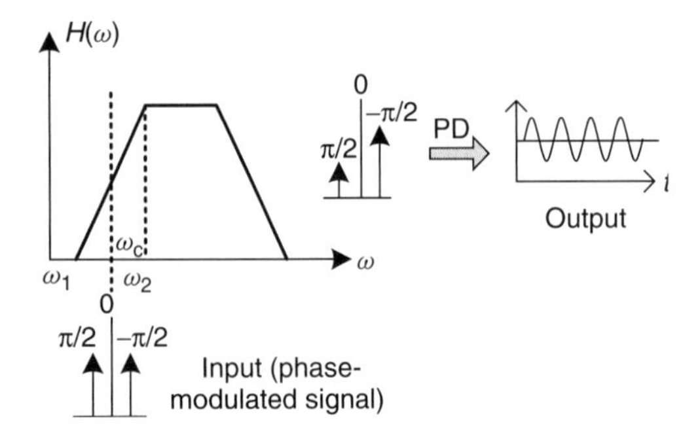

FIGURE 8.18 A first-order microwave differentiator performed by optical phase modulation and phase modulation to intensity modulation conversion in an OBPF.

R_1 and R_2 can be made identical by adjusting the losses of the optical paths via optical attenuators. As seen in Equation 8.16, the ACF is directly proportional to the frequency of the input microwave signal and is not dependent on the input optical power and the microwave modulation index. Therefore, based on the value of the ACF, the frequency of the input signal can be obtained by the following expression:

$$f = \kappa \cdot \text{ACF}, \tag{8.17}$$

where

$$\kappa = \frac{1}{2\pi K^2(\omega_c - \omega_1)}.$$

It should be noted that κ contains the information of the laser wavelength. Therefore, the wavelength drift of the laser source would contribute to the measurement error. For instance, the wavelength of the laser used in our experiment can shift within 5 pm (~600 MHz) in 1-h observation and $\omega_c - \omega_1 \approx 40$ GHz, which gives a measurement error of $\Delta f \approx 0.015f$. When $f = 20$ GHz, the measurement error can be 300 MHz. One possible way to reduce the influence of the wavelength drift is to have a large $\omega_c - \omega_1$, making the relative wavelength drift smaller, or to perform a calibration before the measurement because the wavelength drift is a slowly varying process.

The concept was demonstrated by a proof-of-concept experiment (38). The schematic representation of the microwave frequency measurement system is shown in Figure 8.19a. The polarization-modulated signal from the PolM is split into two branches. In the upper branch, PM–IM conversion is performed at a polarization beam splitter (PBS), which serves as a polarizer. In the lower branch, one of the phase-modulated signals is selected by another PBS and sent to the OBPF to implement a first-order photonic microwave differentiation. The transmission spectrum of the OBPF is shown in Figure 8.19b. The left slope from 1549.52 to 1549.91 nm is almost linear. The time delay variation in this range is less than 3 ps, which is small and can be neglected for a measurement range less than 25 GHz.

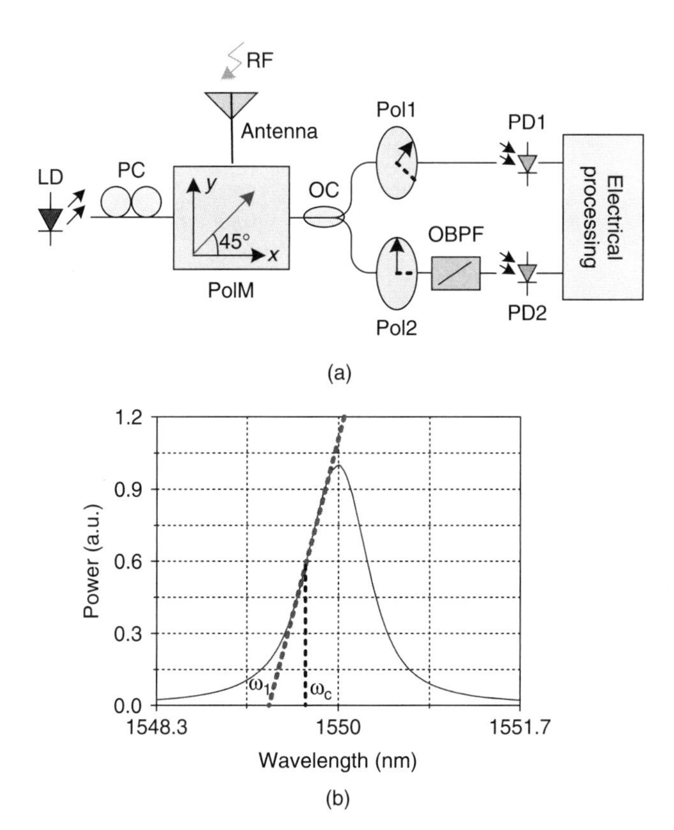

FIGURE 8.19 (a) Schematic representation of the microwave frequency measurement system based on a photonic microwave differentiator. (b) Transmission spectrum of the OBPF used in the experiment.

Figure 8.20a shows the detected microwave powers in the upper and lower branches. The input microwave power is set at 0 dBm. The ACF is almost directly proportional to the frequency of the input microwave signal, which allows the determination of an unknown input frequency from 0 to 25 GHz. Figure 8.20b shows the measurement error versus the frequency of the input microwave signal. In the frequency range from 0.5 to 18 GHz, the measurement error is within ±0.2 GHz. The measurement error is increased to ±0.4 GHz in the frequency range from 20 to 25 GHz, which is due to the imperfections in the electrical devices used in the experiment. In this frequency range, the electrical devices have large insertion losses. The system can be employed to evaluate the carrier frequency of a DSB+C-pulsed microwave signal. The measurement results are shown as the solid squares in Figure 8.20a. When the carrier frequency of the pulsed microwave signal (500-Mb/s NRZ signal) is tuned from 3 to 18 GHz, the measurement error is within ±0.4 GHz.

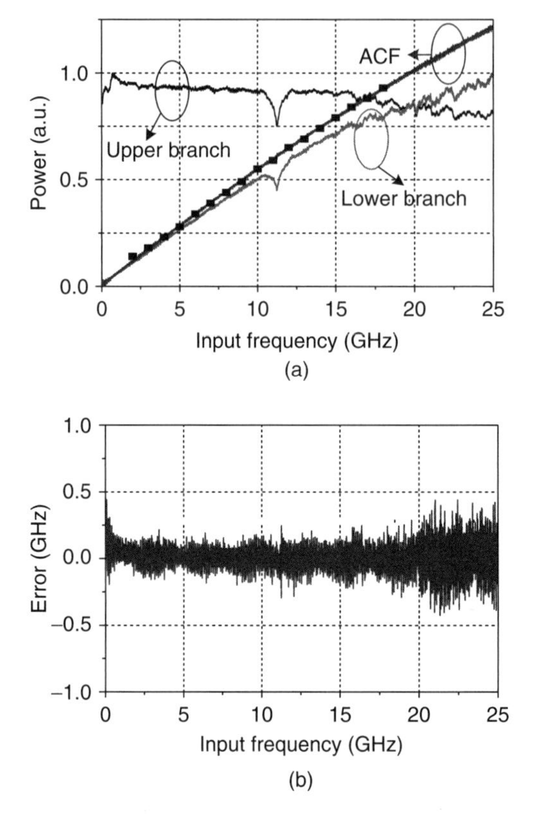

FIGURE 8.20 (a) Measured microwave powers for the signals in the upper and lower branches and the calculated amplitude comparison function (solid squares are the measured results for the pulsed microwave signal). (b) Measurement error versus input microwave frequency.

8.3.5 Optical Power Fading Using Optical Filters

The major problem associated with the IFM by monitoring microwave power is that it requires high frequency PD, microwave transmission lines, and microwave detector, to detect, transmit, and process the microwave signal. If the optical power instead of the microwave power is monitored for microwave frequency measurement, the system will be easier to implement by using low frequency devices at a lower cost because only DC optical powers are measured.

Figure 8.21a illustrates the schematic representation of a typical microwave frequency measurement system based on optical power monitoring. Two optical carriers from two LDs are combined at a multiplexer and sent to an MZM through a polarization controller. The MZM is biased such that the optical carriers at the output of the MZM are completely suppressed. An optical filter with a sinusoidal spectral response is placed at the output of the MZM. The wavelengths of the two optical carriers are set at one peak and one valley of the filter spectral response,

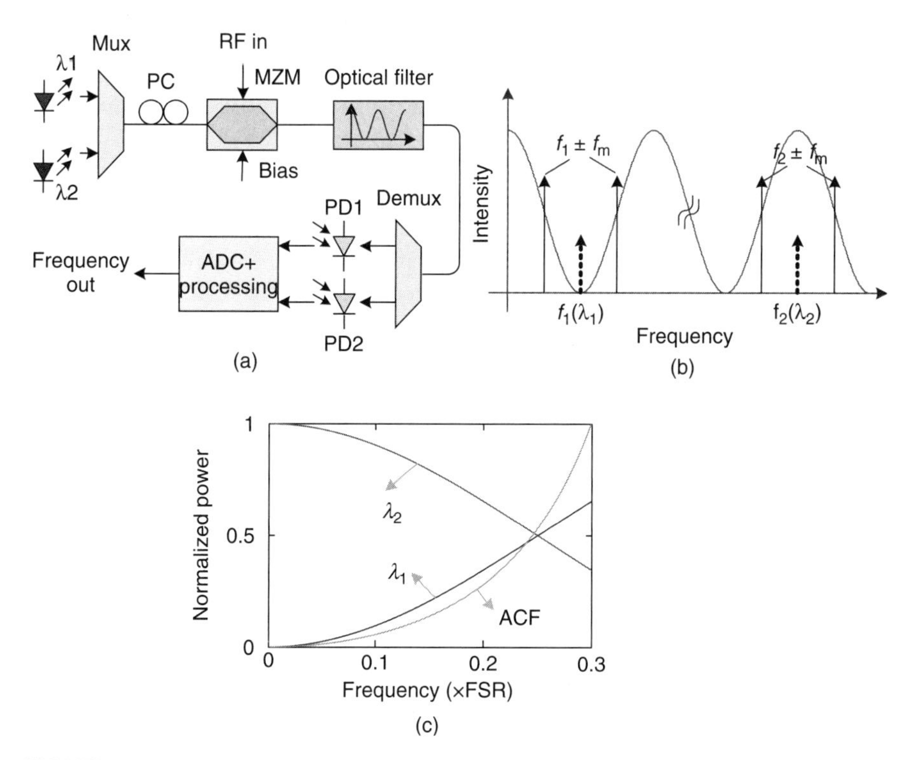

FIGURE 8.21 Microwave frequency measurement system based on optical power monitoring. (a) Schematic diagram. (b) Operation principle. The wavelengths of the two optical carriers are selected such that one carrier is located at one peak and the other is located at the valley of the spectral response. (c) Calculated ACF.

as shown in Figure 8.21b. The carrier-suppressed optical signals are then demultiplexed, with the optical powers measured by two low frequency PDs. The DC currents from the PDs are digitized and compared in a data processing unit. The ratio between the two detected powers, again referred to as *ACF*, can be expressed as (39)

$$\gamma = \frac{1 - \cos(2\pi f/\mathrm{FSR})}{1 + \cos(2\pi f/\mathrm{FSR})} = \tan^2\left(\frac{\pi f}{\mathrm{FSR}}\right), \tag{8.18}$$

The ACF is not dependent on the input optical power and the phase modulation index of the modulator. A calculated ACF is given in Figure 8.21c, which is monotonically increasing in the whole measurement range. This ACF can thus be used to determine the frequency of an unknown input microwave signal. In the experiment reported in Reference 39, a two-tap Sagnac-loop filter (SLF) was used to serve as the sinusoidal filter with an FSR of 50 GHz. A microwave frequency measurement ranging from 1 to 20 GHz with a measurement accuracy of ±0.2 GHz was demonstrated.

The key limitation of this approach is its instability. The measurement error of the system could be significant because of: (i) the bias drift of the MZM, (ii) the wavelength fluctuations of the two laser sources existing in the two-wavelength system, and (iii) the poor stability of the SLF. A sophisticated control circuit is thus needed to stabilize the operation.

The technique in Reference 39 can be improved by replacing the SLF with an optical complementary filter pair (40, 41), which simplifies the system because only a single laser source is required. The measurement error due to power fluctuations of the two laser sources is completely eliminated. Figure 8.22a shows the schematic representation. The key component in the system is the complementary filter pair, which can be implemented by using a length of PMF, in conjunction with two polarization controllers and a PBS, as shown in Figure 8.22b. The polarization direction of a linearly polarized light wave is aligned by the first polarization controller with an angle of θ with respect to the fast axis of the PMF, which is projected to the principal axes with one traveling along the fast axis and the other along the slow axis. Owing to the birefringence of the PMF, a difference in time delay between the two orthogonally polarized light waves is generated. A second polarization controller is connected after the PMF to adjust the two light waves to have an angle of 45° with respect to the axes of the PBS. Two

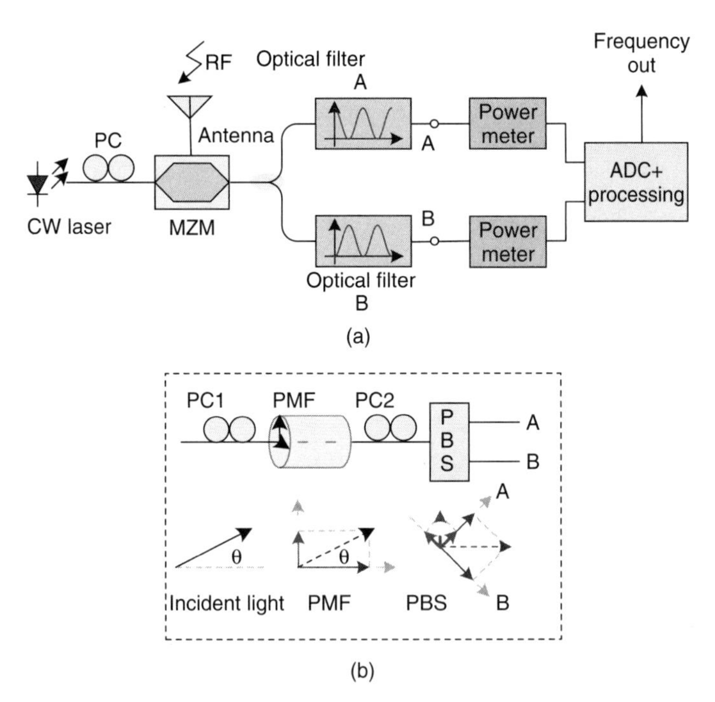

FIGURE 8.22 A simplified microwave frequency measurement system based on optical power monitoring. (a) Schematic representation of the measurement system. (b) Diagram of the complementary filter pair.

complementary transmission responses at ports A and B are obtained. The ratio of the two transmission responses can be expressed as

$$\gamma = \frac{1 - m\cos(2\pi f/\mathrm{FSR})}{1 + m\cos(2\pi f/\mathrm{FSR})}, \tag{8.19}$$

where m is the relative peak-to-notch contrast ratio of the complementary filters, which is related to the incident angle θ by $m = 2 \cdot \tan\theta/(1 + \tan^2\theta)$. In an experiment, the FSR of the complementary filters was 66 GHz. The measurement range was related to the relative peak-to-notch contrast ratio. When $m = 0.6$, 0.825, and 0.99, the estimated measurement errors were less than 0.2 GHz, within a measurement range of 1–26 GHz, 2–24 GHz, and 2–20 GHz, respectively, so a smaller m could provide a larger measurement range.

8.4 OTHER METHODS FOR FREQUENCY MEASUREMENT

In addition to the techniques discussed above, numerous other solutions have also been proposed. In the following sections, we discuss four techniques, including microwave frequency measurement: (i) using a Fabry–Perot scanning receiver, (ii) based on photonic Hilbert transform, (iii) using a monolithically integrated EDG, and (iv) based on frequency-to-time mapping.

8.4.1 Fabry–Perot Scanning Receiver

A Fabry–Perot scanning receiver is a high resolution optical spectrum analyzer, which can be implemented using a piezoelectrically scanned fiber Fabry–Perot filter (FFPF) driven by a ramp voltage (42). With the FFPF scanning over the FSR, the optical spectrum with a high resolution (~90 MHz) can be measured. Microwave frequency is thus derived from the optical spectrum. The measurement requires only low bandwidth postprocessing electronics to perform microwave signal processing, but the piezoelectric scanning of the FFPF plate separation is a slow and nonlinear process, which may not meet the near-real-time measurement requirement.

8.4.2 Photonic Hilbert Transform

The concept of microwave frequency measurement based on Hilbert transform was originally demonstrated using pure electronics. A schematic representation to show the electronic approach is given in Figure 8.23. Similar to the IFM based on power monitoring, the system is constructed to obtain an ACF in which the frequency of the input microwave signal is only a function of the amplitude data. The input unknown microwave signal is divided into two equal parts and then introduced into two interferometers, that is, module 1 and module 2. In each module, the microwave signal is further divided equally into two arms that have a time delay difference of τg. The output signals from the two arms are then multiplied and DC

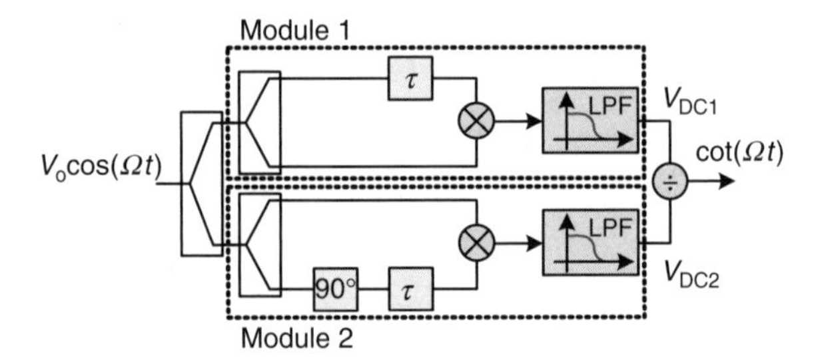

FIGURE 8.23 Block diagram of the IFM system based on Hilbert transform.

filtered. A 90° phase shift is introduced in one arm of module 2, so that the output of module 2 has a sine response, while that of module 1 has a cosine response. The 90° phase shift is realized using a Hilbert transformer. The ratio between the two frequency responses can be written as

$$\gamma = \frac{\frac{1}{8}V_0^2\cos(2\pi f\tau)}{\frac{1}{8}V_0^2\sin(2\pi f\tau)} = \cot(2\pi f\tau). \tag{8.20}$$

As seen in Equation 8.20, the ACF is only a function of microwave frequency. As the signals at the outputs of the two channels are at low frequency band, this system can be implemented at low cost while still enabling broadband frequency measurement; however, practical electrical implementation of such a system requires broadband microwave delay lines and microwave mixers.

The problem will be solved if the system is implemented using photonic method. Figure 8.24 shows the photonic implementation of a microwave frequency measurement system based on Hilbert transform. The microwave signal with its frequency to be measured is converted to the optical domain at an MZM (MZM1). Three optical wavelengths at λ_0, λ_1, and λ_2 are employed to implement a two-tap (λ_1 and λ_2) photonic microwave filter with tap coefficients of $(-1, 1)$ and a reference tap (λ_0). The modulated optical signals at λ_0, λ_1, and λ_2 are reflected by an

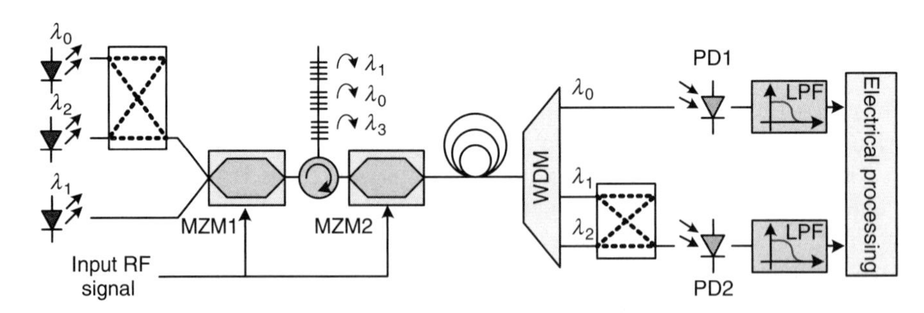

FIGURE 8.24 Photonic implementation of the microwave frequency measurement using Hilbert transform.

optical delay line with three FBGs with their center wavelengths corresponding to λ_0, λ_1, and λ_2. The time delays are determined by the physical locations of the FBGs. The time-delayed signals are then mixed at a second MZM (MZM2). To achieve a broad band 90° phase shift, a Hilbert transformer is employed, which is implemented by the two-tap photonic microwave filter (56). The approach was evaluated experimentally. A frequency measurement ranging from 1 to 10 GHz with a measurement error less than 0.2 GHz was demonstrated. The advantages of the technique are that the time delays were achieved using optical delay lines and mixing was achieved using an MZM, which avoid the use of broadband microwave delay lines and microwave mixers. However, as numerous electrical and optical devices were employed, the system was still complicated.

8.4.3 Monolithically Integrated EDG

The idea to implement IFM based on a monolithically integrated EDG (57) is to monitor the wavelength difference between one optical sideband and the optical carrier of an optical microwave signal. Usually, one can tune the temperature of an EDG chip to change the center wavelengths of the passband. When the detected light intensity reaches the maximum value by adjusting the EDG chip temperature, the center wavelength of the EDG is the same as that of the injected optical signal. By recording this temperature, the exact wavelength of the input signal can be obtained. Knowing the wavelengths of one sideband and the optical carrier, the frequency of the optical microwave signal can be calculated. An EDG-based frequency measurement system based on the above principle can obtain a measurement resolution of better than 1 pm, corresponding to ~120 MHz in the 1550-nm band. As the optical carrier and the other sideband may contribute to the received power in the EDG-based measurement system, a carrier-suppressed modulation is required and a notch filter is employed to filter out the other sideband, which would place a limitation to the measurement range. The measurement would also sensitive to the bias drift of the MZM. A proof-of-concept experiment reports a measurement range from 1 to 15 GHz with a measurement error less than 200 MHz (44). As the EDG can be a multichannel grating, the system can be used to measure the frequencies of multiple microwave signals (Fig. 8.25).

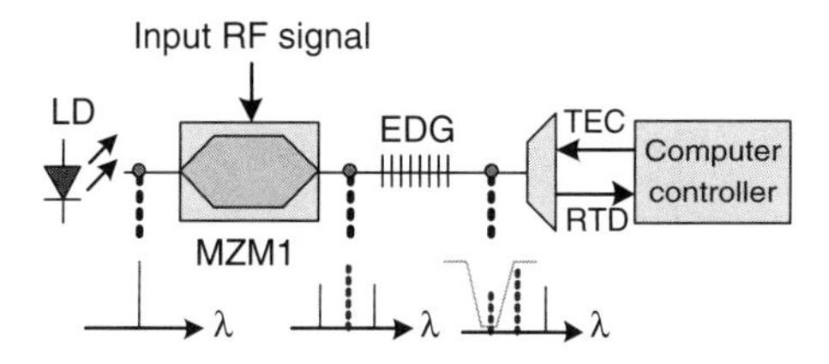

FIGURE 8.25 Photonic microwave frequency measurement using a monolithically integrated EDG.

8.4.4 Incoherent Frequency-to-Time Mapping

IFM can also be realized based on incoherent frequency-to-time mapping, which provides advantages such as fast frequency measurement because the measurement is performed in the time domain. In addition, the entire spectrum in the frequency band of interest is shown; therefore, the approach is suitable for frequency measurement of simultaneously multiple signals. Figure 8.26 shows an IFM system architecture, in which the input signals with different carrier frequencies are mapped to the time domain with different time delays. The mapping operation is achieved by utilizing chromatic dispersive element (58, 59). From Figure 8.26, we can see the input signals are modulated on an optical carrier at an MZM that is biased at the minimum transmission point to form the carrier-suppressed DSB optical signal. A step-recovery stage in the form of a high speed optical on–off switch is used to gate the optical signal to generate a short pulse train. The pulsed carrier-suppressed DSB optical signal is then propagated through a dispersive element and sent to a PD. When propagating through the dispersive element, each optical signal will propagate at a different relative group velocity. This process would map each coupled pair of sidebands to a distinct time delay centered around the imaginary axis of symmetry provided by the wavelength of the optical carrier, as illustrated in Figure 8.26. If the time delays are measured, the frequencies of the multiple microwave signals are obtained.

To accurately estimate the time delays, several issues must be considered. First, the gated signal, that is, the pulsed signal after the step-recovery stage, should have a fast rise time to act as time measurement reference points. Second, the bandwidth of the step-recovery stage and the PD should be with enough bandwidth to replicate the fast gating rise time. Third, the signal processing module should have fast speed to accurately measure the time delays. In a proof-of-concept experiment,

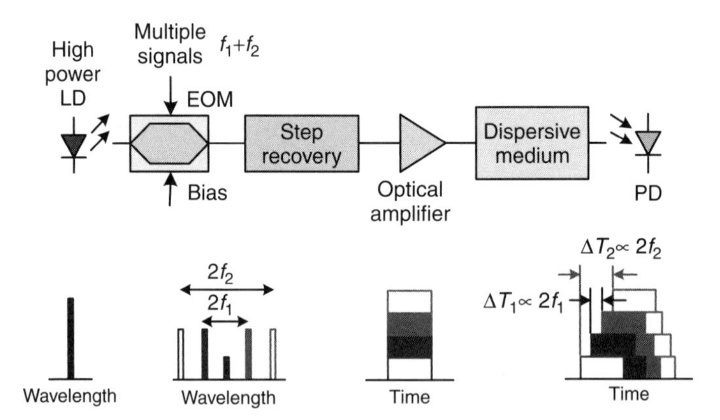

FIGURE 8.26 Frequency measurement of multiple microwave signals based on frequency-to-time mapping in a dispersive element. The measurement of two signals with frequencies at f_1 and f_2 are shown.

two frequencies at 20 and 40 GHz were retrieved using a dispersive element with a value of dispersion of 1000 ps/nm with a measurement error of ~1.6 GHz (45).

8.5 CHALLENGES AND FUTURE PROSPECTS

Although different photonic techniques to achieve IFM have been proposed, there are many challenges in implementing them for practical applications.

The IFM system based on the optical channelizers can estimate the frequencies of simultaneous multiple signals. However, it requires specially designed devices and PD arrays, which is complex and costly. The measurement range (typically <20 GHz) and resolution (typically >1 GHz) are also limited by the fabrication techniques for the optical channelizers, which may not meet the requirement for applications where a wide frequency measurement range and a high measurement resolution are required.

The techniques based on power monitoring are more promising for frequency measurement of a single-channel signal, as it can cover a large frequency measurement range (>20 GHz), and can provide better accuracy (<200 MHz). In addition, the system is much simpler with a much lower cost. Although monitoring the microwave power is an efficient way to estimate the microwave frequency, there are several issues that still need to be addressed. For instance, the PDs should have a wide bandwidth and a flat response. A wideband DC block is also needed to block the DC component. In addition, a wideband microwave power sensor with flat response in the whole measurement range, for example, DLVA, is required to make the energy of the high frequency microwave signal detectable to low frequency receivers. The use of these electrical devices will limit the upper bound of IFM to several tens of gigahertz. The ripples in the frequency responses of the devices would also deteriorate the measurement accuracy. Moreover, there is a trade-off between the measurement range and the accuracy, which is undesirable for applications in which both a large measurement range and a high accuracy are needed.

The techniques based on the monitoring of the optical powers only need low frequency devices. But the MZMs must be biased at the minimum transmission point to suppress the optical carrier, which would suffer severely from the bias drift. The bias drift of the MZMs will make the system unstable; in other words, a sophisticated control circuit is needed to stabilize the operation. Thus to ensure acceptable measurement accuracy, an LD with high wavelength stability is required, as the measurement is highly dependent on the center wavelength of the optical carrier. In addition, the optical filter implemented using a Sagnac loop or an asymmetric Mach–Zehnder interferometer is easily disturbed by environmental variations.

From a practical point of view, EW environment is always filled with spectral clutters. Therefore, it is required that an IFM system can estimate the frequencies of multiple signals. Although IFM based on optical channelizers may have the capability, its measurement range and accuracy are very limited. IFM based on incoherent frequency-to-time mapping also has the ability to perform IFM for multiple signals, but the accuracy is very low or a high speed sampling oscilloscope is required. IFM based on power monitoring can provide broadband and high accuracy frequency measurement, but it is applicable only to single-frequency signals. When multiple signals are received with one signal that has a much higher power, the approach may provide an accurate frequency measurement for the strongest signal. However, the system will report an erroneous frequency if the multiple signals have similar powers.

We would like to point out that in addition to frequency information of a microwave signal, it is essential to have knowledge of other parameters of a microwave signal. For example, in a radar system, the emitted signals are usually in the form of a pulse train. Four other important parameters such as the pulse amplitude (PA), PW, TOA, and angle of arrival (AOA) (2) are also important for warfare applications. Accurate PA information can be used for inverse gain jamming, where the jammer power is inversely proportional to the PA measured. The PW can be used to coarsely identify the type of radars (e.g., weapon radars always have a short PW). The TOA provides the information of the pulse repetition rate, which can be used to generate matched jamming pulses with the same pulse repetition rate. It can also be used to predict the arrival time of the next pulse for a jammer. The AOA gives the direction of a hostile radar. If some of the four parameters can be measured together with the frequency information then the measurement system would provide much detailed information that is important for practical applications.

8.6 CONCLUSION

An overview about the photonic IFM techniques developed in the last two decades was presented. The photonic techniques for IFM have demonstrated remarkable advantages over their electronic counterparts, such as near-real-time measurement, large measurement bandwidth, immunity to electromagnetic interference, low loss, and small size. Although significant progress has been made in the last few years for IFM based on photonics, the techniques have limitations in terms of accuracy and dynamic range. In addition, efforts are desired to design a system that can identify simultaneously the frequencies of multiple signals. To obtain high dynamic range and high performance, the current activities in microwave photonics and silicon photonics would have an important impact on the development and implementation of future photonics IFM systems.

REFERENCES

1. Schleher DC. *Electronic Warfare in the Information Age*. Norwood (MA): Artech House; 1999.
2. Tsui JBY. *Microwave Receivers with Electronic Warfare Applications*. New York: John Wiley and Sons; 1986.

3. Tsui JBY. *Digital Techniques for Wideband Receiver*. 2nd ed. Norwood (MA): Artech House; 2001.
4. Earp CW. Frequency indicating cathode ray oscilloscope. US patent 2,434,914. 1948.
5. James WG. Instantaneous frequency measurement system. US patent 3,939,411. 1976.
6. Tsui JBY, Schrick GH. Instantaneous frequency measurement (IFM) receiver with capability to separate CW and pulsed signals. US patent 4,194,206. 1980.
7. Cuckson H, Curtis PD. Microwave instantaneous frequency measurement apparatus. US patent 4,414,505. 1983.
8. Tsui JBY, Shaw RL. Simultaneous signal detection for IFM receivers by transient detection. US patent 4,547,727. 1985.
9. Asendorf RH, Fitzpatrick JP, Graves BW. Automatic frequency identifier for radar signals. US patent 4,612,545. 1986.
10. Tsui JBY. Instantaneous frequency measurement receiver with digital processing. US patent 4,633,516. 1986.
11. Gagnon A, McMillan M, Gale PM. Method and apparatus for simultaneous instantaneous signal frequency measurement. US patent 4,791,360. 1988.
12. Gale PM, McMillan M, Gagnon A. Apparatus for measuring the frequency of microwave signals. US patent 4,859,934. 1989.
13. Tsui JBY, McCormick WS. Instantaneous frequency measurement (IFM) receiver with only two delay lines. US patent 4,963,816. 1990.
14. Sanderson RB, Tsui JBY. Instantaneous frequency measurement receiver with bandwidth improvement through phase shifted sampling of real signals. US patent 5,109,188. 1992.
15. Tsui JBY, Sharpin DL. Frequency measurement receiver with bandwidth improvement through synchronized phase shifted sampling. US patent 5,198,748. 1993.
16. Choate DB, Tsui JBY. Frequency measurement receiver with means to resolve two folding frequencies of complex signals. US patent 5,323,103. 1994.
17. Ward CR. Hybrid instantaneous frequency measurement compressive receiver apparatus and method. US patent 5,424,631. 1995.
18. Schmidt RO. Simultaneous signals IFM receiver using plural delay line correlators. US patent 5,440,228. 1995.
19. Regev Z. Method for instantaneous frequency measurement. US patent 6,433,531. 2002.
20. Tsui JBY, Lopata SM, Ward CR. Wideband digital IFM receiver. US patent 7,206,707. 2007.
21. Tsui JBY. *Microwave Receivers and Related Components*. Avionics Laboratory, Air Force Wright Aeronautical Laboratories. Los Alton (CA): Peninsula, Publishing; 1985.
22. Wide Band Systems, Inc. Specifications IFM receives 0.5–18.0GHz. Technical Bulletin. Available at http://widebandsystems.com/pdf/ifmspecs.pdf.
23. Lee CH. *Microwave Photonics*. Boca Raton (FL): CRC Press; 2006.
24. Heaton JM, Watson CD, Jones SB, Bourke MM, Boyne CM, Smith GW, Wight DR. 16-channel (1- to 16-GHz) microwave spectrum analyzer device based on a phased array of GaAs/AlGaAs electro-optic waveguide delay lines. Proc SPIE 1998;3278:245–251.
25. Wang WS, Davis RL, Jung TJ, Lodenkamper R, Lembo LJ, Brock JC, Wu MC. Characterization of a coherent optical RF channelizer based on a diffraction grating. IEEE Trans Microw Theory Tech 2001;49:1996–2001.

26. Hunter DB, Edvell LG, Englund MA. Wideband microwave photonic channelised receiver. Proceedings of MWP 2005, Korea; 2005. p 249–252.
27. Winnall ST, Lindsay AC, Austin MW, Canning J, Mitchell A. A microwave channelizer and spectroscope based on an integrated optical Bragg-grating Fabry–Perot and integrated hybrid fresnel lens system. IEEE Trans Microw Theory Tech 2006;54:868–872.
28. Nguyen LVT, Hunter DB. A photonic technique for microwave frequency measurement. IEEE Photon Technol Lett 2006;18:1188–1190.
29. Zou XH, Yao JP. An optical approach to microwave frequency measurement with adjustable measurement range and resolution. IEEE Photon Technol Lett 2008;20:1989–1991.
30. Zou XH, Yao JP. Microwave frequency measurement with improved measurement range and resolution. Electron Lett 2009;45:497–498.
31. Zhang XM, Chi H, Zhang XM, Zheng SL, Jin XF, Yao JP. Instantaneous microwave frequency measurement using an optical phase modulator. IEEE Microw Wireless Compon Lett 2009;19:422–424.
32. Attygalle M, Hunter DB. Improved photonic technique for broadband radio-frequency measurement. IEEE Photon Technol Lett 2009;21:206–208.
33. Zhou JQ, Fu SN, Aditya S, Shum PP, Lin CL. Instantaneous microwave frequency measurement using photonic technique. IEEE Photon Technol Lett 2009;21:1069–1071.
34. Zou XH, Pan SL, Yao JP. Instantaneous microwave frequency measurement with improved measurement range and resolution based on simultaneous phase modulation and intensity modulation. J Lightw Technol 2009;27:5314–5320.
35. Pan SL, Yao JP. Instantaneous microwave frequency measurement using a photonic microwave filter pair. IEEE Photon Technol Lett 2010;22:1437–1439.
36. Zhou JQ, Fu SN, Shum PP, Aditya S, Xia L, Li JQ, Sun XQ, Xu K. Photonic measurement of microwave frequency based on phase modulation. Opt Express 2009;17:7217–7221.
37. Li JQ, Fu SN, Xu K, Zhou JQ, Shum P, Wu J, Lin JT. Photonic-assisted microwave frequency measurement with higher resolution and tunable range. Opt Lett 2009;34:743–745.
38. Pan SL, Yao JP. Instantaneous photonic microwave frequency measurement with a maximized measurement range. Proceedings of MWP 2009 Oct 14–16; Valencia, Spain, Paper Fr 4.3.
39. Chi H, Zou XH, Yao JP. An approach to the measurement of microwave frequency based on optical power monitoring. IEEE Photon Technol Lett 2008;20:1249–1251.
40. Zou XH, Chi H, Yao JP. Microwave frequency measurement based on optical power monitoring using a complementary optical filter pair. IEEE Trans Microw Theory Tech 2009;57:505–511.
41. Drummond MV, Monteiro P, Nogueira RN. Photonic RF instantaneous frequency measurement system by means of a polarization-domain interferometer. Opt Express 2009;17:5433–5438.
42. Winnall ST, Lindsay AC. A Fabry–Perot scanning receiver for microwave signal processing. IEEE Trans Microw Theory Tech 1999;47:1385–1390.
43. Emami H, Sarkhosh N, Bui LA, Mitchell A. Amplitude independent RF instantaneous frequency measurement system using photonic Hilbert transform. Opt Express 2008;16:13707–13712.

44. Guo HL, Xiao GZ, Mrad N, Yao JP. Measurement of microwave frequency using a monolithically integrated scannable echelle diffractive grating. IEEE Photon Technol Lett 2009;21:45–47.
45. Nguyen LVT. Microwave Photonic technique for frequency measurement of simultaneous signals. IEEE Photon Technol Lett 2009;21:642–644.
46. Agrawal GP. *Fiber-Optic Communication Systems*. 3rd ed. New York: John Wiley and Sons; 2002.
47. Smith GH, Novak D, Ahmed Z. Overcoming chromatic-dispersion effects in fiber-wireless systems incorporating external modulators. IEEE Trans Microw Theory Tech 1997;45:1410–1415.
48. Zeng F, Yao JP. Investigation of phase-modulator-based all-optical bandpass microwave filter. J Lightw Technol 2005;23:1721–1728.
49. Bull JD, Jaeger NA, Kato H, Fairburn M, Reid A, Ghanipour P. 40-GHz electro-optic polarization modulator for fiber optic communications systems. Proc SPIE 2004;5577:133–143.
50. Capmany J, Ortega B, Pastor D. A tutorial on microwave photonic filters. J Lightw Technol 2006;24:201–229.
51. Wang Q, Yao JP. Multitap photonic microwave filters with arbitrary positive and negative coefficients using a polarization modulator and an optical polarizer. IEEE Photon Technol Lett 2008;20:78–80.
52. Yao JP, Wang Q. Photonic microwave bandpass filter with negative coefficients using a polarization modulator. IEEE Photon Technol Lett 2007;19:644–646.
53. Wang Q, Yao JP. Switchable optical UWB monocycle and doublet generation using a reconfigurable photonic microwave delay-line filter. Opt Express 2007;15:14667–14672.
54. Yao JP. Microwave photonics. J Lightw Technol 2009;27:314–335.
55. Yao JP, Zeng F, Wang Q. Photonic generation of ultrawideband signals. J Lightw Technol 2007;25:3219–3235.
56. Emami H, Sarkhosh N, Bui LA, Mitchell A. Wideband RF photonic in-phase and quadrature-phase generation. Opt Lett 2008;33:98–100.
57. Chenben P. Wavelength dispersive planar waveguide devices: echelle gratings and arrayed waveguide gratings. In: Calvo ML, Laksminarayanan V, editors. *Optical Waveguides: From Theory to Applied Technologies*. Boca Raton (FL): CRC Press; 2007, Chapter 5. p 173–230.
58. Saperstien RE, Alic N, Pasasenko D, Rokitski R, Fainman Y. Time-domain waveform processing by chromatic dispersion for temporal shaping of optical pulses. J Opt Soc Am B 2005;22:2427–2436.
59. Azana J, Muriel MA. Real-time optical spectrum analysis based on the time-space duality in chirped fiber grating. IEEE J Quantum Electron 2000;36:517–526.

Index

Photonic Sensing: Principles and Applications for Safety and Security Monitoring, First Edition.
Edited by Gaozhi Xiao and Wojtek J. Bock.

WILEY SERIES IN MICROWAVE AND OPTICAL ENGINEERING

KAI CHANG, Editor
Texas A&M University

FIBER-OPTIC COMMUNICATION SYSTEMS, Fourth Edition • *Govind P. Agrawal*

ASYMMETRIC PASSIVE COMPONENTS IN MICROWAVE INTEGRATED CIRCUITS • *Hee-Ran Ahn*

COHERENT OPTICAL COMMUNICATIONS SYSTEMS • *Silvello Betti, Giancarlo De Marchis, and Eugenio Iannone*

PHASED ARRAY ANTENNAS: FLOQUET ANALYSIS, SYNTHESIS, BFNs, AND ACTIVE ARRAY SYSTEMS • *Arun K. Bhattacharyya*

HIGH-FREQUENCY ELECTROMAGNETIC TECHNIQUES: RECENT ADVANCES AND APPLICATIONS • *Asoke K. Bhattacharyya*

RADIO PROPAGATION AND ADAPTIVE ANTENNAS FOR WIRELESS COMMUNICATION LINKS: TERRESTRIAL, ATMOSPHERIC, AND IONOSPHERIC • *Nathan Blaunstein and Christos G. Christodoulou*

COMPUTATIONAL METHODS FOR ELECTROMAGNETICS AND MICROWAVES • *Richard C. Booton, Jr.*

ELECTROMAGNETIC SHIELDING • *Salvatore Celozzi, Rodolfo Araneo, and Giampiero Lovat*

MICROWAVE RING CIRCUITS AND ANTENNAS • *Kai Chang*

MICROWAVE SOLID-STATE CIRCUITS AND APPLICATIONS • *Kai Chang*

RF AND MICROWAVE WIRELESS SYSTEMS • *Kai Chang*

RF AND MICROWAVE CIRCUIT AND COMPONENT DESIGN FOR WIRELESS SYSTEMS • *Kai Chang, Inder Bahl, and Vijay Nair*

MICROWAVE RING CIRCUITS AND RELATED STRUCTURES, Second Edition • *Kai Chang and Lung-Hwa Hsieh*

MULTIRESOLUTION TIME DOMAIN SCHEME FOR ELECTROMAGNETIC ENGINEERING • *Yinchao Chen, Qunsheng Cao, and Raj Mittra*

DIODE LASERS AND PHOTONIC INTEGRATED CIRCUITS, Second Edition • *Larry Coldren, Scott Corzine, and Milan Masanovic*

EM DETECTION OF CONCEALED TARGETS • *David J. Daniels*

RADIO FREQUENCY CIRCUIT DESIGN • *W. Alan Davis and Krishna Agarwal*

RADIO FREQUENCY CIRCUIT DESIGN, Second Edition • *W. Alan Davis*

MULTICONDUCTOR TRANSMISSION-LINE STRUCTURES: MODAL ANALYSIS TECHNIQUES • *J. A. Brandão Faria*

PHASED ARRAY-BASED SYSTEMS AND APPLICATIONS • *Nick Fourikis*

SOLAR CELLS AND THEIR APPLICATIONS, Second Edition • *Lewis M. Fraas and Larry D. Partain*

FUNDAMENTALS OF MICROWAVE TRANSMISSION LINES • *Jon C. Freeman*

OPTICAL SEMICONDUCTOR DEVICES • *Mitsuo Fukuda*

MICROSTRIP CIRCUITS • *Fred Gardiol*

HIGH-SPEED VLSI INTERCONNECTIONS, Second Edition • *Ashok K. Goel*

FUNDAMENTALS OF WAVELETS: THEORY, ALGORITHMS, AND APPLICATIONS, Second Edition • *Jaideva C. Goswami and Andrew K. Chan*

HIGH-FREQUENCY ANALOG INTEGRATED CIRCUIT DESIGN • *Ravender Goyal (ed.)*

RF AND MICROWAVE TRANSMITTER DESIGN • *Andrei Grebennikov*

ANALYSIS AND DESIGN OF INTEGRATED CIRCUIT ANTENNA MODULES • *K. C. Gupta and Peter S. Hall*

PHASED ARRAY ANTENNAS, Second Edition • *R. C. Hansen*

STRIPLINE CIRCULATORS • *Joseph Helszajn*

THE STRIPLINE CIRCULATOR: THEORY AND PRACTICE • *Joseph Helszajn*

LOCALIZED WAVES • *Hugo E. Hernández-Figueroa, Michel Zamboni-Rached, and Erasmo Recami (eds.)*

MICROSTRIP FILTERS FOR RF/MICROWAVE APPLICATIONS, Second Edition • *Jia-Sheng Hong*

MICROWAVE APPROACH TO HIGHLY IRREGULAR FIBER OPTICS • *Huang Hung-Chia*

NONLINEAR OPTICAL COMMUNICATION NETWORKS • *Eugenio Iannone, Francesco Matera, Antonio Mecozzi, and Marina Settembre*

FINITE ELEMENT SOFTWARE FOR MICROWAVE ENGINEERING • *Tatsuo Itoh, Giuseppe Pelosi, and Peter P. Silvester (eds.)*

INFRARED TECHNOLOGY: APPLICATIONS TO ELECTROOPTICS, PHOTONIC DEVICES, AND SENSORS • *A. R. Jha*

SUPERCONDUCTOR TECHNOLOGY: APPLICATIONS TO MICROWAVE, ELECTRO-OPTICS, ELECTRICAL MACHINES, AND PROPULSION SYSTEMS • *A. R. Jha*

TIME AND FREQUENCY DOMAIN SOLUTIONS OF EM PROBLEMS USING INTEGTRAL EQUATIONS AND A HYBRID METHODOLOGY • *B. H. Jung, T. K. Sarkar, S. W. Ting, Y. Zhang, Z. Mei, Z. Ji, M. Yuan, A. De, M. Salazar-Palma, and S. M. Rao*

OPTICAL COMPUTING: AN INTRODUCTION • *M. A. Karim and A. S. S. Awwal*

INTRODUCTION TO ELECTROMAGNETIC AND MICROWAVE ENGINEERING • *Paul R. Karmel, Gabriel D. Colef, and Raymond L. Camisa*

MILLIMETER WAVE OPTICAL DIELECTRIC INTEGRATED GUIDES AND CIRCUITS • *Shiban K. Koul*

ADVANCED INTEGRATED COMMUNICATION MICROSYSTEMS • *Joy Laskar, Sudipto Chakraborty, Manos Tentzeris, Franklin Bien, and Anh-Vu Pham*

MICROWAVE DEVICES, CIRCUITS AND THEIR INTERACTION • *Charles A. Lee and G. Conrad Dalman*

ADVANCES IN MICROSTRIP AND PRINTED ANTENNAS • *Kai-Fong Lee and Wei Chen (eds.)*

SPHEROIDAL WAVE FUNCTIONS IN ELECTROMAGNETIC THEORY • *Le-Wei Li, Xiao-Kang Kang, and Mook-Seng Leong*

COMPACT MULTIFUNCTIONAL ANTENNAS FOR WIRELESS SYSTEMS • *Eng Hock Lim and Kwok Wa Leung*

ARITHMETIC AND LOGIC IN COMPUTER SYSTEMS • *Mi Lu*

OPTICAL FILTER DESIGN AND ANALYSIS: A SIGNAL PROCESSING APPROACH • *Christi K. Madsen and Jian H. Zhao*

THEORY AND PRACTICE OF INFRARED TECHNOLOGY FOR NONDESTRUCTIVE TESTING • *Xavier P. V. Maldague*

METAMATERIALS WITH NEGATIVE PARAMETERS: THEORY, DESIGN, AND MICROWAVE APPLICATIONS • Ricardo Marqués, Ferran Martín, and Mario Sorolla

OPTOELECTRONIC PACKAGING • *A. R. Mickelson, N. R. Basavanhally, and Y. C. Lee (eds.)*

OPTICAL CHARACTER RECOGNITION • *Shunji Mori, Hirobumi Nishida, and Hiromitsu Yamada*

ANTENNAS FOR RADAR AND COMMUNICATIONS: A POLARIMETRIC APPROACH • *Harold Mott*

INTEGRATED ACTIVE ANTENNAS AND SPATIAL POWER COMBINING • *Julio A. Navarro and Kai Chang*

ANALYSIS METHODS FOR RF, MICROWAVE, AND MILLIMETER-WAVE PLANAR TRANSMISSION LINE STRUCTURES • *Cam Nguyen*

LASER DIODES AND THEIR APPLICATIONS TO COMMUNICATIONS AND INFORMATION PROCESSING • *Takahiro Numai*

FREQUENCY CONTROL OF SEMICONDUCTOR LASERS • *Motoichi Ohtsu (ed.)*

INVERSE SYNTHETIC APERTURE RADAR IMAGING WITH MATLAB ALGORITHMS • *Caner Özdemir*

SILICA OPTICAL FIBER TECHNOLOGY FOR DEVICE AND COMPONENTS: DESIGN, FABRICATION, AND INTERNATIONAL STANDARDS • *Un-Chul Paek and Kyunghwan Oh*

WAVELETS IN ELECTROMAGNETICS AND DEVICE MODELING • *George W. Pan*

OPTICAL SWITCHING • Georgios Papadimitriou, Chrisoula Papazoglou, and Andreas S. Pomportsis

MICROWAVE IMAGING • *Matteo Pastorino*

ANALYSIS OF MULTICONDUCTOR TRANSMISSION LINES • *Clayton R. Paul*

INTRODUCTION TO ELECTROMAGNETIC COMPATIBILITY, Second Edition • *Clayton R. Paul*

ADAPTIVE OPTICS FOR VISION SCIENCE: PRINCIPLES, PRACTICES, DESIGN AND APPLICATIONS • *Jason Porter, Hope Queener, Julianna Lin, Karen Thorn, and Abdul Awwal (eds.)*

ELECTROMAGNETIC OPTIMIZATION BY GENETIC ALGORITHMS • *Yahya Rahmat-Samii and Eric Michielssen (eds.)*

INTRODUCTION TO HIGH-SPEED ELECTRONICS AND OPTOELECTRONICS • *Leonard M. Riaziat*

NEW FRONTIERS IN MEDICAL DEVICE TECHNOLOGY • *Arye Rosen and Harel Rosen (eds.)*

ELECTROMAGNETIC PROPAGATION IN MULTI-MODE RANDOM MEDIA • *Harrison E. Rowe*

ELECTROMAGNETIC PROPAGATION IN ONE-DIMENSIONAL RANDOM MEDIA • *Harrison E. Rowe*

HISTORY OF WIRELESS • *Tapan K. Sarkar, Robert J. Mailloux, Arthur A. Oliner, Magdalena Salazar-Palma, and Dipak L. Sengupta*

PHYSICS OF MULTIANTENNA SYSTEMS AND BROADBAND PROCESSING • *Tapan K. Sarkar, Magdalena Salazar-Palma, and Eric L. Mokole*

SMART ANTENNAS • *Tapan K. Sarkar, Michael C. Wicks, Magdalena Salazar-Palma, and Robert J. Bonneau*

NONLINEAR OPTICS • *E. G. Sauter*

APPLIED ELECTROMAGNETICS AND ELECTROMAGNETIC COMPATIBILITY • *Dipak L. Sengupta and Valdis V. Liepa*

COPLANAR WAVEGUIDE CIRCUITS, COMPONENTS, AND SYSTEMS • *Rainee N. Simons*

ELECTROMAGNETIC FIELDS IN UNCONVENTIONAL MATERIALS AND STRUCTURES • *Onkar N. Singh and Akhlesh Lakhtakia (eds.)*

ANALYSIS AND DESIGN OF AUTONOMOUS MICROWAVE CIRCUITS • *Almudena Suárez*

ELECTRON BEAMS AND MICROWAVE VACUUM ELECTRONICS • *Shulim E. Tsimring*

FUNDAMENTALS OF GLOBAL POSITIONING SYSTEM RECEIVERS: A SOFTWARE APPROACH, Second Edition • *James Bao-yen Tsui*

SUBSURFACE SENSING • *Ahmet S. Turk, A. Koksal Hocaoglu, and Alexey A. Vertiy (eds.)*

RF/MICROWAVE INTERACTION WITH BIOLOGICAL TISSUES • *André Vander Vorst, Arye Rosen, and Youji Kotsuka*

InP-BASED MATERIALS AND DEVICES: PHYSICS AND TECHNOLOGY • *Osamu Wada and Hideki Hasegawa (eds.)*

COMPACT AND BROADBAND MICROSTRIP ANTENNAS • *Kin-Lu Wong*

DESIGN OF NONPLANAR MICROSTRIP ANTENNAS AND TRANSMISSION LINES • *Kin-Lu Wong*

PLANAR ANTENNAS FOR WIRELESS COMMUNICATIONS • *Kin-Lu Wong*

FREQUENCY SELECTIVE SURFACE AND GRID ARRAY • *T. K. Wu (ed.)*

PHOTONIC SENSING: PRINCIPLES AND APPLICATIONS FOR SAFETY AND SECURITY MONITORING • *Gaozhi Xiao and Wojtek J. Bock (eds.)*

ACTIVE AND QUASI-OPTICAL ARRAYS FOR SOLID-STATE POWER COMBINING • *Robert A. York and Zoya B. Popovic (eds.)*

OPTICAL SIGNAL PROCESSING, COMPUTING AND NEURAL NETWORKS • *Francis T. S. Yu and Suganda Jutamulia*

ELECTROMAGNETIC SIMULATION TECHNIQUES BASED ON THE FDTD METHOD • *Wenhua Yu, Xiaoling Yang, Yongjun Liu, and Raj Mittra*

SiGe, GaAs, AND InP HETEROJUNCTION BIPOLAR TRANSISTORS • *Jiann Yuan*

PARALLEL SOLUTION OF INTEGRAL EQUATION-BASED EM PROBLEMS • *Yu Zhang and Tapan K. Sarkar*

ELECTRODYNAMICS OF SOLIDS AND MICROWAVE SUPERCONDUCTIVITY • *Shu-Ang Zhou*

MICROWAVE BANDPASS FILTERS FOR WIDEBAND COMMUNICATIONS • *Lei Zhu, Sheng Sun, and Rui Li*